General Theory of
Irregular Curves

Mathematics and Its Applications (*Soviet Series*)

Volume 29

A. D. ALEXANDROV and YU. G. RESHETNYAK

Institute of Mathematics, Siberian Branch of the USSR Academy of Sciences, Novosibirsk

General Theory of Irregular Curves

Kluwer Academic Publishers

Dordrecht / Boston / London

Library of Congress Cataloging in Publication Data

Aleksandrov, A. D. (Aleksandr Danilovich), 1912-
    General theory of irregular curves / A.D. Alexandrov and Yu. G.
Reshetnyak.
        p.   cm. -- (Mathematics and its applications. Soviet series)
    Bibliography: p.
    Includes index.
    ISBN 9027728119
    1. Curves on surfaces.   I. Reshetniak, IUrii Grigor'evich.
II. Title.   III. Title: Irregular curves.   IV. Series: Mathematics
and its applications (Kluwer Academic Publishers). Soviet series.
QA643.A37 1989
516.3'6--dc19   88-23241
                                                                CIP

Published by Kluwer Academic Publishers,
P.O. Box, 17, 3300 AA Dordrecht, The Netherlands

Kluwer Academic Publishers incorporates the publishing programmes
of D. Reidel, Martinus Nijhoff, Dr. W. Junk and MTP Press

Sold and distributed in the U.S.A. and Canada
by Kluwer Academic Publishers,
101 Philip Drive, Norwell, MA 02061, U.S.A.

In all other countries, sold and distributed
by Kluwer Academic Publishers Group,
P.O. Box 322, 3300 AH Dordrecht, The Netherlands

Translated from the Russian by L. Ya. Yuzina

printed on acid free paper

Table of Contents

Series Editor's Preface   ix

Introduction   1

Chapter I: General Notion of a Curve   5

1.1.   Definition of a Curve   5
1.2.   Normal Parametrization of a Curve   15
1.3.   Chains on a Curve and the Notion of an Inscribed Polygonal Line   16
1.4.   Distance Between Curves and Curve Convergence   18
1.5.   On a Non-Parametric Definition of the Notion of a Curve   23

Chapter II: Length of a Curve   29

2.1.   Definition of a Curve Length and its Basic Properties   29
2.2.   Rectifiable Curves in Euclidean Spaces   33
2.3.   Rectifiable Curves in Lipshitz Manifolds   40

Chapter III: Tangent and the Class of One–Sidedly Smooth Curves   43

3.1.   Definition and Basic Properties of One-Sidedly Smooth Curves   43
3.2.   Projection Criterion of the Existence of a Tangent in the Strong Sense   47
3.3.   Characterizing One-Sidedly Smooth Curves with Contingencies   49
3.4.   One-Sidedly Smooth Functions   54
3.5.   Notion of c-Correspondence. Indicatrix of Tangents of a Curve   55
3.6.   One-Sidedly Smooth Curves in Differentiable Manifolds   63

Chapter IV: Some Facts of Integral Geometry   75

4.1.   Manifold G_k^n of k-Dimensional Directions in V^n 75
4.2.   Imbedding of G_k^n into a Euclidean Space   80
4.3.   Existence of Invariant Measure of G_k^n 85
4.4.   Invariant Measure in G_k^n and Integral. Uniqueness of an Invariant Measure   88
4.5.   Some Relations for Integrals Relative to the Invariant Measure in G_k^n 93
4.6.   Some Specific Subsets of G_k^n 96
4.7.   Length of a Spherical Curve as an Integral of the Function Equal to the Number of Intersection Points   101
4.8.   Length of a Curve as an Integral of Lengths of its Projections   105

4.9.   Generalization of Theorems on the Mean Number of the Points of
       Intersection and Other Problems   109

Chapter V: Turn or Integral Curvature of a Curve   118

5.1.   Definition of a Turn. Basic Properties of Curves of a Finite Turn   118
5.2.   Definition of a Turn of a Curve by Contingencies   127
5.3.   Turn of a Regular Curve   132
5.4.   Analytical Criterion of Finiteness of a Curve Turn   134
5.5.   Basic Integro-Geometrical Theorem on a Curve Turn   139
5.6.   Some Estimates and Theorems on a Limiting Transition   144
5.7.   Turn of a Curve as a Limit of the Sum of Angles Between the Secants 148
5.8.   Exact Estimates of the Length of a Curve   151
5.9.   Convergence with a Turn   160
5.10   Turn of a Plane Curve   164

Chapter VI: Theory of a Turn on an n-Dimensional
Sphere   175

6.1.   Auxiliary Results   175
6.2.   Integro-Geometrical Theorem on Angles and its Corrolaries   184
6.3.   Definition and Basic Properties of Spherical Curves of a Finite
       Geodesic Turn   191
6.4.   Definition of a Geodesic Turn by Means of Tangents   197
6.5.   Curves on a Two-Dimensional Sphere   203

Chapter VII: Osculating Planes and Class of Curves with
an Osculating Plane in the Strong Sense   208

7.1.   Notion of an Osculating Plane   208
7.2.   Osculating Plane of a Plane Curve   211
7.3.   Properties of Curves with an Osculating Plane in the Strong Sense 214

Chapter VIII: Torsion of a Curve in a Three-
Dimensional Euclidean Space   217

8.1.   Torsion of a Plane Curve   217
8.2.   Curves of a Finite Complete Torsion   243
8.3.   Complete Two-Dimensional Indicatrix of a Curve of a Finite
       Complete Torsion   249
8.4.   Continuity and Additivity of Absolute Torsion   254
8.5.   Definition of an Absolute Torsion Through Triple Chains and
       Paratingences   255
8.6.   Right-Hand and Left-Hand Indices of a Point. Complete Torsion
       of a Curve   257

Chapter IX: Frenet Formulas and Theorems on Natural Parametrization   268

9.1. Frenet Formulas   268
9.2. Theorems on Natural Parametrization   275

Chapter X: Some Additional Remarks   281

References   285

Index   287

'Et moi, ..., si j'avait su comment en revenir,
je n'y serais point allé.'

        Jules Verne

The series is divergent; therefore we may be
able to do something with it.

        O. Heaviside

One service mathematics has rendered the
human race. It has put common sense back
where it belongs, on the topmost shelf next
to the dusty canister labelled 'discarded non-
sense'.

        Eric T. Bell

Mathematics is a tool for thought. A highly necessary tool in a world where both feedback and non-linearities abound. Similarly, all kinds of parts of mathematics serve as tools for other parts and for other sciences.

Applying a simple rewriting rule to the quote on the right above one finds such statements as: 'One service topology has rendered mathematical physics ...'; 'One service logic has rendered computer science ...'; 'One service category theory has rendered mathematics ...'. All arguably true. And all statements obtainable this way form part of the raison d'être of this series.

This series, *Mathematics and Its Applications*, started in 1977. Now that over one hundred volumes have appeared it seems opportune to reexamine its scope. At the time I wrote

> "Growing specialization and diversification have brought a host of monographs and textbooks on increasingly specialized topics. However, the 'tree' of knowledge of mathematics and related fields does not grow only by putting forth new branches. It also happens, quite often in fact, that branches which were thought to be completely disparate are suddenly seen to be related. Further, the kind and level of sophistication of mathematics applied in various sciences has changed drastically in recent years: measure theory is used (non-trivially) in regional and theoretical economics; algebraic geometry interacts with physics; the Minkowsky lemma, coding theory and the structure of water meet one another in packing and covering theory; quantum fields, crystal defects and mathematical programming profit from homotopy theory; Lie algebras are relevant to filtering; and prediction and electrical engineering can use Stein spaces. And in addition to this there are such new emerging subdisciplines as 'experimental mathematics', 'CFD', 'completely integrable systems', 'chaos, synergetics and large-scale order', which are almost impossible to fit into the existing classification schemes. They draw upon widely different sections of mathematics."

By and large, all this still applies today. It is still true that at first sight mathematics seems rather fragmented and that to find, see, and exploit the deeper underlying interrelations more effort is needed and so are books that can help mathematicians and scientists do so. Accordingly MIA will continue to try to make such books available.

If anything, the description I gave in 1977 is now an understatement. To the examples of interaction areas one should add string theory where Riemann surfaces, algebraic geometry, modular functions, knots, quantum field theory, Kac-Moody algebras, monstrous moonshine (and more) all come together. And to the examples of things which can be usefully applied let me add the topic 'finite geometry'; a combination of words which sounds like it might not even exist, let alone be applicable. And yet it is being applied: to statistics via designs, to radar/sonar detection arrays (via finite projective planes), and to bus connections of VLSI chips (via difference sets). There seems to be no part of (so-called pure) mathematics that is not in immediate danger of being applied. And, accordingly, the applied mathematician needs to be aware of much more. Besides analysis and numerics, the traditional workhorses, he may need all kinds of combinatorics, algebra, probability, and so on.

In addition, the applied scientist needs to cope increasingly with the nonlinear world and the extra mathematical sophistication that this requires. For that is where the rewards are. Linear models are honest and a bit sad and depressing: proportional efforts and results. It is in the non-linear world that infinitesimal inputs may result in macroscopic outputs (or vice versa). To appreciate what I am hinting at: if electronics were linear we would have no fun with transistors and computers; we would have no TV; in fact you would not be reading these lines.

There is also no safety in ignoring such outlandish things as nonstandard analysis, superspace and anticommuting integration, p-adic and ultrametric space. All three have applications in both electrical engineering and physics. Once, complex numbers were equally outlandish, but they frequently proved the shortest path between 'real' results. Similarly, the first two topics named have already provided a number of 'wormhole' paths. There is no telling where all this is leading - fortunately.

Thus the original scope of the series, which for various (sound) reasons now comprises five sub-series: white (Japan), yellow (China), red (USSR), blue (Eastern Europe), and green (everything else), still applies. It has been enlarged a bit to include books treating of the tools from one subdiscipline which are used in others. Thus the series still aims at books dealing with:

- a central concept which plays an important role in several different mathematical and/or scientific specialization areas;
- new applications of the results and ideas from one area of scientific endeavour into another;
- influences which the results, problems and concepts of one field of enquiry have, and have had, on the development of another.

Certainly, a 'long' time ago, such as say 30 years, it used to be a general opinion that for applications at least, regular curves, e.g. piece-wise differentiable ones, were the only ones of real importance. Even at that time it was known that this was really not the case. Irregular curves (and worse) arise very naturally, e.g. as solutions of optimality problems or as boundaries of domains of attraction. Nowadays, with fractals and the like firmly established as a describing and analysing tool, it is quite generally accepted that an irregular idealized limit may well be a much better object to consider than a piece-wise smooth approximation (granted that such exist, which may not be the case).

The problem then, of course, arises that for irregular curves a great deal of machinery from differential geometry is no longer available: such things as curve length, torsion, curvature, natural parametrizations, ...

This unique book is precisely concerned with defining and analysing such notions for irregular curves in such a way that for regular curves the old notions reappear.

The shortest path between two truths in the real domain passes through the complex domain.

J. Hadamard

La physique ne nous donne pas seulement l'occasion de résoudre des problèmes ... elle nous fait pressentir la solution.

H. Poincaré

Never lend books, for no one ever returns them; the only books I have in my library are books that other folk have lent me.

Anatole France

The function of an expert is not to be more right than other people, but to be wrong for more sophisticated reasons.

David Butler

Bussum, March 1989

Michiel Hazewinkel

Introduction

One of the initial chapters of differential geometry is the theory of curves. In a classical form the theory of curves introduces such notions as the length of a curve, which is a function of its arc, the notion of a tangent and osculatory plane of a curve, and defines certain numerical characteristics, such as curvature and torsion as functions of a curve point. It should be remarked that differential geometry commonly studies only the curves obeying certain conditions of regularity. These conditions are imposed by the requirement that the apparatus of differential calculus be applied, but they are hardly justified in a geometrical sense. Moreover, the classical differential-geometrical method often does not work, even in the cases when we deal with regular curves. For instance, a plane curve $y = x^3$ is regular, and analytical, too. At the same time, viewing it as a spatial curve, we see that the differential-geometrical theory of spatial curves cannot be applied to the curve in question, since at the point $x = 0$ the first two derivates of the radius-vector of the curve turn to zero.

The present book is devoted to the development of the scheme suggested by A. D. Alexandrov in 1946 [1] (and later in [2.3]) of constructing a general theory of irregular curves, which is analogous to the theory of curves in differential geometry, but which would be free from the restrictions of the classical differential-geometrical theory of curves given above. A. D. Alexandrov introduced various definitions of a turn and of a total torsion of a curve (in the sense of M. Frechet [8]), proved (for the case of a turn) their equivalence, and established some estimates and theorems on a limiting transition, associated with the use of the notion of a turn. A. D. Alexandrov's theory was further developed and modified by Yu. G. Reshetnyak by way of employing the method of integral geometry.

The peculiarity of the theory under discussion lies in the fact that certain quantities, termed an integral curvature (or a turn) and an integral (or total) torsion of a curve, are introduced for arbitrary curves. In the case when a curve is regular, its turn is equal to an integral from a common curvature of a curve with respect to the lenght of the arc, while the total torsion is equal to an integral from a common torsion with respect to the lenght of the arc. The notions of a turn and an integral torsion are first defined here for the simplest curves in the geometrical sense, which are polygonal lines. A polygonal line is an ordered system of straight line segments, i.e., links of a polygonal line, subsequently connecting certain points A_0 and A_1, A_1 and A_2, ..., A_{m-1} and A_m, which are vertices of a polygonal line. For an arbitrary

1

curve both a turn and an integral torsion are defined as the limits of the turns and of the integral torsions, respectively, of the polygonal lines inscribed into a curve under the condition that the vertices of the polygonal lines condense on the curve without limit.

Let A_0, A_1, ..., A_m be sequential vertices of a polygonal line. An angle between the vectors $a_i = A_{i-1}A_i$ and $a_{i+1} = A_iA_{i+1}$ is termed a turn at the vertex A_i, while a turn of the polygonal line is a sum of the turns at all its vertices. Each trio (A_{i-1}, A_i, A_{i+1}) of the vertices of a polygonal line defines a certain plane P_i. This plane is oriented in such a way that a pair of vectors (a_i, a_{i+1}) is a right-hand pair on it. An angle between two neighbouring planes P_i and P_{i+1} is termed a torsion on the link a_{i+1} of the polygonal line. A torsion of a polygonal line (or, more exactly, an integral torsion) is a sum of the torsions on its links. A torsion on the link a_{i+1} has a sign which is defined by the orientation of the trio of vectors (a_i, a_{i+1}, a_{i+2}) in space (it is positive if this trio is right-handed, and negative if it is left-handed). A sum of the absolute values of the torsions on the link of a polygonal line is termed an absolute torsion of a polygonal line. A turn of a curve is introduced in such a way that for any curve it is defined without any additional assumptions on its structure: it equals, by definition, an exact upper boundary of the turns of the polygonal lines inscribed into a curve. The value of a turn equal to ∞ is allowed. One can easily distinguish a class of curves of a finite turn.

As far as a torsion is concerned, the situation is as follows: first the notion of an absolute torsion of a curve (or, more exactly, of an absolute integral torsion of a curve) is introduced, which is defined for any curve without any additional assumptions. An absolute torsion is defined as a certain upper limit. (Unlike a turn, it cannot be defined as an exact upper boundary of the absolute torsions of an inscribed polygonal line. Such a definition would be incorrect since, as applied to polygonal lines, it will give a value different from that given by the first definition.) A class of curves of a finite absolute torsion is singled out. For the curves of this class the notion of an integral torsion is defined.

In this book use is made of the integro-geometrical identities proved by I. Fari [6, 7] and, somewhat later, by J. Milnor [19, 20]. Integral geometry was being quite intensively developed in the 1930s by the school of W. Blaschke (see L. Santalo [33]). As will be shown later, there exists a simple relation between the equalities derived by I. Fari and J. Milnor and the identities of a integral geometry given by W. Blaschke and L. Santalo. The integro-geometrical relations established by Fari and Milnor comprise a certain apparatus which makes it possible to reduce the investigation of arbitrary curves to the study of some simplest curves. Thus, for instance, one can limit oneself to considering the simplest case of the curves lying in one straight line. In the case when a turn of such a curve is finite, it is a polygonal line with a finite number of links. An angle between two consequent links of such a polygonal line is equal to 0 or π. At the vertices at which an angle between the links is π, a 'finite load' of a turn, which is equal to that angle (i.e. to π) is concentrated.

The integro-geometrical relations derived by Fari and Milnor, in particular, made it possible to prove the equivalency of different definitions of a complete torsion, and no other way to prove this is known; to prove some new theorems on the approximation of a turn and a complete torsion, to simplify the deduction of certain estimates and so on. The possibility of such an application of the relations in question was established by Yu. G. Reshetnyak [24, 26, 27]. The notion of the length of a curve was studied as long ago as the last century by Jordan, who gave an exhaustive characterization of the class of curves for which the notion of a length has profound geometrical sense. The length of a curve, as well as the theory of the area of a surface, are objects of investigation in the theory of functions of a real variable. Perhaps it is this circumstance that explains why the important class of curves of a finite turn attracted the attention of geometricians comparatively late. The curve of this class can, in particular, be characterized as those curves, the unit vector of the tangent of which is a function of a bounded variation. Curves of a finite turn on a plane were first considered by Radon [23] in connection with the study of the theory of potential on a plane as far back as 1919. As a subject of the theory of geometry, however, but not as an auxiliary apparatus, curves of a finite turn were first considered by A. D. Alexandrov. Namely, the notion of a curve turn is basically used in studies in geometry 'in the large', referring to the directions suggested in the papers by A. D. Alexandrov and his pupils (see, for example, [5, 12, 21, 22, 25, 30-32]). One of the basic results of the theory of convex surfaces is the Liberman theorem [12] which proves that the shortest curve on a convex surface is a curve of a finite turn in space. Later various generalizations of this theorem were developed (see, in particular, [5]). The notion of a turn of a curve and the properties of curves of a finite turn were extensively used by A. V. Pogorelov [21, 22] in his well-known studies on the problem of a unique definitness of convex surfaces in a three-dimensional space. In relation with this fact the book studies in detail the properties of the curves of a finite turn; the case of curves in a n-dimensional Euclidean space are considered at the same time. It is, in particular, established that any curve of a finite turn is rectifiable and at every point it has a left-handed and a right-handed tangent. In this case a limit from the left (right) of the tangents at an arbitrary point of a curve is a left-handed (respectively, right-handed) tangent at this point. Estimates of the length of a curve through its turn are discovered, necessary theorems on convergence and approximation of a turn of a curve are established, various methods of defining a turn are investigated and their equivalence to the initial method is proved.

In the part of the book devoted to torsion only curves in a three-dimensional Euclidean space are considered. This is associated with the fact that in the three-dimensional case a turn and an integral torsion, which are considered as functions of the lenght of an arc, define the uniqueness of a curve to the accuracy of motion.

The concluding part of the book presents theorems on the natural equation, analogous to the corresponding theorems of the classical theorems of curves. In its properties the class of curves of a finite absolute torsion is in many respects analogous to the class of curves of a finite turn, and in all cases when for a torsion there exists an analog of a certain property of a turn, this analog is established in this book. (It should be remarked that such an analogy does not always exist. For instance, a curve of a finite absolute torsion also has a finite turn, but no estimate of a turn throughout the absolute complete torsion is possible.)

When studying the properties of curves associated with torsion, an essential role is played by the methods based on the use of the integro-geometrical relations mentioned above. Their use makes it possible to reduce the study of the properties of a torsion of an arbitrary curve to the case of a curve lying in one plane. A plane curve of a finite absolute torsion consists of a finite number of locally convex arcs. At the points where the arcs with oppositely directed convexities osculate there arises a 'point load' of the torsion, equal to π.

The denotations used hereafter coincide with convention. A certain point O is assumed fixed in space, and the points of the space are identified with their radius-vectors with respect to this point.

The results obtained in Chapter I and partially in Chapter II are valid, and the definitions presented there preserve their sense, not only for curves in a common three-dimensional space, but also for more general spaces, for instance, for arbitrary complete metric spaces.

The theory presented in this book is an example of a combination of geometrical considerations with the classical technique of the theory of functions of a real variable. The basic results are obtained with the theory of the Lebesgue integral (theorems on a limiting transition under the sign of an integral, measurable functions and their properties, etc.) combined with certain considerations referring, in essence, to elementary geometry.

The authors take this opportunity to thank Yekaterina Grigoryevna Reshetnyak whose kind help contributed greatly to this book publication.

General Notion of a Curve

1.1. Definition of a Curve

1.1.1. The notion of a curve in the sense considered here was introduced by M. Frechet, and the definition given below is equivalent to that given by M. Frechet. Here we are going to dwell in detail on the definition of a curve with the aim of clarifying certain peculiarities that are important while discussing the theory of curves, and of presenting the definition of a curve in a more geometrical form as compared to the classical definition by M. Frechet.

We assume the notions of topological and metric space are known (see, for example, [11]).

The notion of a curve is complex and different areas of geometry ascribe different senses to it. The definition accepted in this book originates with Jordan. A set of points in the metric space M is called a Jordan continuum if it is a continuous image of the segment $[0, 1]$ of a real axis. We could try to define a curve as a Jordan continuum, but in many cases this definition of a curve is, however, insufficient. For instance, sometimes it is necessary to consider a segment of a straight line as two different curves and sometimes as the same segment but run through twice - in the direct and reverse directions. In other words, it is important to know not only the points of the set, but also a certain order in which they are run through. The order of running through the set A can be given by choosing a certain continuous mapping $x(t)$ on A of the segment $[0, 1]$ of a real axis. This consideration, naturally, results in the idea of determining a curve as a continuous function whose values are the points of the space, and which is set on a segment of a real axis. This definition is, in certain respects, also expedient, as it is necessary sometimes to consider various parametrizations of 'one and the same' curve.

Let $x(t)$ and $y(u)$ be two continuous mappings of the segment $[0, 1]$ onto a certain set A. If, at monotonous changes in the parameters t and u, the points $x(t)$ and $y(u)$ run through one and the same points of the set A, and in the same order, then the parameters u, obviously, must be a non-decreasing function of t. These considerations lead us to the definition of the curve given by M. Frechet [8]. According to Frechet, a curve is a continuous mapping of a segment of the straight line into the space, considered to the accuracy of

5

monotonic transformations of the parameter. In this case, if we limit ourselves only by the transformations expressed with non-decreasing functions, then
we come to the notion of an oriented curve. Hereafter we are going to consider only oriented curves.

Different variants of the definition by Frechet can be found elsewhere (see,
for instance, [3.10]). We shall give the definition close to that given by
Frechet himself. Now, after the preliminary considerations discussed above,
let us pass to exact definitions.

Let us arbitrarily select a topological space M. A parametrized curve, or a
path in M is any continuous mapping

$$f:[a, b] \rightarrow M$$

where $[a, b]$ is a (closed) segment of the set of all real numbers $\mathbb{R}$. The compact set $f([a, b])$ is called a support of the path f.

Let there be a path $f:[a, b] \rightarrow M$, and τ is a connected component of the set
$f^{-1}(x)$. It is clear that τ is either a one-point set, or a certain segment
$[\alpha, \beta] \subset [a, b]$. The point $x \in M$ is called a support of the point X of the
path f, τ is the totality of the parameter values corresponding to the point
Z. Let $X_1 = (\tau_1, x_1)$ and $X_2 = (\tau_2, x_2)$ be two arbitrary points of the path f.
Let us assume that X_1 precedes X_2 or that X_2 follows X_1, and write $X_1 < X_2$ if
for any $t_1 \in \tau_1$ and any $t_2 \in \tau_2$ we always have $t_1 < t_2$. An ordered set of the
points of the path $f:[a, b] \rightarrow M$ will be denoted through the symbol $\tilde{f}$. Any
value $t \in [a, b]$ defines a certain point of the path $f:[a, b] \rightarrow M$, i.e. the
point $\{\tau, f(t)\}$, where τ is the connected component of the set $f^{-1}[f(t)]$, to
which the given value of t belongs. The point of the parametrized curve
$f:[a, b] \rightarrow M$ which is thus defined through the number $t \in [a, v]$ will be
further denoted by $f[t]$. If X_1 and X_2 are the points of the path $f:[a, b] \rightarrow M$
and $X_1 < X_2$, then we shall say that X_1 lies to the left of X_2 and, correspondingly, that X_2 is located to the right of X_1.

Now let us establish the terminology. The set S will be called ordered if in
S there is set a relation between its elements, denoted through the symbol $<$,
such that (a) if $x < y$, then $x \neq y$, and (b) if $x < y$ and $y < z$, then $x < z$.
The ordered set S is called linearly ordered if in it for any elements x and y
one of the following three relations hold: (1) $x < y$, (2) $x = y$, (3) $y < x$. If
x, y are the elements of an ordered set, then the statement '$y > x$' can, according to the definition, be considered equivalent to the statement '$x < y$'.

Let S be an ordered set, and let $<$ be the relation of the order in it. The
expression: $x \leqslant y$, where $x, y \in S$ means that either $x < y$ or $x = y$. In this
case we shall also write $y \geqslant x$, therefore the expressions $x \leqslant y$ and $y \geqslant x$ are
equivalent. Let $a, b \in S$ be such that $a < b$. In this case (a, b) denotes the
totality of all $x \in S$ such that $a < x < b$. Then, by analogy with the case of
the set of real numbers $\mathbb{R}$, let us set $[a, b) = (a, b) \cup \{a\}$, $(a, b] = (a, b) \cup
\{b\}$ and, finally, $[a, b] = (a, b) \cup \{a\} \cup \{b\}$. The ordered sets S_1 and S_2 are

called similar (anti-similar) if there exists a one-to-one mapping φ of the set S_1 onto the set S_2, such that for any $x, y \in S_1$ it results from $x < y$ that $\varphi(x) < \varphi(y)$ (respectively, it follows from $x < y$ that $\varphi(x) > \varphi(y)$). The mapping φ, which obeys this condition is called a correspondence of similarity (anti-similarity) of the sets S_1 and S_2.

A set of the points of a parametrized curve either consists of a single point or it does not. In the former case the parametrized curve is called degenerate. In the latter case this set is similar to the interval $[0, 1]$ of the real axis. The validity of the latter statement results from Lemma 1.2.1, which will be proved below.

On the set of all parametrized curves in the topological space M let us introduce an equivalence relation.

Let $f : [a, b] \to M$ and $g : [c, d] \to M$ be two arbitrary parametrized curves in the space M. Let us define in which cases we shall consider f and g as being equivalent.

If the path f degenerates into a point, then the path g is considered equivalent to f if and only if the path g is also degenerate and $f([a, b]) = g([a, b])$.

Let us assume that the paths f and g are non-degenerate. Let us set that the path g is equivalent to f, if there exists a mapping φ of the set $\tilde{f}$ onto $\tilde{g}$, such that:

(a) φ is the similarity correspondence of the ordered sets $\tilde{f}$ and $\tilde{g}$;
(b) for any point $X \in \tilde{f}$ its support coincides with the support of its image $\varphi(X) = Y \in \tilde{g}$.

The fact that the paths f and g are equivalent will be denoted as $f \sim g$.

If the paths $f : [a, b] \to M$ and $g : [c, d] \to M$ are equivalent, then any mapping φ obeying the conditions (a) and (b) given above will be called a connecting correspondence of the paths f and g.

The property of equivalence of two parametrized curves, as can easily be verified, is symmetrical, reflective and transitive and, as a result, the totality of all parametrized curves falls into classes of equivalence. As to the parametrized curves belonging to one class, let us assume that they define one and the same curve and call them the parametrizations of this curve. Formally, this means that the curve is a class of equivalence of a set of parametrized curves, which corresponds to a notion of equivalence defined here.

A curve is uniquely defined by giving one of its parametrizations and, thus, we are going to use the following form of notation: $K\{f : [a, b] \to M\}$, where the letter K denotes the given curve, and the expression in braces denotes one of its parametrizations.

Let K be a curve, and let $f : [a, b] \to M$ be its arbitrary parametrization. Let us define through f the function $f^* : [-b, -a] \to M$, setting $f^*(u) = f(-u)$ for $u \in [-b, -a]$. Let us denote by K^* the curve, one of whose parametrizations is the path f^*. The curve K^* is independent of the choice of parametrization f of the curve K. It results from the fact that if $f:[a, b] \to M$ and $g:[c, d] \to M$ are two equivalent paths in M, then the paths of $f^*:u \mapsto f(-u)$, $g^*:v \mapsto g(-v)$ are also equivalent. Let us assume that the curve K^* is obtained by changing the orientation of the curve K. It is obvious that through changing the orientation of the curve K^* we obtain the initial curve K.

In the literature one can sometimes find another definition of a curve which arises when, in the definition given above, the equivalence of the paths is understood in the following way. The paths $f:[a, b] \to M$ and $g:[c, d] \to M$ are equivalent if there is a mapping $\varphi \to \tilde{f}$ onto $\tilde{g}$, such that (a) φ is either the similarity correspondence or the anti-similarity correspondence of the ordered sets $\tilde{f}$ and $\tilde{g}$; (b) for any point $X \in \tilde{f}$ the supports X and $\varphi(X)$ coincide. The only difference from the previous definition of equivalence is that in the latter case the correspondence φ in the condition (a) can be anti-similarity.

Let $f:[a, b] \to M$ be an arbitrary path. Let us denote through K the class of paths which are equivalent to f in the sense of our previous definition, and through $\hat{K}$ the class of paths equivalent to f in the sense of the new definition. In this case, as is easily proved, $\hat{K} = K \cup K^*$.

A curve in the sense of the definition arising under a new formulation of path equivalence will be called a non-oriented curve.

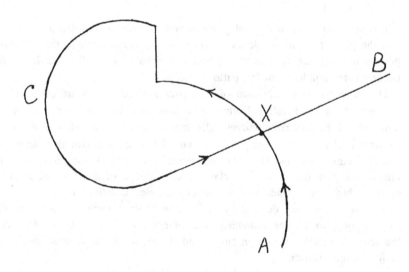

Fig. 1.

1.1.2. Let us now define the notion of a curve point. A point of the curve is not understood as a point of the curve as a set but as something different. For instance, on Figure 1 the point X should be considered as two points of the curve depending in the way this point is approached along the curve - either via the arc AX or the arc CX. A more appropriate term here would be 'a place on the curve'.

LEMMA 1.1.1. *If two parametrized curves are equivalent, then there exists only one similarity correspondence between those points at which the supports of the corresponding points coincide.*

Proof. Let $f:[a, b] \to M$ and $g:[c, d] \to M$ be two equivalent parametrized curves, $\tilde{f}$ be a set of points of the first curve, $\tilde{g}$ be a set of points of the second curve. Let us assume that there exist two different similar correspondences $\varphi_1:\tilde{g} \to \tilde{f}$ and $\varphi_2:\tilde{g} \to \tilde{f}$. Then to a certain point $q \in \tilde{g}$ in the correspondence φ_1 there corresponds a point $p_1 \in \tilde{f}$, and in the correspondence φ_2 a different point p_2. Let $p_2 < p_1$. Let us set $\varphi = \varphi_2 \circ \varphi_1^{-1}$. In this case φ is a mapping of the similarity $\tilde{f}$ onto itself, such that the supports p and $\varphi(p)$ coincide for any $p \in \tilde{f}$ and $\varphi(p_1) = p_2$. Let $p_2 = f(t_2)$, $p_1 = f(t_1)$. As far as $p_1 \neq p_2$, the function f is not constant in the segment $[t_2, t_1]$ and, hence, there exists t' such that $f(t') \neq f(t_1) = f(t_2)$. Let p_1' be a point corresponding to the value of the parameter t', $p_1' = f(t)$. Let us define the sequence of the points (p_m) and (p_m'), setting $\forall m \; p_{m+1} = \varphi(p_m)$, $p_{m+1}' = \varphi(p_m')$. We have: $p_2 < p_1' < p_1$. From this, by way of induction, we get $\forall m \; p_{m+1} < p_m' < p_m$. In this case the supports of all points p_m coincide with the support A of the point p_1, while the supports of the points p_m' coincide with the support B of the point p_1', $A \neq B$. Let $p_m = f(t_m)$, $p_m' = f(t_m')$. In this case $t_{m+1} < t_m' < t_m$ at every m. The sequences $(t_m)_{m \in N}$ and $(t_m')_{m \in N}$ converge to a certain limit t_0. Due to the continuity of f at $m \to \infty \; f(t_m) \to f(t_0)$, $f(t_m') \to f(t_0)$. This, however, contradicts the fact that at every $m \; f(t_m') = B \neq A = f(t_m)$. The contradiction obtained proves the lemma.

Lemma 1.1.1 yields the fact that every point of an arbitrary parametrization of the curve defines a single point in any other parametrization, namely, the point that corresponds to it due to the similar correspondence discussed in the definition of equivalence. As to these points, let us assume that they define one and the same point of the curve, or, in other words, they are the images of the same point of the curve. A point of the curve is totally defined by the indication of the parametrization point, and, therefore, we can write: the point $f(t)$ of the curve K and so on, identifying here a point of the curve with the point of its parametrization.

All points in different parametrizations of a given curve, which represents the given point P of the curve have one and the same support. It will be called the support of the point P and be denoted through the symbol $|P|$.

Formally, a point of a curve is a function P which puts in correspondence to every parametrization a certain point of this parametrization, in which case if $f:[a, b] \to M$ and $g:[c, d] \to M$ are two parametrizations, then $P(f)$ and $P(g)$ correspond to one another at such a similarity mapping of the set of the points of one parametrization onto the set of the points of the another one, which is required by the definition of equivalency.

The points of a curve are ordered if they are considered to be located in the same order as the parametrization points representing them. When writing the order of the points of a curve, we shall use the same terminology and the same notations as we did in the case of parametrized curves.

The definitions so far formulated do not rule out curves consisting of a single point, which will be called degenerate.

Let K be a non-degenerate curve. Among its points there is, obviously, the utmost left point and the utmost right one. We shall call them the end points of the curve, the first point being the beginning, and the second one the end. The points of the curve which are not end points are referred to as interior points.

The totality of all points of the curve located between the two given points are called the curve arcs. An arc is called closed or open depending on whether the utmost end points are ascribed to it. By analogy with the segments of a real axis, we shall denote the arcs with the end points X and Y in the following way: the closed arc through $[X, Y]$, the open one through (XY), and the semi-open one through $[XY)$ and $(XY]$. If the curve parametrization is given, then any closed arc of this curve is totally defined by giving the segment $[\alpha, \beta]$, for which $f(\alpha)$ and $f(\beta)$ are the end points of the considered curve. A closed arc of the curve will, as a rule, be identified with the curve having the following parametrization:

$$f:t \in [\alpha, \beta] \mapsto f(t) \in M$$

Let K be an arbitrary curve, A be its beginning, B its end, and X its arbitrary point. The left (right) semi-neighbourhood of the point X is any semi-open arc of type $(YX]$ ($[XY)$, respectively), where $Y < X$ ($Y > X$). The neighbourhood of an interior point of the curve is an arbitrary open arc containing this point. The neighbourhood of the point A which is the beginning of the curve is any right-hand semi-neighbourhood of this point. Analogously, the neighbourhood of the point B is its left-hand semi-neighbourhood.

Here follows the proposition which will be referred to as the Borel lemma.

LEMMA 1.1.2. *Then there exists a finite sequence of the points* $X_1, X_2, \ldots,$ X_m, *such that the neighbourhoods* $U(X_1), U(X_2), \ldots, U(X_m)$ *contain all the points of the curve.*

Lemma 1.1.2 is readily deduced from a common Borel lemma for a segment of line by going over to the curve parametrization.

Let (Y_m), $m = 1, 2, \ldots$, be an arbitrary sequence of the curve points. Let us establish that the sequence (Y_m) converges along the curve K to the point X

of the curve K, if for any neighbourhood of the point X there is an m_0, such that for $m \geqslant m_0$, the point Y_m belongs to this neighbourhood. If in this case, for all m, $Y_m < X$ ($Y_m > X$), then we say that $Y_m \to X$ from the left (from the right, respectively).

It can be easily shown that if the points (Y_m) converge to the point X along the curve, then their supports converge to the support of X. The inverse statement is not valid, which can be proved by the simplest examples (see, for instance, Figure 2).

The set of points of the space M, at each of the points of which at least one point of the curve K lies, is called the support of the curve and is denoted through $|K|$. A support of any curve is a continuous image of a closed segment of a line and is, therefore, a compact set.

Let us assume that M is a metric space and ρ is its metric. Let K be a curve in the space M. The diameter of the curve K is the least upper boundary d(K) of the distances $\rho(X, Y)$ between its points X and Y. In so far as the support of the curve is a compact set, then there exist the points X_0 and Y_0 of the curve, such that

$$\rho(X_0, Y_0) = \mathrm{d}(K).$$

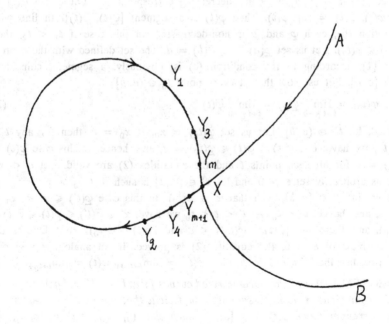

Fig. 2.

A point of the curve is called multiple if there exists another point, different from it, coinciding with the first one by its position in space. The curve is called an arc if it has no multiple points. A non-degenerate curve with its end points coinciding as space points, is called a loop.

1.1.3. In some cases a criterion of equivalence of two parametrized curves proves to be useful. This criterion is involved in Theorem 1.1.1, which will be proved below.

Let us introduce the notion of non-decreasing correspondence. Let there be an segment $[a, b] \subset \mathbb{R}$. Let us assume that to any point $t \in [a, b]$ there corresponds a certain subset $\varphi(t)$ of the set R. Let us say that it defines a non-decreasing correspondence $\varphi(t)$, $t \in [a, b]$, provided the following conditions are met:

(a) for any $t \in [a, b]$ the set $\varphi(t)$ is not empty and is, by itself, a closed interval of the set $\mathbb{R}$ (probably degenerating into a point);

(b) if $t_1 < t_2$, then for any $x \in \varphi(t_1)$ and for any $y \in \varphi(t_2)$ the inequality $x \leqslant y$ holds;

(c) the set

$$\bigcup_{t \in [a,b]} \varphi(t) \tag{1}$$

is a segment.

Let $\varphi(t)$, $t \in [a, b]$ be a non-decreasing correspondence. Let us set $\underline{\varphi}(t) =$ inf $\varphi(t)$, $\overline{\varphi}(t) =$ sup $\varphi(t)$. Then $\varphi(t)$ is a segment $[\underline{\varphi}(t), \overline{\varphi}(t)]$. In line with condition (b), each $\underline{\varphi}$ and $\overline{\varphi}$ is non-decreasing. In this case if $t_1 < t_2$, then $\underline{\varphi}(t_1) \leqslant \overline{\varphi}(t_2)$. Let us set $\underline{\varphi}(a) = c$, $\overline{\varphi}(b) = d$. The set defined with the expression (1), according to the condition (c) is, obviously, a segment, coinciding with $[c, d]$. Let us show that at every point $t_0 \in (a, b)$

$$\underline{\varphi}(t_0) = \lim_{t \to t_0 - 0} \underline{\varphi}(t) = \lim_{t \to t_0 - 0} \overline{\varphi}(t). \tag{2}$$

Indeed, let $t_0 \in (a, b]$. Let us set $\underline{\varphi}(t_0) = x_0$. If $x_0 = c$, then for any $t \in [a, t_0)$ we have: $c \leqslant \underline{\varphi}(t) \leqslant \overline{\varphi}(t) \leqslant \underline{\varphi}(t_0) = c$, and, hence, in this case $\underline{\varphi}(t) = \overline{\varphi}(t) = c$ for all such points t and the equalities (2) are valid. Let $c < x_0$. Let us arbitrarily set $\varepsilon > 0$ and let $x' \in [c, d]$ is such that $x_0 > x' > x_0 - e$. Let us find $t' \in [a, b]$, such that $x' \in \varphi(t_0)$. In this case $\underline{\varphi}(t') \leqslant x' < x_0 = \underline{\varphi}(t_0)$, and, hence, $t' < t_0$. At $t' < t < t_0$ we have: $x' \leqslant \overline{\varphi}(t') \leqslant \underline{\varphi}(t) \leqslant \overline{\varphi}(t) \leqslant \varphi(t_0)$, and, therefore, $|\overline{\varphi}(t) - \underline{\varphi}(t_0)| < \varepsilon$ and $|\underline{\varphi}(t) - \underline{\varphi}(t_0)| < \varepsilon$. Due to the arbitrariness of $\varepsilon > 0$, the equality (2) is proved. In an analogous way one can establish that if $a \leqslant t_0 < b$, then $\overline{\varphi}(t_0) = \lim_{t \to t_0 + 0} \underline{\varphi}(t) = \lim_{t \to t_0 + 0} \overline{\varphi}(t)$.

THEOREM 1.1.1. *For the parametrized curves $f : [a, b] \to M$ and $g : [c, d] \to M$ to be equivalent, it is necessary and sufficient that there be a non-decreasing correspondence $\varphi(t)$, $t \in [a, b]$, such that $\bigcup_{t \in [a,b]} \varphi(t) = [c, d]$, the function g on the interval $\varphi(t)$ being constant at any t, in which case $g[\varphi(t)] = f(t)$. (In other words, for all $u \in \varphi(t)$, $g(u) = f(t)$).*

Proof. Let us assume that the parametrized curves $f:[a, b] \rightarrow M$ and $g:[c, d]$ $\rightarrow M$ be equivalent. Then they define a certain curve K for which they serve as parametrizations. Let us arbitrarily choose $t \in [a, b]$. Let X be a point of the curve K which corresponds in the parametrization of t. Let α be the least and β be the greatest of the values of t', t'', respectively, such that $t' \leqslant t$ $\leqslant t''$ and the function f in the segment $[t', t'']$ is constant. Let $[\gamma, \delta]$ be the totality of the values of parameter u corresponding to the point X in the parametrization of g. Let $\alpha < \beta$. Let us define the function $\tilde{\varphi}(t)$ setting $\tilde{\varphi}(t) = kt + l$ for $t \in [\alpha, \beta]$, where the constants k, l are chosen in such a way that $\tilde{\varphi}(\alpha) = \gamma$, $\tilde{\varphi}(\beta) = \delta$, and let us set $\varphi(t) = \{\tilde{\varphi}(t)\}$ for $t \in [\alpha, \beta]$. It is obvious that $f(t) = g[\varphi(t)]$ for all $t \in [\alpha, \beta]$. If $\alpha = \beta$, we set $\varphi(t) = [\gamma, \delta]$, in which case also $f(t) = g[\varphi(t)]$. In both cases $\bigcup_{t \in [\alpha, \beta]} \varphi(t) = [\gamma, \delta]$. Let us show that the correspondence φ defined in this way is non-decreasing. Let $t_1 < t_2$, X_1 and X_2 be the curve points corresponding to t_1 and t_2 in the parametrization f, $[\alpha_1, \beta_1]$ and $[\alpha_2, \beta_2]$ be the segments formed by all the values of the parameter corresponding to X_1 and X_2 in the parametrization f. If $[\alpha_1, \beta_1]$ $= [\alpha_2, \beta_2] = [\alpha, \beta]$, then $X_1 = X_2$. In this case at $t \in [\alpha, \beta]$ $\varphi(t)$ consists of a single element which equals $kt + l$, where $k, l = \text{const}$, $k > 0$. The condition $t_1 < t_2$ yields that $kt_1 + l < kt_2 + l$, and, hence, in this case the condition (b) of the definition of a non-decreasing correspondence is fulfilled. Let us assume that the segments $[\alpha_1, \beta_1]$ and $[\alpha_2, \beta_2]$ are different. Then $\beta_1 < \alpha_2$, the points X_1 and X_2 of the curve K are different, in which case $X_1 < X_2$. Let $[\gamma_1, \delta_1]$ and $[\gamma_2, \delta_2]$ be the intervals consisting of all the values of the parameter and corresponding to the points X_1 and X_2, respectively, in the parametrization of g. It gives $\delta_1 < \gamma_2$, $\varphi(t_1) \subset [\gamma_1, \delta_1]$, $\varphi(t_2) \subset [\gamma_2, \delta_2]$, which means that the condition (b) of the definition of a non-decreasing correspondence is fulfilled in this case.

Let us show that $\bigcup_{t \in [a, b]} \varphi(t) = [c, d]$. Let us arbitrarily take $u \in [c, d]$ and let X be the point of the curve K, which corresponds this u in the parametrization g. Let $[\alpha, \beta]$ and $[\gamma, \delta]$ be the segments consisting of all the values of the parameter corresponding to X in the parametrizations f and g, respectively. In this case we have: $v \in [\gamma, \delta]$, $[\gamma, \delta] = \bigcup_{t \in [\alpha, \beta]} \varphi(t)$, and, hence, to the allowances made for $\varphi(t) \subset [c, d]$ which is valid at any t we have: $\bigcup_{t \in [a, b]} \varphi(t) = [c, d]$, and, thus, the condition (c) of the definition of a non-decreasing correspondence is also fulfilled.

Now, when the necessity condition of the theorem has been proved, let us prove the sufficiency condition.

Let us assume that there exists a non-decreasing correspondence $\varphi(t)$, $t \in [a, b]$, such that $\bigcup_{t \in [a, b]} \varphi(t) = [c, d]$ and $f(t) = g[\varphi(t)]$ for all $t \in [a, b]$. To the point $f(t)$ of the path f let there be a corresponding point $g(u)$, where $u \in \varphi(t)$, of the path g. Therefore, we shall get a similar mapping of the ordered set $\tilde{f}$ onto the set $\tilde{g}$. In this case to every point from $\tilde{f}$ there will correspond a point from $\tilde{g}$, having the same support. In line with the definition, it means that the paths f and g are equivalent.

The theorem is proved.

Let us now define the notion of a curve image with respect to the continuous mapping of the space wherein the curve is given.

Let M_1 and M_2 be topological spaces, $F:M_1 \to M_2$ be a continuous mapping. Let $f:[a, b] \to M_1$ be an arbitrary path in M_1. Then $F \circ f$ is a certain path in M_2. Let us prove that if the paths $f:[a, b] \to M_1$ and $g:[c, d] \to M_2$ are equivalent, then the paths $F \circ f$ and $F \circ g$ are also equivalent. Let us make use of the equivalency criterion of Theorem 1.1.1. Let us assume that the paths $f:[a, b] \to M_1$ and $g:[c, d] \to M_2$ are equivalent. Then there exists a non-decreasing correspondence $\varphi:[c, d] \to \mathbb{R}$ which maps the segment $[c, d]$ onto $[a, b]$, such that $g(u) = f[\varphi(u)]$ for all $u \in [c,d]$. But in this case, obviously, the relation

$$F[g(u)] = F\{f[\varphi(u)]\}$$

holds for all $u \in [c, d]$. Therefore, the parametrized curves $F \circ g$ and $F \circ f$ are equivalent.

Let us suppose that in the space M_1 there is a curve K, and let $F: M_1 \to M_2$ be a continuous mapping. If f and g are two arbitrary parametrizations of K, then due to what we have proved above, the paths $F \circ f$ and $F \circ g$ are equivalent, i.e. they are the parametrizations of a certain curve in M_2. The mapping F, consequently, transforms (according to the rule $f \mapsto F \circ f$) the totality of all parametrizations of the curve K in M_1 into a set of parametrizations of a certain curve in the space M_2, which will be denoted by the symbol $F(K)$ and called the image of the curve K at the mapping $F:M_1 \to M_2$.

It is obvious that if $F = I:M \to M$ is the identical mapping of M onto itself, then for any curve K in M $I(K) = K$. Let there be continous mappings $F:M_1 \to M_2$, $G:M_2 \to M_3$ given. In this case for any curve K in M_1 we have: $G[F(K)] = (G \circ F)(K)$.

1.1.4. Let us define the notion of a closed curve. First of all, we define a certain relation on a manifold of loops of the space M, i.e., let K_1 and K_2 be two arbitrary loops in M, A_1 and B_1 be the terminal points of K_1, A_2 and B_2 be those of K_2. The loops K_1 and K_2 are called equicomposed provided they either coincide or one can find the points C_1 on K_1 and C_2 on K_2, such that the arc $[A_1 C_1]$, as a curve in M, coincides with the curve $[C_2 B_2]$, and the arc $[C_1 B_1]$ coincides with the curve $[A_2 C_2]$. In other words, K_1 and K_2 are equicomposed if K_2 can be obtained from K_1 by way of cutting K_1 at the point C_1 and glueing the arcs $[A_1 C_1]$ and $[C_1 B_1]$ together by the coinciding spatial points A_1 and B_1. The relation of equicomposition is reflexive, symmetrical and transitive.

Let K_1 and K_2 be two arbitrary loops in M. In this case if they are equicomposed, we shall say that K_1 and K_2 define one and the same closed curve K in M, and that K is obtained from K_1 and K_2 by way of bringing the terminal point into coincidence.

Let K be a closed curve. K' be the loop from which K is obtained by superposing the terminal points, $x:[a, b] \rightarrow M$ be an arbitrary parametrization of K'; the function x is also called a parametrization of K. We have: $x(a) = x(b)$. Let us assume that the function x is periodical (with the period $T = b - a$) continued onto all $\mathbb{R}$, so that $x(t + T) = x(t)$ for all t. The operation of cutting and glueing used in the definition of loops equicomposition is analogous to finding an arbitrary segment $[p, q] \subset \mathbb{R}$, such that $q - p = T$ and considering the function $x(t)$ limitations on the segment $[p, q]$.

1.2. Normal Parametrization of a Curve

1.2.1. The parametrized curve $f:[a, b] \rightarrow M$ is called normal if f has no segment of constancy, different from one-point sets. The parametrization of the curve K is called normal if it is a normal parametrized curve.

LEMMA 1.2.1. *An ordered set of points of any non-degenerate path is identical to the segment* $[0, 1]$ *of a real axis.*

Proof. Let $f:[a, b] \rightarrow M$ be an arbitrary non-degenerate path, $\tilde{f}$ be an ordered set of the points of the path f. The task is to prove that $\tilde{f}$ is similar to the segment $[0, 1]$. The proof rests on the following characteristic property of the sets similar to the segment $[0, 1]$, the proof of which is not given here.

Let S be an ordered set. Let us assume that S has the following properties:

(a) S is linearly ordered;
(b) there exists a countable set $I \subset S$ which is everywhere dense in S, i.e. such that for any $x, y \in S$, where $x < y$, the segment (x, y) contains the elements of the set T;
(c) for any sequence of segments $([a_n, b_n])$, $n = 1, 2, \ldots$, such that $[a_n, b_n] \supset [a_{n+1}, b_{n+1}]$ for all n there exists a point $x \in S$ belonging to all the segments of the sequence;
(d) the set S has both the least and the greatest elements.

In this case S is similar to the segment $[0, 1]$.

It can be easily verified that the set $\tilde{f}$ obeys all the above listed conditions (a) – (d), which results in the lemma.

COROLLARY. *Any non-degenerate curve has normal parametrization.*

Proof. Let $f:[a, b] \rightarrow M$ be an arbitrary parametrization of a non-degenerate curve K. The ordered set $\tilde{f}$ is similar to the segment $[0, 1]$. Let ξ^* be a similarity mapping of the segment $[0, 1]$ onto the ordered set $\tilde{f}$. For any $t \in [0, 1]$ $\xi^*(t)$ is a pair $\{X, \tau\}$, where $X \in S(K)$, and τ is a connected component of the set $f^{-1}(X)$. If we set $\xi(t) = X$, then we define the mapping of the segment $[0, 1]$ onto the set $S(K) = f([a, b])$. We can easily establish that ξ is continuous. Therefore, $\xi:[0, 1] \rightarrow M$ is a parametrized curve in M. It can be readily shown that it is a normal parametrization of the given curve K. The lemma is proved.

LEMMA 1.2.2. *If K is a non-degenerate curve in the metric space M with the metric ρ, then for any of its parametrizations $f:[a, b] \rightarrow M$ by any $\varepsilon > 0$ we can find a normal parametrization $f^*:[a, b] \rightarrow M$, such that at all t $\rho[f(t), f^*(t)] < \varepsilon$.*

Proof. By $\varepsilon > 0$ we find the sequence of the points $t_0 = a < t_1 < t_2 < \cdots < t_m = b$ such that the function $f(t)$ is not constant in any of the segments $[t_{i-1}, t_i]$ and the diameters of arcs $f(t)$, $t_{i-1} \leqslant t \leqslant t_{i+1}$ are less than ε. The parametrized curve $f(t)$, $t \in [t_{i-1}, t_i]$ defines a non-degenerate arc of the curve K. In line with Lemma 1.2.1, this arc has a normal parametrization. Linear transformation of the parameter can result in the fact that the domain of setting this parametrization $f_i(t)$ will be the segment $[t_{i-1}, t_i]$. In $[a, b]$ let us define the function f^*, setting $f^*(t) = f_i(t)$ at $t \in [t_{i-1}, t_i]$, $i = 1, 2, \ldots, m$. The function f^* is the desired normal parametrization of the curve K.

1.3. Chains on a Curve and the Notion of an Inscribed Polygonal Line

1.3.1. Let K be an arbitrary curve in the metric space M. A chain on the curve K is any finite sequence of the points $\xi = \{X_1, X_2, \ldots, X_m\}$ of the curve K, such that $X_1 < X_2 < \cdots < X_m$. The points X_i are called the vertices of the chain. The biggest of the diameters of the arcs into which the curve K is divided by the vertices of the chain ξ is called a module of the chain and is denoted through $\lambda(\xi)$.

In a Euclidean space the curve K is called polygonal if one can find on it a chain $\xi = \{X_0 = A < X_1 < \cdots < X_m = B\}$, where A is the beginning of the curve, B is its end, and such that every arc $[X_{i-1}, X_i]$ is simple and lies in one straight line. The points X_i are called the vertices of the polygonal line and the arcs $[X_{i-1}, X_i]$, $i = 1, 2, \ldots, m$, are called its links.

Let K be an arbitrary curve in an n-dimensional Euclidean space and L be a polygonal line with the vertices $X_0 < X_1 < \cdots < X_m$. We say that the polygonal line L is inscribed into the curve K if on the latter there can be found a chain $\xi = \{Y_0, Y_1, \ldots, Y_m\}$ such that for any i Y_i coincides with X_i as a space point. Such a chain is, generally speaking, not unique. Henceforths, when referring to the polygonal line inscribed in a curve, we shall assume, if the opposite is not the case, that one of the given chains is fixed for the polygonal line in question. The module of ξ is called the module of the inscribed polygonal line L and is designated through $\lambda_K(L)$. The index K will henceforth mostly be omitted in cases when this results in no ambiguity.

Hereafter we are, in particular, going to consider the curves lying on a unit sphere of a Euclidean space, when it is expedient to consider, alongside with common polygonal lines, the so-called spherical polygonal lines. The curve K lying on the sphere S^{n-1} is called a spherical polygonal curve if on it one can find a sequence of points

$$X_0 = A < X_1 < X_2 < \cdots < X_m = B$$

where A is the beginning and B is the end of the curve K, such that each arc $[X_{i-1}, X_i]$ is simple and lies on the semicircle of a certain big circle.

By analogy with what has been written above, we can define the notion of an inscribed spherical polygonal line and that of the module of an inscribed spherical curve, employing the same notations as in the case of common polygonal lines.

With the view of studying curves, various numerical characteristics are ascribed to them. One of the ways of introducing these characteristics is as follows: first we define some functions of the chains formed by the curve points, then, through the limiting transition at $\lambda(\xi) \to 0$ we get the values directly referring to the curve under consideration. Let us define some notions pertaining to the latter stage, i.e., the limiting transition.

Let there be a certain function $f(\xi)$ given on the totality of all chains ξ of the curve K. We say that the number a is the limit of the function $f(\xi)$ at $\lambda(\xi) \to 0$, $a = \lim_{\lambda(\xi) \to 0} f(\xi)$, if for any $\varepsilon > 0$ one can find such $\delta > 0$ that if $\lambda(\xi) < \delta$ then $|f(\xi) - a| < \varepsilon$. We say that $\lim_{\lambda(\xi) \to 0} f(\xi) = \infty$, if for any $\varepsilon > 0$ one can find $\delta > 0$ such that at $\lambda(\xi) < \delta$ $f(\xi) > 1/\varepsilon$. We say that $\lim_{\lambda(\xi) \to 0} f(\xi) = -\infty$, if $\lim_{\lambda(\xi) \to 0} [-f(\xi)] = \infty$.

Let us also define the notions of the upper and lower limits of the function $f(\xi)$ at $\lambda(\xi) \to 0$. For $t = 0$, let $\underset{\sim}{f}(t) = \inf_{\lambda(\xi) < t} f(\xi)$, $\tilde{f}(t) = \sup_{\lambda(\xi) < t} f(\xi)$. It is obvious that $\tilde{f}$ is a non-decreasing, and $\underset{\sim}{f}$ is a non-increasing function of t. Therefore, there exist finite or infinite limits.

$$\tilde{F} = \lim_{t \to 0} \tilde{f}(t), \qquad \underset{\sim}{F} = \lim_{t \to 0} \underset{\sim}{f}(t)$$

which will be called, respectively, the upper and the lower limits of the function $f(\xi)$ at $\lambda(\xi) \to 0$, denoting

$$\tilde{F} = \varliminf_{\lambda(\xi) \to 0} F(\xi), \qquad \underset{\sim}{F} = \varlimsup_{\lambda(\xi) \to 0} f(\xi).$$

For any function $f(\xi)$ the lower limit does not exceed its upper limit, and for the function f to have the limit in the common sense of the word, it is necessary and sufficient that its lower and upper limits coincide. In this case the limit of the function equals their general value.

In the case of curves in a Euclidean space, each chain on a curve defines a certain inscribed polygonal line. Conversely, by the condition, to every inscribed polygonal line there corresponds a certain curve chain. Each function $f(L)$, defined on a set of inscribed polygonal lines, can therefore be considered as the function $\varphi(\xi) = f(L)$, defined on the set of all curve chains, where ξ is the chain formed by the vertices of the polygonal line L. The limit of the function $f(L)$ at $\lambda(L) \to 0$ will be viewed as the limit of the function

$\varphi(\xi)$ at $\lambda(\xi) \to 0$. In an analogous way one can view the upper and lower limits of the function $f(L)$ at $\lambda(L) \to 0$.

1.4. Distance Between Curves and Curve Convergence

1.4.1. Let K and L be two curves in the metric space M with the metric ρ, $f(t)$ and $g(t)$ be their arbitrary parametrizations defined on one and the same segment $[a, b]$. Let us define the quantity $\max_{a \leqslant t \leqslant b} \rho[f(t), g(t)]$. The exact lower boundary of this maximum taken by all the pairs of the parametrizations of the curves K and L defined on the same segment, is called the distance between the curves K and L, and is defined through $\rho(K, L)$.

Let us make some notes before going on to consider the properties of the distance.

If both curves K and L do not degenerate into a point, then, while determining the distance, one can consider only normal parametrizations, since for any parametrization there exists, in line with Lemma 1.2.2, a normal parametrization arbitrarily close to it.

Then, when determining $\rho(K, L)$ one can fix the parametrization of one of the curves K and L, i.e. if $f : [a, b] \to M$ is an arbitrary parametrization of K, then

$$\rho(K, L) = \inf_{g} \left\{ \max_{a \leqslant t \leqslant b} \rho[f(t), g(t)] \right\}$$

where the lower boundary is taken by all the parametrizations g of the curve L, the segment $[a, b]$ serving as the basis for setting them. Indeed, let us arbitrarily fix the parametrization $f(t)$ of the curve K. Let $f_1(t)$, $g_1(t)$ be an arbitrary pair of normal parametrizations of the curves K and L. Then $f(t) = f_1[\varphi(t)]$, where $\varphi(t)$ is a continuous non-decreasing function of the parameter t. The function $g(t) = g_1[\varphi(t)]$ is continuous and is a parametrization of the curve L. At the same time:

$$\max_{a \leqslant t \leqslant b} \rho[f(t), g(t)] = \max_{a \leqslant t \leqslant b} \rho[f_1(t), g_1(t)]$$

which yields the required statement.

It should be remarked that if one of the curves K and L, for instance, the curve K, degenerates into a point, then $\rho(K, L)$ equals the greatest of the distances of the points of the curve L up to this point.

The following theorem shows that the totality of all the curves with the distance determined in such a way is a metric space.

THEOREM 1.4.1. *The distance between the curves has the following properties*:

(1) *for any K and L $\rho(K, L) \geqslant 0$, in which case $\rho(K, L) = 0$ if and only if $K = L$*;

(2) $\forall K, L$ $\rho(K, L) = \rho(L, K)$;

(3) $\forall K, L, Q$ $\rho(K, Q) \leqslant \rho(K, L) + \rho(L, Q)$.

Proof. (1) It is obvious that if $K = L$, then $\rho(K, L) = 0$. Let us assume that the curves K and L are such that $\rho(K, L) = 0$. Let us prove that in this case $K = L$. Let us arbitrarily fix the normal parametrizations $f_0:[a, b] \rightarrow M$ and $g_0:[a, b] \rightarrow M$ of the curves K and L and find the sequence $(g_m(t))$ of the parametrizations of the curve L, such that at $\forall m$ $\max_{a \leqslant t \leqslant b} \rho[f_0(t), g_m(t)] < 1/m$. In view of Theorem 1.1.1 $g_m(t) = g_0[\varphi_m(t)]$, where $\varphi_m(t)$ is a non-decreasing correspondence, mapping the segment $[a, b]$ onto itself. This correspondence is continuous; as to its points of discontinuity they should have corresponded the segments of constancy of the function $g_0(t)$ which is not the case since the $g_0(t)$ parametrization is normal. In view of the known Helli theorem of choice, the sequence $(\varphi_m(t))$ can yield the subsequence $(\varphi_{m_k}(t))$, $m_1 < m_2 < \cdots$, converging to a certain non-decreasing function $\varphi_0(t)$ at all the points of continuity of the latter. For all t, for which $\varphi_{m_k}(t) \rightarrow \varphi_0(t)$ we, obviously, have: $f_0(t) = g_0[\varphi_0(t)]$. Let us prove that φ_0 is continuous. Indeed, let us assume that it is not the case and, for instance, the point t_1, $a \leqslant t_1 \leqslant b$, is the point of discontinuity for the function $\varphi_0(t)$. Let $p = \varphi_0(t_1 - 0)$, $q = \varphi_0(t_1 + 0)$ (at $t_1 = a$ ($t_1 = b$) we set $p - \varphi_0(t_1)$ (and accordingly $q = \varphi_0(t_1)$). Then $p < q$. As far as the parametrization of $g_0(t)$ is normal, the diameter of δ of the arc $\{g_0(t), p \leqslant t \leqslant q\}$ is other than zero. Let us now find the numbers α and β, $\alpha < \beta$, such that $\alpha \leqslant t_1 \leqslant \beta$ and the diameter of the arc $\{f_0(t), \alpha \leqslant t \leqslant \beta\}$ is less than $\delta/4$ and, besides, at $k \rightarrow \infty$ $\varphi_{m_k}(\alpha) \rightarrow \varphi_0(\alpha)$, $\varphi_{m_k}(\beta) \rightarrow \varphi_0(\beta)$. Since $\max_{\alpha \leqslant t \leqslant \beta} \rho(f_0(t), g_0[\varphi_{m_k}(t)]) \rightarrow 0$ at $k \rightarrow \infty$, then at sufficiently large k the diameter of the arc $\{g_0[\varphi_{m_k}(t)], \alpha \leqslant t \leqslant \beta\}$ is less than $3\delta/4$. On the other hand, as can be easily seen, at sufficiently large k the image of the segment $[\alpha, \beta]$ at mapping of the segment $[a, b]$ onto itself, realized by the function $\varphi_{m_k}(t)$, will cover up the segment which is arbitrarily close to the segment $[p, q]$ and, hence, at large k the diameter of the arc $\{g_0[\varphi_{m_k}(t), \alpha \leqslant t \leqslant \beta\}$ will exceed $3\delta/4$. Thus, we have come to a contradiction and, hence, the continuity of the function φ_0 is proved.

Therefore, $f_0(t) = g_0[\varphi_0(t)]$ where $\varphi_0(t)$ is a continuous non-decreasing function, which proves the coincidence of the curves K and L.

The fulfillment of condition (2) results directly from the definition of the distance.

Let us prove property (3). Let K, L and Q be three curves, $f(t)$, $g(t)$ and $h(t)$, $a \leqslant t \leqslant b$, be their arbitrary parametrizations. For any t we have: $\rho[f(t), h(t)] \leqslant \rho[f(t), g(t)] + \rho[g(t), h(t)]$ which yields $\max_t \rho[f(t), h(t)] \leqslant \max_t \rho[f(t), g(t)] + \max_t \rho[g(t), h(t)]$. Let us fix the parametrization g, leaving the parametrizations f and h arbitrary. We have: $\rho(K, Q) \leqslant \max_t \rho[f(t), h(t)]$, which yields

$$\rho(K, Q) \leqslant \max_t \rho[f(t), g(t)] + \max_t \rho[g(t), h(t)].$$

Let us arbitrarily set $\varepsilon > 0$. The parametrizations f and h of the curves K and Q can be chosen in such a way that $\rho[f(t), g(t)] < \rho(K, L) + \varepsilon/2$, $\rho[g(t), h(t)]$

$< \rho(L, Q) + \varepsilon/2$ for all $t \in [a, b]$. It yields: $\rho(K, Q) \leqslant \rho(K, L) + \rho(L, Q) + \varepsilon$. In line with $\varepsilon > 0$ arbitrarity, it leads us to the conclusion that $\rho(K, Q) \leqslant \rho(KL) + \rho(L, Q)$. The theorem is proved.

1.4.2. Let us define the notion of curve convergence. We say that the sequence of the curves (K_m), $m = 1, 2, \ldots$, at $m \to \infty$ converges to the curve K, if $\rho(K_m, K) \to 0$ at $m \to \infty$. The definition of convergence can be formulated in another way. As has been noted earlier, by the definition of the distance one can fix the parametrization of one of the curves. Let $f:[a, b] \to M$ be the parametrization of the curve K. Let us find the parametrization $f_m:[ab] \to M$ of the curve K_m, such that all $t \in [a, b]$ $\rho[f(t), f_m(t)] \leqslant \rho(K, K_m) + 1/m$. At $m \to \infty$ $\rho(K, K_m) \to 0$. This yields the following Lemma.

LEMMA 1.4.1. *For the sequence of curves* (K_m), $m = 1, 2, \ldots$, *to converge to the curve K at $m \to \infty$, it is necessary and sufficient that the curves K and K_m allow the parametrization $f(t)$ and $f_m(t)$, $a \leqslant t \leqslant b$, such that at $m \to \infty$ the functions f_m uniformly converge to f in $[a, b]$.*

The necessity has already been proved, the sufficiency is obvious.

Let M be an arbitrary metric space, and ρ be the metric of this space. The sequence (X_m), $m = 1, 2, \ldots$, of the points of the space M is called converging in itself, if for any $\varepsilon > 0$ one can find a number $\bar{n}$, such that for any $m \geqslant \bar{n}$ and $n \geqslant \bar{n}$ $\rho(X_m, X_n) < \varepsilon$. If the sequence (X_m), $m = 1, 2, \ldots$, is converging in the common sense of the word, then it is converging in itself. The space M is called complete if any sequence in it which converges in itself is converging.

THEOREM 1.4.2. *If the metric space M is complete, then the set of all curves in M, characterized by the metrics defined above, is a complete metric space.*

Proof. Let M be a complete metric space, and (K_m), $m = 1, 2, \ldots$, be a sequence of curves converging in itself in M. The task is to prove that this sequence converges to a certain curve. Let us find an increasing sequence of integer numbers $n_1 < n_2 < \cdots < n_k < \cdots$ such that at $n \geqslant n_k$, $m \geqslant n_k$ $\rho(K_n, K_m) < 2^{-k-2}$. The curves $K_1, K_2, \ldots K_{n_1}$ are arbitrarily parametrized in such a way that the domain of the parameter changes is the segment $[0, 1]$. Let there be a parametrization $f_{n_k}:[0, 1] \to M$ of the curve K_{n_k} given. The parametrization $f_j:[0, 1] \to M$ of the curves K_j, where $n_K < j \leqslant n_{K+1}$ will be constructed in such a way that at all $t \in [0, 1]$ $\rho[f_j(t), f_{n_k}(t)] < 2^{-K-1}$. By way of induction we, therefore, have constructed certain parametrizations for all the curves of the sequence. If $m \geqslant n$, then

$$\rho[f_m(t), f_{n_k}(t)] < 2^{-k} \tag{3}$$

for all $t \in [0, 1]$. Indeed, let $r \geqslant k$ and $n_r < m \leqslant n_{r+1}$. Then we have:

$$\rho[f_{n_k}(t), f_m(t)] \leqslant \rho f_{n_k}(t), f_{n_{k+1}}(t)] + \cdots + \rho[f_{n_{r-1}}(t), f_{n_r}(t)] +$$

$$+ \rho[f_{n_r}(t), f_m(t)] < \frac{1}{2^{k+1}} + \frac{1}{2^{k+2}} + \cdots \frac{1}{2^{r+1}} < \frac{1}{2^k},$$

which is the required proof. It yields that at every t the sequence of points $(f_m(t))$, $m = 1, 2, \ldots$, converges in itself and, hence, has the limit $f(t)$. In line with (3) we have:

$$\rho[f(t), f_{n_k}(t)] \leqslant 2^{-k}$$

which gives

$$\rho[f(t), f_m(t)] \leqslant 2^{-k+1}$$

at $m \geqslant n_k$. The function f_m, therefore, uniformly converge to f. Thus, the function f is continuous and defines in the space a certain curve K, to which, in accordance with Lemma 1.4.1, the curves K_m converge at $m \to \infty$. The theorem is thus proved.

In a metric space the set A is termed relatively compact, if a converging subsequence can be obtained from any sequence of its points. The following theorem involves a criterion of a relative compactness of a family of curves.

Let A be an arbitrary family of curves in the space M. The curves of the family A are considered to be uniformly divisible if for any $\varepsilon > 0$ one can find a positive integer n, such that any curve $K \in A$ can be subdivided into n arcs, with their diameters less than ε.

THEOREM 1.4.3 (M. Frechet). *For the family of curves A in a complete metric space M to be relatively compact, it is necessary and sufficient that these curves be uniformly divisible and that there be a relatively compact set R, which contains all the curves of the family.*

Proof. Let us first remark that any converging sequence of curves (K_m), $m = 1, 2, \ldots$, obeys the conditions of the theorem. Indeed, let $f:[0, 1] \to M$ and $f_m:[0, 1] \to M$ be parametrizations of the curves K and K_m, such that at $m \to \infty$ $f_m(t) \to f(t)$ uniformly in $[0, 1]$. Let us take $\varepsilon > 0$ and find a sequence of points $t_0 = 0 < t_1 < t_2 < \cdots t_n = 1$, such that the diameter of the arc $[f(t_{i-1})f(t_i)]$ of the curve K is less than $\varepsilon/3$, $i = 1, 2, \ldots, n$. In view of the uniform convergence of the functions $f_m(t)$ to $f(t)$ there is an integer m_0, such that at $m \geqslant m_0$ $\rho[f_m(t), f(t)] < \varepsilon/3$ at all $t \in [0, 1]$. At such m the arc diameter $\{f_m(t), t_{i-1} \leqslant t \leqslant t_i\}$, $i = 1, 2, \ldots, n$, is less than ε and, hence, at $m \geqslant m_0$ the curve K_m is subdivided into n arcs with diameters less than ε. For $m \leqslant m_0$ let us subdivide K_m into the arcs of the diameter less than ε in an arbitrary way. Let n_m be the number of the arcs. If $n(\varepsilon)$ is the largest of the numbers $n_1, n_2, \ldots, n_{m_0}$, n, then each curve of the sequence can be divided into $n(\varepsilon)$ arcs with a diameter less than ε, and, hence, in view of $\varepsilon > 0$ being arbitrarity, the curves $K_1, K_2, \ldots, K_m \ldots$, are uniformly divisible. Uniform convergence of the functions $f_m(t)$ to $f(t)$ allows one to deduce that the set $S = \bigcup_{m=1}^{\infty} |K_m|$ is relatively compact.

Let the family of the curves A be relatively compact. If the curves of this family were not uniformly divisible, then for a certain $\varepsilon > 0$ there could be found a sequence (K_m), $m = 1, 2, \ldots, \forall m \ K_m \in A$, such that it would be impossible to divide the curve K_m into fewer than m arcs of diameter less than ε. In view of the relative compactness of A one can derive a converging subsequence from (K_m). Its curves K_{m_k}, due to its construction, are not uniformly divisible, which contradicts what we have proved above. The contradiction obtained demonstrates that the curves of the family A are uniformly divisible. Let R be the union of supporters of all the curves included in A. Let us assume that R is not a relatively compact set. In this case there is a sequence (X_m), $m = 1, 2, \ldots$, in it, which contains no converging subsequence. Let K_m be a curve of the family A, passing through the point X_m. Let us choose from the sequence (K_m) a converging subsequence $K_{n_1} < K_{n_2} < \cdots < K_{n_m} < \cdots$. The set

$$R' = \bigcup_{m=1}^{\infty} |K_{n_m}|$$

is relatively compact and, hence, from the sequence (X_{n_k}), $k = 1, 2, \ldots$, one can extract a converging subsequence, which contradicts the construction of the initial sequence (X_m), $m = 1, 2, \ldots$. The necessity of the theorem's condition is thus proved.

Let us prove the sufficiency. Let us say that the sequence (K_m), $m = 1, 2, \ldots$, of the curves ε-converges if there exists a number m_0, such that $m \geqslant m_0$, $n \geqslant m_0 \ \rho(K_m, K_n) < \varepsilon$. Let the curves of the family A be uniformly divisible and be contained in a relatively compact set R. Let us choose an arbitrary sequence (K_m), $m = 1, 2, \ldots$, of the curves from A and show that a converging subsequence can be extracted from it. To this end, let us first demonstrate that for any $\varepsilon > 0$ from (K_m) one can find an ε-converging subsequence. Let us find an n which corresponds, in line with the definition of uniform divisibility, to the number $\varepsilon/3$, and divide all the curves from A into n arcs of diameter less than $\varepsilon/3$. Let us now parametrize the curves from A in such a way that the domain of the changes of parameter t is the segment $[0, 1]$, and to the points of division there correspond the values $t = k/n$, $k = 0, 1, \ldots, n$. Let $f_m : [0, 1] \rightarrow M$ be a parametrization of the curve K_m. Let us choose from (K_m) a subsequence (K_{m_i}), such that at $m_i \rightarrow \infty$ the points $X_{m_i}(k/n)$, $k = 0, 1, \ldots, n$, converge, which is possible since the curves are contained in a relatively compact set R. Let m^* be such that at $m_i > m^* \ \rho[f_{m_i}(k/n), f_{m_j}(k/n)] < \varepsilon/3$ for any $k = 0, 1, 2, \ldots, n$. Let us arbitrarily set $t \in [0, 1]$. If we can find a k such that $k/n \leqslant t < (k+1)/n$, then at $m_i \geqslant m^*$, $m_j \geqslant m^*$

$$\rho[f_{m_j}(t), f_{m_i}(t)] \leqslant \rho[f_{m_j}(t), f_{m_j}(\tfrac{k}{n})] + \rho[f_{m_j}(\tfrac{k}{n}), f_{m_i}(\tfrac{k}{n})] +$$
$$+ \ \rho[f_{m_i}(\tfrac{k}{n}), f_{m_i}(t)] < \frac{\varepsilon}{3} + \frac{\varepsilon}{3} + \frac{\varepsilon}{3} = \varepsilon,$$

and hence $\rho(K_{m_i}, K_{m_j}) < \varepsilon$, i.e. K_{m_i} is an ε-converging sequence.

For $\nu = 1, 2, \ldots$ let us now set $\varepsilon_\nu = 1/\nu$. Let us extract from (K_m) and ε_1-

converging subsequence, and let this subsequence be K_{11}, K_{12}, ..., K_{1m}, ...
Let us now extract from the latter an ε_2-converging subsequence K_{21}, K_{22}, ...,
K_{2m}, ... From the subsequence just obtained we can extract an ε_3-converging
subsequence K_{31}, K_{32}, ..., K_{3m}, ..., and so on. The diagonal sequence K_{11},
K_{22}, ..., K_{mm}, ..., obviously converges in itself and, hence, in view of Theorem 1.4.2, it converges. The theorem is thus proved.

LEMMA 1.4.2. *Let K be a curve in the metric space M, (K_m), $m = 1, 2, ...,$ be a
sequence of curves in M. Let us assume that for any $\varepsilon > 0$ there is m_0, such
that at $m \geqslant m_0$ on the curves K and K_m one can construct the sequences of
points: X_1, X_2, ..., X_k on the curve K and X'_1, X'_2, ..., X'_k on the curve K_m,
such that $\rho(X_i, X'_i) < \varepsilon$ for all i and the diameters of the arcs into which the
curves K and K_m are divided by the points are less than ε. In this case at $m
\to \infty$ the curves K_m converge to the curve K.*

    Proof. The definition of m_0 yields that at $m \geqslant m_0$ $\rho(K_m, K) < 3\varepsilon$ and hence
$\lim_{m \to \infty} \rho(K_m, K) = 0$, which is the required proof.

COROLLARY. *Let K be a curve in an n-dimensional Euclidean space, and let
L_m, $m = 1, 2, ...,$ be an arbitrary sequence of the polygonal lines inscribed
in it. In this case, if at $m \to \infty$ $\lambda(L_m) \to 0$, then the polygonal lines L_m
converge to the curve K.*

1.5. On a Non–Parametric Definition of the Notion of a Curve

1.5.1. The definition by M. Frechet [8] given above (see also [3] and [10]) is
based on the notion of a curve parametrization. In line with this definition,
a curve in the metric space M is a continuous mapping into M of a segment
$[a, b]$ of the real axis, considered to the accuracy of monotonic transformations of the segment $[a, b]$.

    In this section we are going to show that the notion of a curve can be de-
fined without employing the notion of parametrization.

    Let M be an arbitrary metric space with the metric ρ. A chain in M is any
finite sequence ξ of compact sets of the space X_1, X_2, ..., X_p, numbered in a
definite order and such that for all $i = 1, 2, ..., p - 1$ $X_i \cap X_j \neq \varnothing$. The
largest of the diameters of the sets X_i will be called the chain module denoted through $\lambda(\xi)$.

    Let ξ_1 and ξ_2 be two arbitrary chains in M,

$$\xi_1 = \{X_1, X_2, ..., X_p\}, \xi_2 = \{Y_1, Y_2, ..., Y_q\}.$$

Let us say that the chain ξ_2 is inscribed into the chain ξ_1, using the notation $\xi_2 \succ \xi_1$, if there is such a sequence of the indices $i_0 = 1 < i_1 < i_2 <
\cdots < i_p = q + 1$, such that

$$\forall s = 1, 2, ..., p \quad X_s = \bigcup_{j = i_{s-1}}^{i_s - 1} Y_i.$$

A set Z of the chains is called a system of chains if it obeys the following conditions:

(1) if $\xi_1 \in Z$ and $\xi_2 \in Z$, then there exists a chain $\xi_3 \in Z$ inscribed in ξ_1 and in ξ_2;

(2) for any $\varepsilon > 0$ there exists a chain $\xi \in Z$, the module of which is less than ε.

A system of chains is called complete is it obeys one more condition:

(3) if the chain ξ_1 is inscribed into ξ_2 and $\xi_1 \in Z$, then $\xi_2 \in Z$ as well.

Let Z be an arbitrary system of chains. Let us denote by $\widetilde{Z}$ the set of all the chains in each of which there is at least one of the chains of the system Z inscribed. It can be easily seen that $\widetilde{Z}$ is a complete system of chains. Let us call $\widetilde{Z}$ a completion of Z.

The chain system Z is called closed if it obeys the following condition:

(4) there exists no chain system Z' different from Z, such that $Z \subset Z'$.

Let $f:[a, b] \longrightarrow M$ be an arbitrary parametrized curve in the space M. Let $t_0 = a \leqslant t_1 \leqslant \cdots \leqslant t_p = b$ be a sequence of points in the segment $[a, b]$. Let us denote by X_k the set $f([t_{k-1}, t_k])$. A sequence of the sets $(X_1, X_2, \ldots, X_p)$ forms a certain chain. We shall say that this chain is generated by a parametrized curve f through the system of points $t_0, t_1, \ldots, t_m$. Let Z be an arbitrary system of chains. If each of the chains of this system is generated by the given parametrized curve $f:[a, b] \longrightarrow M$, then the chain system Z is assumed to obey the parametrized curve f.

THEOREM 1.5.1. *For any system of chains Z there exists a parametrized curve $f:[a, b] \longrightarrow M$, to which the given system of chains obeys.*

If a system of chains obeys one of two equivalent parametrized curves, then it also obeys the second one. The reverse statement is also valid if for two parametrized curves there exists a system of chains, obeying each of them, then these parametrized curves are equivalent.

Proof. Let Z be an arbitrary system of chains. Let us prove that there exists a parametrized curve generating this system. Let us construct a sequence of chains (ξ_m), $m = 1, 2, \ldots$, of the system Z, such that ξ_{m+1} is inscribed into ξ_m at every m and $\lambda(\xi_m) \longrightarrow 0$ at $m \longrightarrow \infty$. Such a sequence can be constructed by induction with respect to m. The chain ξ_1 can be chosen arbitrarily. Let us suppose that the chain ξ_m has been constructed. Let us find a chain ξ_m', such that $\lambda(\xi_m) < 1/(m + 1)$. As ξ_{m+1}, let us choose an arbitrary chain inscribed in both ξ_m and ξ_m'.

Let us construct a sequence of divisions (η_m), $m = 1, 2, \ldots$, of the segment $[0, 1]$ of the real axis in such a way that the following conditions be fulfilled. The division η_m is formed by the points $t_0^m = 0 < t_1^m < \cdots < t_{k_m}^m = 1$, where k_m is the number of elements of the chain ξ_m. Let us correspond to the segment $[t_{i-1}^m, t_i^m]$, $i = 1, 2, \ldots, k_m$, an i-th set of the chain ξ_m. At every m

the division η_{m+1} is obtained by subdividing the segments of the division η_m in exactly the same way as the chain ξ_{m+1} has been obtained from the chain ξ_m. Now let $f_m(t)$ be a function such that, in the segment $[t_{i-1}^m, t_i^m)$ (at $i = k_m$ in the segment $[t_{i-1}^m, t_i^m]$), f_m is constant and coincides with one of the points of the set of the chain ξ_m which corresponds to the segment $[t_{i-1}^m, t_i^m]$. The function f_m is defined at every m. It should be remarked that at $k \geqslant m$ and $n \geqslant m$ the points $f_k(t)$ and $f_n(t)$ are the points of one and the same set of the chain ξ_m, and, hence, for such k and n $\rho[f_k(t), f_n(t)] < \lambda(\xi_m)$. As far as $\lambda(\xi_m) \to 0$ at $m \to \infty$, the sequence $f_m(t)$ uniformly converges to a certain function $f_0(t)$, $0 \leqslant t \leqslant 1$ at $m \to \infty$.

The function f_0 is continuous. Indeed, if t' and t'' belong to one segment $[t_{i-1}^m, t_i^m)$, then $\rho[f_n(t'), f_n(t'')] \leqslant \lambda(\xi_m)$ at any $n \geqslant m$. If t' and t'' belong to two neighbouring segments of this type, then $\rho[f_n(t'), f_n(t'')] \leqslant 2\lambda(\xi_m)$, since, by the definition, the neighbouring sets of the chains have common points. Therefore, if $|t' - t''| < \delta$, where δ is the length of the least of the segments $[t_{i-1}^m, t_i^m]$, then $\rho[f_n(t'), f_n(t'')] \leqslant 2\lambda(\xi_m)$ at any $n \geqslant m$ and, hence, $\rho[f_0(t'), f_0(t'')] \leqslant 2\lambda(\xi_m)$.

Let us prove that for any i the set B_i^m of the points $f_0(t)$, such that $t_{i-1}^m \leqslant t \leqslant t_i^m$ coincides with the corresponding set of the chain ξ_m. Indeed, each point $f_0(t)$ for $t \in [t_{i-1}^m, t_i^m]$ is a limit of the sequence of the points of the set X_i^m and, hence, belongs to X_i^m, since the set X_i^m is closed. Therefore, $B_i^m \subset X_i^m$. Now let x be an arbitrary point of the set X_i^m. For any $n \geqslant m$ we can find a set C_n of the chain ξ_n which contains the point x. According to the construction of the function $f_n(t)$ there is a point $t_n \in [t_{i-1}^m, t_i^m]$, such that $f_n(t_n) \in C_n$. At $n \to \infty$ $f_n(t_n) \to x$. From among the points (t_n) let us select a subsequence (t_{n_k}) converging to a certain point t_0. The point $f_{n_k}(t_{n_k}) \to f_0(t_0)$ and, hence, $x = f_0(t_0)$. It yields $X_i^m \subset B_i^m$ and consequently $X_i^m = B_i^m$.

Therefore, each of the chains ξ_m obeys the parametrized curve $f_0 : [0, 1] \to M$.

Now let ξ be an arbitrary chain of the system Z, $\xi = \{X_1, X_2,..., X_p\}$. Through ξ'_m let us denote a chain inscribed into ξ and ξ_m. Let $Y_{k_i+1}^m$, $Y_{k_i+2}^m,..., Y_{k_{i+1}}^m$ be the sets of the chain, a union of which yields the element X_i of the chain ξ. For $Y_{k_i+1}^m$ and $Y_{k_{i+1}}^m$ let us find the containing them elements of the chain ξ_m. Let $[\overline{\alpha}_i^m, \overline{\beta}_i^m]$ and $[\alpha_{i+1}^m, \beta_{i+1}^m]$ be the segments in the segment $[0, 1]$, the images of which in the space M by the mapping f_0, are the given elements of the chain ξ_m. The construction shows that the segment $[\alpha_i^m, \beta_i^m]$ is located to the left from the segment $[\overline{\alpha}_i^m, \overline{\beta}_i^m]$, adjoining it (or coinciding with it), while the segment $[\overline{\alpha}_i^m, \overline{\beta}_i^m]$ is located to the left from or coincides with the segment $[\alpha_{i+1}^m, \beta_{i+1}^m]$. Let us set $t_i^m = \alpha_i^m$ at $i \neq p$, $t_i^m = \beta_i^m$ at $i = p$. The image of the segment $[t_i^m, t_{i+1}^m]$ when mapping f_0 is a closed set X_i^m. In this case the symmetrical difference $(X_i^{(m)} \backslash X_i) \cup (X_i \backslash X_i^{(m)})$ is a sum of not more than two sets with diameters less than $\lambda(\xi_m)$. Let us construct a subsequence $m_1 < m_2 < \cdots < m_k \cdots$, such that at $k \to \infty$ there exists the limit

$$\lim_{k \to \infty} t_i^{m_k} = t_i, \quad i = 1, 2, \ldots, p.$$

It is easily seen that $0 = t_0 \leqslant t_1 \leqslant \cdots \leqslant t_{p-1} \leqslant t_p = 1$, in which case the image of the segment $[t_i, t_{i+1}]$, when mapping f_0, is the set X_i. Therefore we have proved that the chain ξ is generated by the parametrized curve $f : [0, 1] \to M$.

Thus we have constructed the parametrized curve which the system of chains Z obeys.

Let $f : [a, b] \to M$ and $g : [c, d] \to M$ be two equivalent parametrized curves. Let us assume that the system of chains Z obeys the parametrized curve f. Let us prove that in this case it also obeys the parametrized curve g. Indeed, to each of the sequences $a = t_0 \leqslant t_1 \leqslant \cdots \leqslant t_m = b$ there corresponds the sequence $c = u_0 \leqslant u_1 \leqslant \cdots u_m = d$. In this case the set of the points x of type $x = f(t)$, $t_{i-1} \leqslant t \leqslant t_i$, coincides with the set of all points $y = g(u)$, $u_{i-1} \leqslant u \leqslant u_i$, where $i = 1, 2, \ldots, m$. Each chain, generated by way of parametrization f, is thus also generated by way of parametrization g, which proves that the system of chains Z also obeys the parametrized curve $g : [c, d] \to M$.

Let $f : [a, b] \to M$ and $g : [c, d] \to M$ be two parametrized curves. Let us assume that there exists a system of chains Z which obeys each of the above parametrized curves. Let us prove that in this case these curves are equivalent.

Let us take $\varepsilon > 0$ and find the chain $\xi \in Z$, such that $\lambda(\xi) < \varepsilon$. Let X_i, $i = 1, 2, \ldots, m$, be a set of this chain. In line with the definition, X_i coincides with the totality of all $x = f(t)$, where $t_{i-1} \leqslant t \leqslant t_i$ and with the totality of all $y = g(u)$, where $u_{i-1} \leqslant u \leqslant u_i$. In this case $a = t_0 \leqslant t_1 \leqslant \cdots \leqslant t_m = b$ and $c = u_0 \leqslant u_1 \leqslant \cdots u_m = d$. Now, let $\varphi(t)$ be a function defined in the segment $[a, b]$, linear in each of the segments $[t_{i-1}, t_i]$, and such that $\varphi(t_i) = u_i$, $i = 0, 1, 2, \ldots, n$. If for any i $t_{i-1} = t_i$, and $u_{i-1} \neq u_i$, then the function φ at the point t_i is considered to assume all the values from the segment $[u_{i-1}, u_i]$. Let us set $f^*(t) = g[\varphi(t)]$. The function f^*, as can be seen, is equivalent to g. At the same time, at every t $\rho[f^*(t), f(t)] < \varepsilon$, as at all t the points $f^*(t)$ and $f(t)$ belong to the same set of the chain ξ, while, from the condition, $\lambda(\xi) < \varepsilon$. Due to the fact that $\varepsilon > 0$ is arbitrary, we see that the distance between the curves K and L, having the parametrizations f and g, respectively, is equal to zero, i.e., these curves coincide and, hence, f is equivalent to g which is the required proof.

Each parametrized curve $f : [a, b] \to M$ has only the complete closed system of chains obeying it. Such a system is the totality of all chains generated by the parametrized curve $f : [a, b] \to M$. Indeed, it is obvious that Z is a complete system. Let us prove that it is closed. Let $Z' \supset Z$ be an arbitrary system of chains containing Z. Let $g : [c, d] \to M$ be the parametrized curve, which the system of chains Z' obeys. As far as $Z \subset Z'$, then, obviously, the system of chains Z obeys g. Thus, by Theorem 1.5.1 we see that the curves f and g are equivalent and, hence, the system of chains Z, which obeys g, also obeys f. Consequently, in line with the definition of Z, $Z \supset Z'$ and, thus, $Z = Z'$, which proves the system Z is complete.

There is therefore a one-to-one correspondence between the totality of all classes of equivalence in a set of parametrized curves and the set of all complete closed chains, which fact can serve the basis for a new definition of a curve, when it is viewed as a complete closed system of chains.

1.5.2. Without giving the corresponding proofs, let us demonstrate the way of defining the basic notations referring to a curve on the basis of the above definition of a curve.

First of all, let us consider the problem of defining the notion of a point of the curve. Let there be a system of chains Z. A place of the system is any function $\tau(\xi)$ of the chains of the system, which corresponds to every chain a certain element of it and has the following property: if $\xi \succ \eta$, then $\tau(\xi) \subset \tau(\eta)$.* In this case the inclusion is understood in the sense that $\tau(\xi)$ is one of those sets of the chain ξ, by union of which $\tau(\eta)$ is obtained.

Let an arc of the chain ξ be any chain formed by all the sets of the chain ξ the numbers of which are included between any two numbers k and l. The diameter of a chain is that of the union of all the sets included in it.

Let $\tau_1(\xi)$ and $\tau_2(\xi)$ be two places of the system Z. Let us assume that these places coincide if for any $\varepsilon > 0$ one can find a chain $\xi \in Z$, which includes the arc containing the elements $t_1(\xi)$ and $\tau_2(\xi)$ and having the diameter less than ε.

Let $\tau_1(\xi)$ and $\tau_2(\xi)$ be two places of the system. Let us say that $\tau_1(\xi) < \tau_2(\xi)$, if $\tau_1(\xi)$ and $\tau_2(\xi)$ do not coincide, and that at any $\xi \in Z$ the set $\tau_1(\xi)$ has a lower number than $\tau_2(\xi)$ within the chain ξ.

If $\tau(\xi)$ is the place of the system of chains, then the intersection of all the sets $\tau(\xi)$, $\xi \in Z$, consists of a single point which will be termed the place support.

Let $f:[0, 1] \to M$ be any normal parametrization of the curve defined by the system of chains Z. Each chain ξ of the system Z is defined by a certain sequence $t_0 = 0 \leqslant t_1 \leqslant t_2 \leqslant \cdots \leqslant t_m = 1$ of the points of the segment $[0, 1]$. Let X_i be a set of all $x = f(t)$, where $t_{i-1} \leqslant t \leqslant t_i$. Let us choose an arbitrary $t \in [0, 1]$, and let $\tau_t(\xi) = X_i$, where i is the least of the numbers i, such that $t > t_{i-1}$ ($t \geqslant t_{i-1}$, if $t = 0$). The function $\tau_t(\xi)$, as is easily proved, is the place of the system, the support of which is the point $f(t)$. Let us say that the place $\tau_t(\xi)$ is generated by the point $f(t)$ of the curve f. For each point there exists a place of the system generated by this point. In this case, if $t_1 < t_2$, then $\tau_{t_1}(\xi) < \tau_{t_2}(\xi)$. It can be then demonstrated that any place of the system is generated by a certain point of the curve, in which case two places coincide if and only if the corresponding points of the curve do.

---

* It should be remarked that an element of the chain is a set denoted by a certain index, so that the two elements of the chain are considered different if they have different numbers, even if they coincide as sets.

What has been stated leads to the conclusion that if a curve is to be defined as a system of chains, then the points of the curve are to be defined as the places of this system. In this case the order of the points of the curve, as well as the notions of neighbourhood and semi-neighbourhood are obviously defined. A definition of the notion of an inscribed polygonal line also presents no difficulty. Let us dwell on the notion of the distance between the curves.

First of all, let us define a sort of a distance between the sets. Let A and B be arbitrary closed sets. If $h > 0$ is a real number, then through $A(A, h)$ let us denote the totality of all points x, the distance of which to the set A does not exceed h. A deviation of the set A from B is the exact lower boundary $l(A, B)$ of the numbers h, such that $A \subset S(B, h)$. It can be proved that the deviation has the following properties:

(1) $l(A, B) \geqslant 0$ and $l(A, B) = 0$ if and only if $A \subset B$;
(2) $l(A, B) + l(B, C) \geqslant l(A, C)$.

It should be noted that, generally speaking, $l(A, B) \neq l(B, A)$.

Now, let $\xi = \{A_1, A_2, \ldots, A_m\}$ and $\eta = \{B_1, B_2, \ldots, B_m\}$ be two chains consisting of the same number of elements. A deviation of the chain ξ from the chain η is the quantity $l(\xi, \eta)$ which is equal to the largest of the numbers $l(A_i, B_i)$, $i = 1, 2, \ldots, m$. Let $\tilde{K}$ and $\tilde{L}$ be two systems of chains. The distance between the systems $\tilde{K}$ and $\tilde{L}$ will be the exact lower boundary $\rho(\tilde{K}, \tilde{L})$ of the number ρ, such that for every chain $\xi \in \tilde{K}$ there exists a chain $\eta \in \tilde{L}$, the deviation of which from ξ, $l(\xi, \eta) \leqslant \rho$, and for every chain $\eta \in \tilde{L}$ there exists a chain $\xi \in \tilde{K}$, the deviation of which from η, $l(\eta, \xi) \leqslant \rho$. We suggest the reader the idea to prove the fact that if K and L are the curves defined by the systems $\tilde{K}$ and $\tilde{L}$, then the distance $\rho(\tilde{K}, \tilde{L})$ coincides with that between the curves in the sense defined in section 1.4.

CHAPTER II

Length of a Curve

2.1. Definition of a Curve Length and its Basic Properties

2.1.1. Consider an arbitrary metric space M with the metric ρ. Let K be a curve in the space, and let $\xi = (X_1, X_2, \ldots, X_m)$ be an arbitrary chain of the curve points, i.e., a finite sequence of the points of K, such that $X_1 \leqslant X_2 \leqslant X_m$. Let us set

$$s(\xi) = \sum_{i=1}^{m=1} \rho(X_i, X_{i+1}).$$

The least upper boundary of the quantity $s(\xi)$ on the set of all chains of the curve K is called a length of the curve K and is denoted as $s(K)$. The curve K is termed rectifiable if its length is finite.

THEOREM 2.1.1 (Length additivity). *If $K = [AB]$ is an arbitrary curve, and C is its point, then*

$$s(AB) = s(AC) + s(CB).$$

Proof. Let ξ be an arbitrary chain of the points of the arc $[AC]$, η be a chain of the points of the arc $[CB]$. Combining the chains ξ and η, we get the chain ζ of the curve K points. We obviously have $s(K) \geqslant \rho(\zeta) \geqslant s(\xi) + s(\eta)$. Since the chains ξ and η on the arcs $[AC]$ and $[CB]$ are arbitrarily taken, then

$$s(K) \geqslant \sup_{\zeta} s(\xi) + \sup_{\eta} s(\eta) = s(AC) + s(CB). \tag{1}$$

Let ζ be an arbitrary chain of the points of the curve K. Let us add to it the points A, C and B as new vertices. Then we get a new chain ζ', for which, obviously, $s(\zeta') \geqslant s(\zeta)$. With the point C the chain ζ' is subdivided into two chains ξ and η, where ξ is formed by the vertices of the chain ζ' lying on the arc $[AC]$ (the point C included), and η consists of those vertices of the chain ζ' which lie on $[CB]$ (the point C included again). We have:

$$s(\zeta) \leqslant s(\zeta') = s(\xi) + s(\eta) \leqslant s(AC) + s(CB).$$

From this, due to the arbitrariness of ζ, we get

$$s(K) = \sup_{\zeta} s(\zeta) \leqslant s(AC) + s(CB). \tag{2}$$

Inequalities (1) and (2) obviously yield the equality of the theorem.

THEOREM 2.1.2 (Lower semi-continuity of length). *If the curves (K_m) converge to the curve K at $m \to \infty$, then $s(K) \leqslant \lim_{m \to \infty} s(K_m)$.*

Proof. Let ξ be an arbitrary chain of the points of the curve K, and let $X_0 \leqslant X_1 \leqslant \cdots \leqslant X_p$ be its vertices. For every m let us find on the curve K_m a chain ξ_m with the vertices $X_0^{(m)} \leqslant X_1^{(m)} \leqslant \cdots \leqslant X_p^{(m)}$, in such a way that $X_k^{(m)} \to X_k$ at $m \to \infty$ for all k. In this case, as can easily be seen, $s(\xi_m) \to s(\xi)$ at $m \to \infty$. At every m we have: $s(K_m) \geqslant s(\xi_m)$, which yields

$$\underline{\lim}_{m \to \infty} s(K_m) \geqslant \lim_{m \to \infty} s(\xi_m) = s(\xi).$$

In so far as the chain ξ of the curve K has been arbitrarily chosen, we come to the conclusion that

$$\underline{\lim}_{m \to \infty} s(K_m) \geqslant \sup_{\xi} s(\xi) = s(K)$$

which is the required proof.

COROLLARY. *Let K be a curve in an n-dimensional space E^n, and let (L_m), $m = 1, 2, \ldots$, be a sequence of the polygonal lines inscribed in it, such that $\lambda(L_m) \to 0$ at $m \to \infty$. In this case $s(L_m) \to s(K)$ at $m \to \infty$.*

Proof. The polygonal lines L_m converge to the curve K. Therefore,

$$s(K) \leqslant \underline{\lim}_{m \to \infty} s(L_m). \tag{3}$$

At every m, obviously, $s(L_m) = s(\xi_m)$, where ξ_m is a chain of the curve K points, formed by the vertices of the polygonal line L_m. It yields $s(L_m) \leqslant s(K)$ at every m, and, therefore,

$$\overline{\lim}_{m \to \infty} s(L_m) \leqslant s(K). \tag{4}$$

Inequalities (3) and (4) afford $s(K) = \lim_{m \to \infty} s(L_m)$, which is the required proof.

THEOREM 2.1.3 (Length continuity). *Let K be a rectifiable curve. Then by any $\varepsilon > 0$ we can find a $\delta > 0$, such that any arc of the curve K with its diameter less than δ has a length less than ε.*

Proof. Let us assume, inversely to the required proof, that the theorem is not valid. In this case for any $\varepsilon > 0$ there is a sequence of the arcs $([X_m, Y_m])$ of the curve K, the diameters of which tend to zero, in which case at all m $s(X_m, Y_m) \geqslant \varepsilon$. Without reducing the generality we can suppose that at $m \to \infty$ the points X_m and Y_m converge to a certain point X_0 of the curve K. Theorem 2.1.1 gives: $s(X_m, Y_m) \leqslant s(X_0 X_m) + s(X_0 Y_m)$. Let us prove that at $m \to \infty$ $d(X_0 X_m) \to 0$. Let us divide the sequence (X_m) into two sequences, referring to one of them all the points X_m lying to the left of X_0, and to the other one all the points lying to the right of X_0. In this case for the points of the first subsequence we have: $s(A X_0) \geqslant s(A X_m)$ and, hence, $s(A X_0) \geqslant \overline{\lim}_{m \to \infty} (A X_m)$. The arcs $A X_m$ converge to the arc $A X_0$ and, hence, in line with Theorem 2.1.2,

$s(AX_0) \leqslant \underline{\lim}_{m\to\infty} s(AX_m)$. Comparing the inequalities obtained, we see that for the first subsequence the limit $\lim_{m\to\infty} s(AX_m)$, does exist and is equal to $s(AX_0)$. Therefore, $s(X_0 X_m) = s(AX_0) - s(AX_m) \to 0$ at $m \to \infty$. In an analogous manner we can establish that for the second subsequence we have $s(X_0 X_m) \to 0$. Thus, $s(X_0 X_m) \to 0$ for the whole sequence (X_m). For $s(X_0 Y_m)$ the considerations are the same. It follows from what was proved above that $s(X_m, Y_m) \to 0$, which contradicts the fact that $s(X_m Y_m) \geqslant \varepsilon > 0$ for all m. The contradiction obtained proves the theorem.

COROLLARY. *If K is a rectifiable curve, X is a point of the curve and the point Y tends to X along the curve, then the length of the arc [XY] tends to zero.*

Let $f:[a, b] \to M$ be an arbitrary parametrization of the rectifiable curve K, $s(t)$ be the length of its arc corresponding to the values of the parameter from the segment $[a, t]$. According to Theorem 2.1.1 and the corollary to Theorem 2.1.3 the quantity $s(t)$ is a non-decreasing continuous function of the parameter t. The length of any curve, as can be seen, is not less than its diameter, i.e., the maximum length of the straight line connecting two points of the curve. Therefore, the intervals of the function f constancy coincide with those of the function s. Let $t = t(s)$ be a function inverse to the function $s(t)$. In line with what has been said above, the function $x(s) = f[t(s)]$ is a parametrization of the curve K, and we thus obtain the following theorem.

THEOREM 2.1.4. *Any rectifiable curve allows parametrization $x(s)$, $0 \leqslant s \leqslant l$, wherein the parameter s is the length of a curve arc, counted off from the origin, i.e., $s = s[x(0)x(s)]$.*

THEOREM 2.1.5. *Assume that the metric space M is complete. In this case any set of curves in the space M, such that all curves of the set are contained in one relatively compact set of the space M, and their lengths do not exceed one number $L < \infty$, is relatively compact.*

Proof. The theorem is an obvious corollary to Theorem 1.4.3 since the curves of the totality considered are uniformly divisible. Indeed, if L is the upper boundary of the lengths of the curves of the totality, then as $n(\varepsilon)$ corresponding to $\varepsilon > 0$ we can choose the number $[l/\varepsilon] + 1$. (Here $[x]$ is an integer part of the number x.)

THEOREM 2.1.6. *Let M_1 and M_2 be metric spaces, $F:M_1 \to M_2$ be a continuous mapping, and let K be a curve in M_1. Let us assume that there exists a number $H < \infty$, such that for any two points x', $x'' \in |K|$ the inequality*

$$\rho_{M_2}[F(x'), F(x'')] \leqslant H\rho_{M_1}(x', x'')$$

is valid. Then

$$s[F(K)] \leqslant Hs(K).$$

REMARK. The condition of the theorem is, in particular, fulfilled in the inequality

$$\rho_{M_2}[F(x'), F(x'')] \leqslant H\rho_{M_1}(x', x'')$$

and is valid for any $x', x'' \in M_1$.

Proof. Let $\eta = \{Y_0 \leqslant Y_1 \leqslant \cdots \leqslant Y_m\}$ be an arbitrary chain of the points of the curve $F(K)$. Then there is a chain $\xi = \{X_0 \leqslant X_1 \leqslant \cdots X_m\}$ of the curve K points, such that $f(X_i) = Y_i$. We have:

$$s(\eta) = \sum_{i=1}^{m} \rho_{M_2}(Y_{i-1}, Y_i) \leqslant H \sum_{i=1}^{m} \rho_{M_1}(X_{i-1}, X_i) = Hs(\xi) \leqslant Hs(K).$$

Due to arbitrarity of the chain η of the curve $F(K)$ points, we get

$$s[F(K)] = \sup_{\eta} s(\eta) \leqslant Hs(K)$$

which is the required proof.

COROLLARY. Let the mapping $\varphi: M_1 \to M_2$ of the metric space M_1 into M_2 be an isometry, i.e., such that for any $x, y \in M_1$ $\rho_{M_2}[\varphi(x), \varphi(y)] = \rho_{M_1}(x, y)$. Then for any curve K in M_1 $s[\varphi(K)] = s(K)$.

Proof. Let $\varphi: M_1 \to M_2$ be an isometry. Then φ is one-to-one. Let us set $\varphi(M_1) = N$. The mapping $\varphi^{-1}: N \to M_1$ is obviously also an isometry. Let K be an arbitrary curve in M_1. Then the theorem gives $s[\varphi(K)] \leqslant s(K)$. Applying the theorem to the curve $K_1 = \varphi(K)$ and the mapping $\varphi_1 = \varphi^{-1}$, we get $s(K) = s[\varphi_1(K_1)] \leqslant s(K_1) = s[\varphi(K)]$. By a correlation of the inequalities obtained we come to the conclusion that $s(K) = s[\varphi(K)]$, which is the required proof.

In particular, we find that the length of a curve in E^n remains constant in the process of motion of the space E^n.

THEOREM 2.1.7. A curve length in the metric space M remains constant when changing the curve's orientation, i.e., $s(K) = s(K^*)$ whatever the curve K is.

Proof. Let $\varphi:[a, b] \to M$ be some normal parametrization of the curve K. In this case the function $\varphi^*: u \mapsto \varphi(-u)$, $u \in [-b, -a]$ is a parametrization of the curve K^*. Let $\xi = \{X_0 \leqslant X_1 \leqslant \cdots X_m\}$ be an arbitrary chain of the points of the curve K, and let $t_0 \leqslant t_1 \leqslant \cdots \leqslant t_m$, be the values of the parameter t corresponding to the vertices of the chain ξ, $u_i = -t_i$, $Y_i = \varphi^*(u_{m-i})$. Obviously, $Y_0 \leqslant Y_1 \leqslant \cdots \leqslant Y_m$. Let η be a chain with the vertices Y_i, $i = 0, 1, \ldots, m$, of the points of the curve K^*. The point Y_i coincides spatially with the point X_{m-i} and, hence, $\rho(Y_{i-1}, Y_i) = \rho(X_{m-i}, X_{m-i+1})$. This yields

$$s(K^*) \geqslant s(\eta) = \sum_{i=1}^{m} \rho(Y_{i-1}, Y_i) = \sum_{i=1}^{m} \rho(X_{m-i}, X_{m-i+1}) = \sum_{i=1}^{m} \rho(X_{m-i}, X_i) = s(\xi)$$

and, since the chain ξ has been arbitrarily chosen, it affords $s(K) = \sup_{\xi} s(\xi) \leqslant s(K^*)$. In so far as $(K^*)^* = K$, then what has been proved above demonstrates also that $s(K^*) \leqslant s(K) = s[(K^*)^*]$, from which we have $s(K) = s(K^*)$, which is the required proof.

2.2. Rectifiable Curves in Euclidean Spaces

2.2.1. Let us introduce some notations. The symbol E^n denotes an n-dimensional Euclidean space. Let X and Y be two arbitrary points of E^n. Then $[XY]$ denotes a segment with the ends X and Y, $|XY|$ is the distance between X and Y, i.e., the length of the segment $[XY]$. To the space E^n there corresponds in a natural way a certain vector space, i.e., the space of vectors of the space E^n. This vector space is denoted by the symbol V^n. Each ordered pair (X, Y) of the points of the space E^n uniquely defines the vector $u \in V^n$, which is denoted by the symbol XY. Let us set $|u| = |XY|$. The symbol $\mathbb{R}^n$ will denote an n-dimensional arithmetic vector Euclidean space, i.e., a set the elements of which are the ordered systems $x = (x_1, x_2, \ldots, x_n)$ from n real numbers. In $\mathbb{R}^n$ the operations of addition of the elements and of multiplying an element by a number are defined in a known way. For the vectors $x = (x_1, x_2, \ldots, x_n)$, $y = (y_1, y_2, \ldots, y_n)$ from $\mathbb{R}^n$ $<x, y>$ denotes their scalar product, i.e., $<x, y> = \sum_{i=1}^{n} x_i y_i$. Finally, the length of the vector $x \in \mathbb{R}^n$ is henceforth denoted by the symbol $|x|$. We have:

$$|x| = \sqrt{<x, x>} = \sqrt{\sum_{i=1}^{n} x_i^2}.$$

A Cartesian orthogonal coordinate system in E^n is any isometric mapping κ of the space E^n onto $\mathbb{R}^n$. Let $\kappa : E^n \longrightarrow \mathbb{R}^n$ be a Cartesian orthogonal system of coordinates in E^n, $X \in E^n$ and $\kappa(X) = (x_1, x_2, \ldots, x_n) \in \mathbb{R}^n$. The numbers $x_1, x_2, \ldots, x_n$ are called the coordinates of the point x in the coordinate system κ. The coordinates of the vector $u \in V''$ in the Cartesian orthogonal system of coordinates $\kappa : E^n \longrightarrow \mathbb{R}^n$ are defined in a known way.

The symbol Ω^{n-1} will henceforth denote an $(n-1)$-dimensional spherical space, i.e., the totality of all unit vectors of the space E^n. A big circumference or a circumference of a big circle is any cross-section of the sphere Ω^{n-1} by a two-dimensional plane passing through its center. For any two different points $u, v \in \Omega^{n-1}$ such that $u + v \neq 0$ there exists only one big circumference Γ, passing through those u and v. The points u and v divide the circumference Γ into two arcs, one of which has a length less than π. This arc will be referred to as the shortest arc on the sphere Ω^{n-1} connecting the points u and v, which are called polar if $-u = v$. If u and v are polar points of the sphere, then they are connected with an infinite set of the shortest arcs on Ω^{n-1}, each of which is a semi-circumference and has a length equal to π.

Hereafter we are going to consider some other geometrical objects related to the space E^n, with the corresponding terminology and notations introduced in proper places.

2.2.2. Let us suppose that in E^n there is a certain fixed Cartesian orthogonal system of coordinates. Let K be a curve in E^n, $X : [a, b] \longrightarrow E^n$ be its arbitrary parametrization, $x_i(t)$, $i = 1, 2, \ldots, n$, be the coordinates of the point $X(t)$, $r(t)$ be the radius-vector of the point $X(t)$ with respect to the coordinate origin.

THEOREM 2.2.1 (Jordan). *For the curve K to be rectifiable it is necessary and sufficient that the functions $x_1(t)$, $i = 1, 2, \ldots, n$, be functions of a limited variation.*

A proof of the theorem may be found, for instance, in [10].

Let us of dwell the problem of the expression of the length of a rectifiable curve through the functions $x_1(t)$, $i = 1, 2, \ldots, n$, and of the expression of the radius of the vector $r(t)$ through its derivative. As is known, any continuous function of bounded variation $f:[a, b] \to \mathbb{R}$ is equal to the sum $\varphi + \sigma$, where the function φ is absolutely continuous, while σ is a so-called singular function, the derivative of which $\sigma'(t) = 0$ for almost all t. The functions φ and σ are defined, in this presentation, to the accuracy of the constant addend C. The function σ is referred to as a function of peculiarities for f. Any function $f:[a, b] \to \mathbb{R}$ of bounded variation is differentiable almost everywhere in $[a, b]$ and the following formula for the restoration of the function by its derivative is valid:

$$f(t) - f(a) = \int_a^t f'(u)\mathrm{d}u + \sigma(t) - \sigma(a).$$

THEOREM 2.2.2 (Lebesgue). *Let $\{X(t), a \leqslant t \leqslant b\}$ be a rectifiable curve in E^n; $s(t)$ is the length of the arc $|X(a)X(t)|$ of a given curve. The function s is absolutely continuous if and only if each of the functions x_i is absolutely continuous. In the case when the functions x_i are absolutely continuous, the equality*

$$s(t) = \int_a^t \sqrt{\sum_{i=1}^n [x_i'(t)]^2}\ \mathrm{d}t \qquad (5)$$

is valid.

If as a parameter for a curve we assume the length of the arc s read off from the origin, the coordinates $x_i(s)$, $i = 1, 2, \ldots, n$, of the point $X(s)$ of the curve are absolutely continuous functions. Indeed, in this case for any $s_1, s_2 \in [a, b]$ the inequalities

$$|x_i(s_1) - x_i(s_2)| \leqslant |X(s_1) - X(s_2)| \leqslant |s_1 - s_2|$$

are fulfilled, which shows that the functions x_i obey the Lipshitz condition and, hence, are absolutely continuous.

The conditions under which equality (5) is valid do exist, in particular, if the functions x_i are continuously differentiable in $[a, b]$. In the case of an arbitrary rectifiable curve K, if $X(t)$ is its arbitrary parametrization, formula (5), generally speaking, does not hold, and the integral on the right-hand side is not, in general, equal to the curve length. In this case the following statement is valid.

THEOREM 2.2.3 (Lebesgue). *If $\sigma(t)$ is the function of singularities of the function $s(t)$, then*

$$s(K) = \int_a^b \sqrt{\sum_{i=1}^n [x_i'(t)]^2} \, dt + \sigma(b) - \sigma(a). \tag{6}$$

Finally, let us comment on the following known result, referring to the notion of a curve's length.

THEOREM 2.2.4. *Let K be a rectifiable curve in E^n and let $\{X(s),\ 0 \leqslant s \leqslant l\}$ be its parametrization, where the parameter s is an arc length. Then for nearly all $s \in [0, l]$ the vector-function $r(s)$ is differentiable. In this case $|r'(s)| = 1$ for almost all s, and for any segment $[s_1, s_2] \subset [0, l]$ the equality*

$$r(s_2) - r(s_1) = \int_{s_1}^{s_2} r'(t) \, dt$$

holds.

Let $X(t)$, $0 \leqslant t \leqslant 1$, be a parametrization of the curve K. Let us say that t is a reduced length for any $t_1, t_2 \in [0, 1]$, $t_1 < t_2$, the arc length $[X(t_1)\ X(t_2)]$ of the curve K equals $s(K)(t_2 - t_1)$. It is obvious that any curve allows the parametrization wherein a parameter is the reduced length. Namely, let $Y(s)$, $0 \leqslant s \leqslant l = s(K)$, be a parametrization wherein s is the arc length. The function $X(t) = Y[s(K)t]$ obviously gives the required parametrization of the curve K. The parametrization $X(t)$, $0 \leqslant t \leqslant 1$, of the curve K is called reduced if the parameter t is a reduced length.

If $X(t)$, $0 \leqslant t \leqslant 1$, is the parametrization of the curve K where the parameter t is a reduced length, then the functions $x_i(t)$ are absolutely continuous and for almost all $t \in [0, 1]$

$$|r'(t)| = s(K).$$

2.2.3. The following theorem demonstrates that the properties of length established in Section 2.1 are, in a known sense, characteristic of it in the case of curves in a Euclidean space.

THEOREM 2.2.5. *On a set of all continuous curves of the space E^n there is a certain non-negative function $f(K)$ (the values $f(K) = \infty$ are allowed), possessing the following properties:*

(1) *if the curves K and L are congruent, i.e., if they can be superposed by the motion of E^n, then $f(K) = f(L)$;*

(2) *if A and B are the end points of the curve K, and C is any of its points, then $f(K) = f([AC]) + f([CB])$;*

(3) *if the curves (K_m) converge to the curve K at $m \to \infty$, then $f(K) \leqslant \underline{\lim}_{m \to \infty} f(k_m)$;*

(4) *there exists a non-degenerating to a point curve K, for which $0 < f(K) < \infty$.*

In this case for any curve K $f(K) = Cs(K)$, where C is constant, $0 < C < \infty$.

Proof. First, let the curve K be a segment of a straight line. Let us assume that $f(K) < \infty$. As far as any segment can be covered with a finite number of segments equal to K, then for any segment L $f(L) < \infty$. Due to condition (1) $f(K)$ depends only on the length of the segment K, $f(K) = \varphi[s(K)]$. In this case the function $\varphi(s)$ is defined and non-negative for any $s \geqslant 0$, and, due to condition (2), for any s_1 and s_2 the equality $\varphi(s_1) + \varphi(s_2) = \varphi(s_1 + s_2)$ is valid. It affords $\varphi(s) = Cs$, where $0 \leqslant C \leqslant \infty$.*

Therefore, we have shown that if there exists a straight line segment K, such that $f(K) < \infty$, then $f(L) = Cs(L)$ for any segment L where $0 \leqslant C < \infty$. It is obvious from what has been proved that in this case $f(K) = Cs(K)$ for any polygonal line K. let K be an arbitrary curve, and let (K_m) be a sequence of the polygonal lines inscribed in it. In this case $\lim_{m \to \infty} f(K_m)$ exists and is equal to zero, if $C = 0$, and $Cs(K)$ at $C > 0$. In line with (3) we conclude that if $C = 0$ then for any curve K $f(K) = 0$, if $C > 0$ then $f(K) \leqslant Cs(K)$. According to condition (4) we deduce that the case when $C = 0$ is impossible and, hence, $C > 0$.

Let us select a ray l having its origin O in the space E^n. Let K be an arbitrary non-degenerate curve in E^n, and let (L_m), $m = 1, 2, \ldots$, be a sequence of the polygonal lines inscribed in the curve K and sharing general end points with it, such that $\lambda(L_m) \to 0$ at $m \to \infty$. On the ray l let us, beginning from the point O, lay down mutually non-overlapping segments $A_0^m A_1^m$, $A_1^m A_2^m$, $\ldots$, $A_{k_m-1}^m A_{k_m}^m (A_0^m = 0)$, which are equal to the consequent links of the polygonal line L_m. The curve K is divided into arcs by the vertices of the polygonal line L_m. Let us transfer the i-th arc of the curve K in the space in such a way that its terminal points coincide with the points A_{i-1}^m and A_i^m. The transferred arcs form a certain curve K_m. From (1) and (2) we have: $f(K_m) = f(K)$. At $m \to \infty$ $s(L_m) \to s(K)$. If we set $A_{k_m}^m = B_m$, then we get $|OB_m| = s(L_m)$ and, hence, $|OB_m| \to s(K)$ at $m \to \infty$. As a result, by $m \to \infty$ the point B_m is either tending to a certain point B_0, such that $|OB_0| = s(K)$, or it is going to infinity; the latter case, obviously, occurs when $s(K) = \infty$. In the second case any segment OB of the ray l is a limit of a certain sequence of the arcs K'_m of the curves K_m, and, hence, in line with (3)

$$f(OB) \leqslant \varliminf_{m \to \infty} f(K'_m) \leqslant \varliminf_{m \to \infty} f(K_m) = f(K). \tag{7}$$

In the first case we analogously get

---

* Indeed, due to its non-negativity and additivity, the function $\varphi(s)$ is non-decreasing. For any integer positive m and n $\varphi(ms/n) = \varphi(s)m/n$. As a result, if for a certain $s_0 > 0$ $\varphi(s_0) = 0$, then for all values of s $\varphi(s) = 0$, so that in this case $\varphi(s) = Cs$, where C equals zero. Let, for a certain $s_0 \neq 0$, $0 < \varphi(s_0)$. Then at $0 < s < s_0/m$ $0 \leqslant \varphi(s) \leqslant \varphi(s_0)/m$ (m is a natural number), which affords for any s_1 and s_2 $s_1 \geqslant 0$, $s_2 \geqslant 0$, such that $|s_1 - s_2| < s_0/m$ $|\varphi(s_1) - \varphi(s_2)| \leqslant \varphi(s_0)/m$. Hence, $\varphi(s)$ is continuous and therefore $\varphi(s) = Cs$, in line with the famous Cauchy theorem.

$$f(OB_0) \leqslant \lim_{m \to \infty} f(K_m) = f(K).\tag{8}$$

Let us assume that the curve K is such that $0 < f(K) < \infty$. Inequalities (7) and (8) allow us to conclude that $f(OB) < \infty$ for a certain segment OB of the ray l (in the case when $s(K) = \infty$ for any segment, and when $s(K) < \infty$ for the segment OB_0). We therefore come to the conclusion that there exists a straight segment of K, such that $f(K)$ is finite. What has been proved above yields that, for any segment L $f(L) = Cs(L)$, where $0 < C < \infty$, and for any curve K

$$f(K) \leqslant Cs(K).\tag{9}$$

Inequalities (7) lead us to the conclusion that if $s(K) = \infty$, then for any segment OB on the ray l

$$C|OB| = f(OB) \leqslant f(K)$$

and, since $C > 0$, it affords $f(K) = \infty \geqslant Cs(K)$. If $s(K) < \infty$, then, applying (8), we get

$$Cs(K) = C|OB_0| = f(OB_0) \leqslant f_0(K).$$

It follows from what has been proved above that

$$f(K) = Cs(K)$$

where $0 < C < \infty$, $C = \text{const}$ for any curve K in E^n. The theorem is thus proved.

By way of conclusion let us note that none of the conditions (1) - (4) of Theorem 2.2.5 can be neglected. The validity of this statement is proved in the examples cited below.

1. Let an orthogonal system of Cartesian coordinates be defined in a given space. Let $\varphi:[a, b] \to E^n$ be a parametrization of the curve K, and let $x_i(t)$, $i = 1, 2, \ldots, n$, be the coordinates of the point $\varphi(t)$. Let us set $f(K) = V_a^b$ ar $x_1(t)$. The value of $f(K)$ defined in such a way is independent of the choice of the curve parametrization. It obviously obeys conditions (2), (3) and (4), but does not obey condition (1).

2. Settings $f(K) = 1$ for any curve K, we get a function obeying all the conditions but condition (2).

3. The functions $f_1(K) \equiv 0$ and $f_2(K) \equiv \infty$ obey conditions (1), (2) and (3) but do not obey condition (2).

4. If $s(K) = \infty$, then we set $f(K) = \infty$. If $s(K) < \infty$, then let the parameter s be an arc length, and let $\varphi_k(s)$ be the function equal to 1/2 if a certain neighbourhood of the point $X(s)$ is a straight segment, and equal to 1 in the opposite case. Let us assume $f(K) = \int_0^{s(K)} \varphi_k(s) ds$. The function $f(K)$ obeys condition (1), (2) and (4), and condition (3) is not fulfilled for it. Indeed, let, for instance, K be an arc of the circumference and (L_m) be a sequence of polygonal lines inscribed in it, such that $\lambda(L_m) \to 0$. For every polygonal line L_m

$$f(L_m) = \tfrac{1}{2} s(L_m), \lim_{m \to \infty} f(L_m) = \tfrac{1}{2} s(K) < s(K) = f(K)$$

and condition (3) is thus not fulfilled.

REMARK 1. In the formulation of the theorem it is enough to demand that the function $f(K)$ be defined only for the curves of a certain class Σ which contains all the polygonal lines and such that the operation of curve rectifying used in the proof does not result in withdrawal from this class. It is, in particular, enough to demand that $f(K)$ be defined in order that all the curves are rectifiable.

REMARK 2. Theorem 2.2.5 is valid for the curves in an arbitrary Riemann space characterized with a sufficiently rich group of motions, i.e., such that the following condition is met. In the space M there exists a number $\delta > 0$ and a geodesic line l, such that for any pair X, Y of the space points and for any segment $[X'Y']$ of the geodesic line l, such that its length is $\rho(X, Y) < \delta$, we can find a spatial motion transferring the point X to the point X', and the point Y to the point Y'. This condition is, for instance, met in the spaces of constant curvature; the Theorem 2.2.5, therefore, is valid for such spaces, too.

2.2.4. Let $K, K_1, K_2, \ldots, K_m, \ldots$ be rectifiable curves and, by $m \to \infty$, $K_m \to K$. We say that the curves K_m converge to K with the length of the curve, if, at $m \to \infty$ $s(K_m) \to s(K)$.

Let us prove some theorems on convergence to the length for the case of curves in the space E^n.

LEMMA 2.2.1. *Let $K = AB$ and $K_m = A_mB_m$, in which case, at $m \to \infty$, the curves K_m converge to the length of the curve K, C and C_m are the points on the curves K, K_m such that $A_mC_m \to AC$, $C_mB_m \to CB$. In this case at $m \to \infty$*

$$s(A_mC_m) \to s(AC) \quad and \quad s(C_mB_m) \to s(CB).$$

Proof. We have: $s(AC) \leqslant \underline{\lim}_{m \to \infty} s(A_mC_m)$, $s(CB) \leqslant \underline{\lim}_{m \to \infty} s(C_mB_m)$. Let us arbitrarily set $\varepsilon > 0$ and, using it, find a number m_1, such that, at every $m \geqslant m_1$, $s(A_mC_m) > s(AC) - \varepsilon/2$, $s(C_mB_m) > s(CB) - \varepsilon/2$. In so far as $s(A_mB_m) \to s(AB)$ at $m \to \infty$, then there is an m_2 such that at $m \geqslant m_2$ $s(A_mB_m) < s(AB) + \varepsilon/2$. Let us set $m_0 = \max \{m_1, m_2\}$. Then, if $m \geqslant m_0$, then $s(A_mC_m) = s(A_mB_m) - s(B_mC_m) < s(AB) + \varepsilon/2 - S(CB) + \varepsilon/2 = s(AC) + \varepsilon$. Therefore, at all $m \geqslant m_0$ $|s(A_mC_m) - s(AC)| < \varepsilon$. Due to the arbitrary nature of $\varepsilon > 0$ we have thus established that $s(A_mC_m) \to s(AC)$ at $m \to \infty$, and the lemma is thus proved.

Let K be a non-degenerate rectifiable curve, and let $X(t)$, $0 \leqslant t \leqslant 1$, be a parametrization of this curve. Let us assume that the parameter t is a reduced length, i.e., $t = s/s(K)$, where s is the length of the arc $[X(0)X(t)]$ of the curve K. The function $X(t)$, $0 \leqslant t \leqslant 1$ will be referred to as a reduced parametrization of the curve K.

THEOREM 2.2.6. *At $m \to \infty$ let all the rectifiable curves K_m converge to a rec-*

*tifiable curve K_0, in which case $s(K_m) \to s(K)$ and at all m $s(K_m) \neq 0$ and $s(K)$
$\neq 0$. In this case, if $\{X_m(t),\ 0 \leqslant t \leqslant 1\}$, $m = 0, 1, 2, \ldots$, are the reduced
parametrizations of the curves K_m, $m = 0, 1, 2, \ldots$, then the functions X_m
converge uniformly to X_0 at $m \to \infty$.*

Proof. Let M be a constant, such that for all m $s(K_m) \leqslant M$. In this case we
obviously have: $|X_m(t') - X_m(t'')| \leqslant M(t' - t'')$ for any t' and t'' for all m,
which implies that the functions $X_m(t)$ are uniformly continuous. Therefore,
applying the Arzelá theorem [10], from the sequence (X_m) we can derive a sub-
sequence X_{m_k} which converges uniformly to a certain function $\tilde{X}(t)$, $0 \leqslant t \leqslant 1$.
The function $\tilde{X}(t)$, as can easily be seen, is a parametrization of the curve
K_0. Let us prove that this parametrization coincides with $X_0(t)$. Indeed, let
us choose an arbitrary value of t. The point $X(t)$ subdivides the curve K into
two arcs K' and K''. The point $X_{m_k}(t)$ divides the curve K_{m_k} into two arcs K'_{m_k}
and K''_{m_k}. At $k \to \infty$ $K'_{m_k} \to K'$, $K''_{m_k} \to K''$. In line with Lemma 2.2.1, $s(K'_{m_k}) \to$
$s(K')$ and $s(K''_{m_k}) \to s(K'')$. But $s(K'_{m_k}) = ts(K_{m_k})$. Therefore $s(K') = ts(K)$,
which shows that t is a reduced length and, hence, $X_0(t) = X(t)$. Therefore,
from (X_m) we can derive a subsequence (X_{m_k}) which uniformly converges to X_0.

Let us now assume that $X_m(t)$ does not converge to $X(t)$ uniformly. In this
case there exists a subsequence $(\tilde{X}_{m_k})$, such that max $|X_0(t) - \tilde{X}_{m_k}(t)| \geqslant \delta > 0$
at all K. Applying the considerations given above to this subsequence we ob-
tain a contradiction. The theorem is thus proved.

THEOREM 2.2.7 (Radò). *Let K_0 and K_m, $m = 1, 2, \ldots$, be non-degenerate rectifiable curves, and let, for $m \to 0$, $K_m \to K$ and $s(K_m) \to s(K)$. In this case
if $\{x_m(t),\ 0 \leqslant t \leqslant 1\}$, $m = 0, 1, 2, \ldots$, are the parametrizations of curves
K_m, such that t is a reduced length, then at $m \to \infty$ the vector-function $x'_m(t)$
converges in measure to the vector-function $x'(t)$, i.e., for any $\sigma > 0$, the
set measure of the set $E_m(\sigma)$ of those t for which $|x'_m(t) - x'_0(t)| \geqslant \sigma$ tends to
zero at $m \to \infty$.*

Proof. Let L_m be a curve given by the parametrization $y(t) = \frac{1}{2}[x_0(t) + x_m(t)]$. It is obvious that at $m \to \infty$ the curves L_m converge to the curve K. We
have:

$$s(K) + s(K_m) - 2s(L_m) = \int_0^1 (|x'_0(t)| + |x'_m(t)| - |x'(t) + x'_m(t)|)\ dt \geqslant 0.$$

As far as $s(K_m) \to s(K_0)$ and $\underline{\lim}_{m \to \infty} s(L_m) \geqslant s(K_0)$, then

$$\overline{\lim}_{m \to \infty} [s(K_0) + s(K_m) - 2s(L_m)] \leqslant 0.$$

Since the expression $s(K_0) + s(K_m) - 2s(L_m)$ is non-negative at all m, what has
been proved above demonstrates that at $m \to \infty$ it tends to zero. Therefore,

$$\mathcal{I}_m = \int_0^1 (|x'_0(t)| + |x'_m(t)| - |x'_0(t) + x'_m(t)|)\ dt \to 0.$$

For any two non-zero vectors a and b we have:

$$|a|+|b|-|a+b| = \frac{(|a|+|b|)^2-(a+b)^2}{|a|+|b|+|a+b|} \geq \frac{(|a|+|b|)^2-(a+b)^2}{2|a|+2|b|} \tag{10}$$

and, besides,

$$(|a|+|b|)^2-(a+b)^2 = (a-b)^2-(|a|-|b|)^2. \tag{11}$$

Applying relation (10) to the subintegral expression of the integral $\mathcal{I}_m$ and allowing for the fact that $|x_m'(t)| = s(K_m)$, $|x'(t)| = s(K_0)$ for almost all t, we get

$$\mathcal{I}_m \geq \frac{1}{2s(K_m)+2s(K_0)} \int_0^1 \{(|x_0'(t)|+x_m'(t)|)^2-(x'(t)+x_m'(t))^2\} \ dt \geq 0.$$

Allowing for equality (11) we get:

$$\int_0^1 \{(x_m'(t)-x_0'(t))^2-(|x_0'(t)|-|x_m'(t)|)^2\} \ dt \to 0 \quad \text{at} \quad m \to \infty.$$

Since at $m \to \infty$ $|x_m'(t)| - |x_0'(t)| \to 0$ almost everywhere, then

$$\int_0^1 [x_m'(t)-x_0'(t)]^2 \ dt \to 0$$

at $m \to \infty$.

Therefore, at $m \to \infty$ the function $x_m'(t)$ converges to the function $x_0(t)$ in the mean. Consequently, $x_m'(t) \to x_0'(t)$ in measure and the theorem is thus proved. (It should be noted that the convergence in measure is in this case equivalent to convergence in the mean, so in fact no strengthening of the theorem has been achieved.)

2.3. Rectifiable Curves in Lipshitz Manifolds

2.3.1. Let us recall some definitions concerning manifolds. Let $\mathcal{H}$ denote a certain class of mappings $\alpha:G \to \mathbb{R}^n$, where G is an arbitrary open set in $\mathbb{R}^n$. Of most importance for us are the following particular cases of the class $\mathcal{H}$:

(1) $\mathcal{H} = \text{Lip}$ is a class of Lipshitz mappings. We consider $\alpha:G \to \mathbb{R}^n$ to be a Lipshitz mapping if for any compact set $F \subset G$ there exists a number $L = L(F) < \infty$ such that for any $x_1, x_2 \in F$

$$|\alpha(x_1)-\alpha(x_2)| \leq L|x_1-x_2|.$$

(2) $\mathcal{H} = C^k$. We consider $\alpha:G \to \mathbb{R}^n$ to be a mapping of the class C^k, where $k \geq 1$ is an integer number, if α has all partial derivatives of the order k, these derivatives being continuous.

(3) $\mathcal{H} = C^\infty$ and $\mathcal{H} = C^\omega$. We consider α to be a mapping of the class C^∞, if $\alpha \in C^k$ for all k. If $\alpha \in C^\infty$, then we say that $\alpha \in C^\omega$ if α is an analytical function.

The following obvious inclusion occurs: $\text{Lip} \supset C^k \supset C^\infty \supset C^\omega$. (Only the first of them requires some proofs, but its validity can easily be deduced from the known results of the analysis.)

The topological space M is called an n-dimensional manifold if it is a Hausdorfian space and has a countable base and for any point x one can find a neighbourhood U of this point and a topological mapping $\varphi:U \rightarrow \mathbb{R}^n$, such that $\varphi(U)$ is an open set in $\mathbb{R}^n$. Let M be an n-dimensional manifold. Any topological mapping $\varphi:U \rightarrow \mathbb{R}$, where U is an open set in M is referred to as a local system of coordinates or a chart in M. For the point $p \in U$ the coordinates of the point $x = \varphi(p) = (x_1, x_2, \ldots, x_n)$ in $\mathbb{R}^n$ are in this case called the coordinates of the point p with respect to the chart $\varphi:U \rightarrow \mathbb{R}^n$. Let us assume that we are given the arbitrary charts $\varphi_1:U_1 \rightarrow \mathbb{R}^n$ and $\varphi_2:U_2 \rightarrow \mathbb{R}^n$ of the n-dimensional manifold M. The charts φ_1 and φ_2 are called overlapping if $U_1 \cap U_2 \neq \emptyset$. In this case the sets $G_1 = \varphi_1(U_1 \cap U_2)$ and $G_2 = \varphi_2(U_1 \cap U_2)$ in the space $\mathbb{R}^n$ are open, and the topological mappings $\varphi_2 \circ \varphi_1^{-1}:G_1 \rightarrow G_2$, $\varphi_1 \circ \varphi_2^{-1}:G_2 \rightarrow G_1$, which are called the transition function for the considered maps φ_1 and φ_2 are defined.

We say that in an n-dimensional manifold M there is given a structure of the n-dimensional manifold of the class $\mathcal{H}$ or, more briefly, that M is an n-dimensional manifold of the class $\mathcal{H}$, if there is a family $(\varphi_\alpha:U_\alpha \rightarrow \mathbb{R}^n)_{\alpha \in A}$ of the charts in the space M, the domains of the definitions of which cover M, in which case for any two overlapping charts φ_α and φ_β of the given family the corresponding functions of transition belong to the class $\mathcal{H}$. In this case the family of the charts $(\varphi_\alpha:U_\alpha \rightarrow \mathbb{R}^n)_{\alpha \in A}$ is called an atlas setting the structure of the manifold M. The chart $\varphi:U \rightarrow \mathbb{R}^n$ is called admissible if for any α the transition functions $\varphi_\alpha \circ \varphi^{-1}$ and $\varphi \circ \varphi_\alpha^{-1}$ for a pair of chart φ and φ_α is a mapping of the class $\mathcal{H}$. Manifolds of the class Lip will be referred to as Lipshitz's ones.

Let M be a manifold of the class C^k, $k \geqslant 1$, φ_α and φ_β be arbitrary charts of the atlas defining the structure of the manifold of the class C^k in M. The functions $\varphi_\alpha^{-1} \circ \varphi_\beta$ and $\varphi_\beta^{-1} \circ \varphi_\alpha$ are one-to-one, each belonging to the class C^k. This, in particular, yields the fact that the Jacobians of the functions $\varphi_\alpha^{-1} \circ \varphi_\beta$ and $\varphi_\beta^{-1} \circ \varphi_\alpha$ do not become zero. If M is a manifold of the class Lip, then it follows from the fact that the mappings $\theta = \varphi_\alpha^{-1} \circ \varphi_\beta$ and $\theta^{-1} = \varphi_\beta^{-1} \circ \varphi_\alpha$ are both Lipshitz's, and that for any compact F, contained in the domain G of the definitions of θ, there exists a constant $L(F)$, $1 \leqslant L(F) < \infty$, such that for any $x, y \in F$

$$|x - y|/L(F) \leqslant |\theta(x) - \theta(y)| \leqslant L(F)|x - y|.$$

The wider the class $\mathcal{H}$, the greater the class of manifolds introduced by this definition. On any open set $U \subset M$ the structure of an n-dimensional manifold of the class $\mathcal{H}$ is naturally defined. Namely, let $(\varphi_\alpha:U_\alpha \rightarrow \mathbb{R}^n)_{\alpha \in A}$ be an atlas setting the structure of a differentiable manifold of the class $\mathcal{H}$ in M. In this case the restriction of the mappings φ_α onto the sets $U_\alpha \cap U$, obviously, forms an atlas of the class $\mathcal{H}$ in U and defines in U the structure of a manifold of the class $\mathcal{H}$, which is referred to as an induced one.

Let M_1 and M_2 be manifolds of the class $\mathcal{H}$, $n_1 = \dim M_1$, $n_2 = \dim M_2$. Let us say that the mapping $f:M_1 \rightarrow M_2$ belongs to the class $\mathcal{H}$, if for any two admissible charts $\varphi:U_1 \rightarrow \mathbb{R}^{n_1}$ in M_1 and $\psi:V \rightarrow \mathbb{R}^{m_2}$ in M_2 the function $\psi \circ f \circ \varphi^{-1}$ belongs to the class $\mathcal{H}$.

For any Lipshitz manifold one can define a class of curves which can be naturally termed rectifiable. In this case the notion of a curve length remains undefined. If we want to be able to ascribe to any rectifiable curve in the manifold a certain number which can be considered as its length, the manifold must be supplied with some additional structure, i.e., it must be a Finsler space. We are not going here to dwell upon Finsler spaces, since it would lead us far from the subject of the book.

Let M be a Lipshitz manifold, and let K be an arbitrary curve in M. In this case let us assume that the curve K is rectifiable if for any of its points X we can find such allowable local system of coordinates $\varphi:U \rightarrow \mathbb{R}^n$, defined in the neighbourhood U of the point X in M, and an arc $[YZ]$ of the curve K, contained in U, such that $Y < Z, Y = X$ if X is the beginning of the curve, $Z = X$ if X is the end of the curve, $Y < X < Z$ in the remaining cases, and $\varphi([YZ])$ is a rectifiable curve in $\mathbb{R}^n$.

THEOREM 2.3.1. *Let $\varphi:U \rightarrow \mathbb{R}^n$ be an arbitrary chart in an n-dimensional Lipshitz manifold M. In this case, for any rectifiable curve K lying in U, the curve $\varphi(K)$ in $\mathbb{R}^n$ is rectifiable.*

Proof. Let K be a rectifiable curve lying in the domain of definition of an admissible chart $\varphi:U \rightarrow \mathbb{R}^n$. According to the definition, for any point X of the curve K there exists an admissible chart $\theta_x:V_x \rightarrow \mathbb{R}^n$, such that the image of a certain closed neighbourhood of the point X on the curve K is a rectifiable curve in $\mathbb{R}^n$. Applying the Borel theorem we get a certain sequence of the points $X_0 = A < X_1 < \cdots < X_m = B$, where A is the beginning, B is the end of the curve K and the sequence of the admissible chart $\theta_i:V_i \rightarrow \mathbb{R}^n$, such that $[X_{i-1}X_i] \subset V_i$ and $L_i = \theta_i([X_{i-1}X_i])$ is a rectifiable curve in $\mathbb{R}^n$. Without diminishing the generality, we can assume that $V_i \subset U$ for all i. Let $L = \varphi(K)$, $Y_i = \varphi(X_i)$, $L_i' = [Y_{i-1}Y_i]$. Then let $\psi_i = \varphi \circ \theta_i^{-1}$ be a transition function for the maps φ and θ_i. In this case $L_i' = \psi_i(L_i)$. The mapping ψ_i is a Lipshitz's one. The set $|L_i|$ is compact and contained in the domain of the definition of the mapping ψ_i. Therefore, in line with the definition of a Lipshitz mapping, there is a constant $H_i < \infty$, such that for any

$$x', x'' \in |L_i| \quad |\psi_i(x') - \psi_i(x'')| \leqslant H_i|x' - x''|.$$

On the basis of the Theorem we can deduce that the curve $L_i = \varphi(L_i')$ is rectifiable. Therefore, the curve L falls into a finite number of rectifiable arcs, and, hence, L is rectifiable, which is the required proof.

CHAPTER III

Tangent and the Class of One-Sidedly Smooth Curves

3.1. Definition and Basic Properties of One–Sidedly Smooth Curves

3.1.1. The definition of a tangent to a curve given below is essentially the same as that given in courses of differential geometry.

Here, as in the previous chapters, V^n denotes an n-dimensional vector Euclidean space, the elements of which are the vectors of the space E^n, Ω^{n-1} is a unit sphere in E^n, $\Omega^{n-1} = \{x \in E^n \mid |x| = 1\}$.

The straight line l in the space E^n is called directed if there is a certain non-zero vector a which is colinear to l. The vector a is referred to as a directing vector of l. The vector b which is colinear to a is said to have the same direction as l (opposite direction to l) if $b = \lambda a$, where $\lambda > 0$ (respectively, $\lambda < 0$).

Let l_1 and l_2 be two directing straight lines in E^n, a_1 and a_2 be their directing vectors, the quantity $(a_1 \mathbin{\widehat{}} a_2)$ is understood here as the angle between l_1 and l_2.

Let K be an arbitrary curve in E^n, X and Y be its two points not coinciding in the space. Let us draw a straight line XY and orient it in such a way that the vector $\overline{XY}$ have the direction coinciding with that of the straight line in the case when $X < Y$, and the opposite one in the case when $X > Y$. The straight line XY oriented in such a way is called a secant of the curve K and is denoted through $l(X, Y)$. (The sequence of the order of the points X, Y under such a denotation is of no importance, since in all cases $l(X, Y) = l(Y, X)$).

Let K be a curve in E^n, where X is an arbitrary point of this curve. The directed straight line $t = t_r(x)$ (respectively, $t = t_l(x)$) is called a right-hand (respectively, left-hand) tangent at the point X of the curve K, if for any $\varepsilon > 0$ we can find on the curve such a right-hand (left-hand) semi-neighbourhood U of the point X, such that, for any point $Y \in U$ for which there exists a secant $l(X, Y)$, the angle between the straight line t and the secant $l(X, Y)$ is less than ε.

The unit vector $e_l(X)$ ($e_r(X)$) lying on the left-hand (right-hand) tangent at the point X of the curve K, and having the direction of this tangent, will be referred to as the left-hand (right-hand) tangent unit vector at the point X of the curve K.

43

Let X be an internal point of the curve K, such that in it there exist both the right-hand and the left-hand tangent of the curve. The point X is called smooth if both tangents coincide, and angular in the opposite case. The angular point is called a point of return if the angle between the left- and right-hand tangents equals π.

The directed straight line t is called a right-hand (left-hand) tangent in a strong sense at the point X of the curve K if, for any $\varepsilon > 0$, we can find on the curve such a right-hand (left-hand) semi-neighbourhood of the point X that any secant of this semi-neighbourhood forms with the straight line t an angle less than ε.

It is obvious that a tangent in the strong sense is that in the sense of the initial definition. The opposite case is invalid, which can be easily proved with simple examples.

A curve is referred to as one-sidedly smooth if in each of its internal points there exist both left-hand and right-hand tangents in the strong sense, the right-hand tangent existing at the beginning of the curve, and the left-hand one at the end.

Let us recall some properties of one-sidedly smooth curves in E^n.

LEMMA 3.1.1. *If at the point X of the curve there exists a left-hand (right-hand) tangent in the strong sense, then a certain left-hand (right-hand) semi-neighbourhood of the point X is rectifiable and $s(XY)/|XY| \to 1$, when $Y \to X$ from the left (right).*

Proof. At the point X let there exist a left-hand tangent in the strong sense. Let us assume $\varepsilon > 0$, $\varepsilon < \pi/2$. Let $U = (YX)$ be a left-hand semi-neighbourhood of the point X, such that all the secants in it form angles less than ε with $t_l(X)$. The links of any polygonal line inscribed into the arc $[XY]$ comprise angles less than $\varepsilon < \pi/2$ with $t_l(X)$. Therefore, if h is the length of the projection of the segment XY onto $t = t_l(X)$, then the lengths of the polygonal lines inscribed into the arc $[YX]$ are less than $h/\cos \varepsilon$. Therefore, $s(YX) \leqslant h/\cos \varepsilon$. Then, as far as $|XY| \geqslant h$, hence,

$$1 \leqslant \frac{s(YX)}{|XY|} \leqslant \frac{1}{\cos \varepsilon} \, .$$

This proves that the arc $[XY]$ is rectifiable. Due to the fact that $\varepsilon > 0$ is arbitrary, it also affords

$$\frac{s(YX)}{|XY|} \to 1$$

at Y tending to X from the left (right). The Lemma is proved.

In line with Lemma 3.1.1, if the curve K is one-sidedly smooth, then each of its points has a rectifiable neighbourhood, which, following the Borel theorem, results in the following theorem:

THEOREM 3.1.1. *Any one-sidedly smooth curve is rectifiable.*

THEOREM 3.1.2. *Any one-sidedly smooth curve is a union of a finite number of simple arcs.*

Proof. At every point $X \in K$, where K is a one-sidedly smooth curve, there exists a neighbourhood $U = (Y_1 Y_2)$, such that all the secants of each of the arcs $(Y_1 X)$ and (XY_2) form angles less than $\pi/2$ with the corresponding tangents. Any plane which is orthogonal to $t_l(X)$ intersects the arc $(Y_1 X)$ at only one point, since in the opposite case there can be found a secant of the arc $(Y_1 X)$ which is perpendicular to $t_l(X)$. The arcs $[Y_1 X]$ and $[XY_2]$, therefore, have no multiple points. Hence, each point of the curve has a neighbourhood which is a combination of two simple arcs. In keeping with the Borel theorem, a curve can be covered with a finite number of such neighbourhoods. Subdividing each of them into two simple arcs, we subdivide the curve K into a finite number of simple arcs. The theorem is thus proved.

Let us now consider the problem of invariance of the property of the curve so that it has a tangent in the strong sense with respect to transformations of a domain of the space E^n.

Let there be given an open manifold $U \subset E^n$ and the mapping $f : U \rightarrow E^m$. Let us assume that the mapping f is strictly differentiable at the point $x_0 \in U$, provided that there exists a linear mapping $L : V^n \rightarrow V^m$ and a function $\alpha(y, z)$ of the variables $y \in U$, $z \in U$, where $y \neq z$ are such that $\alpha(y, z) \rightarrow 0$ at $y \rightarrow x_0$, $z \rightarrow x_0$ and for any two different points y, $z \in U$ the equality

$$f(y) - f(z) = L(y-z) + |y-z| \alpha(y, z) \qquad (1)$$

holds. (The condition $\alpha(y, z) \rightarrow 0$ at $y \rightarrow x_0$ and $z \rightarrow x_0$ denotes here that for any $\varepsilon > 0$ there is a $\delta > 0$ such that, if $|y-x_0| < \delta$ and $|z-x_0| < \delta$, then $|\alpha(y, z)| < \varepsilon$).

If the function $f : U \rightarrow E^m$ is strictly differentiable at the point x_0, then it is differentiable at this point in the general sense as well.

If the function $f : U \rightarrow E^m$ has in U all the partial derivatives $\partial f / \partial x_1$, $\partial f / \partial x_2, \dots, \partial f / \partial x_n$ (here x_i, $i = 1, 2, \dots, n$, are the Cartesian orthogonal coordinates in E^n) and each of these derivatives is continuous at the point x_0, then f is strictly differentiable at this point. In manuals on mathematical analysis it is usually proved that, under this condition, the function f is differentiable at the point x_0. Obviously, by changing the considerations with which it is established, one can prove the strict differentiability as well, but this will not be discussed here. In particular, if the function $f : U \rightarrow E^m$ belongs to the class C_1, i.e. it has in U all the partial derivatives $\partial f / \partial x_i$, $i = 1, 2, \dots, n$, in which case these derivatives are continuous in the domain U, then f is strictly differentiable at every point of the domain U.

THEOREM 3.1.3. *Let there be a given domain U of the space E^n, a continuous mapping $f : U \rightarrow E^m$ and a curve K lying in U. Let $M = f(K)$. In this case if the* *curve K has at a certain point X a right-hand (left-hand) tangent in the strong sense, and the mapping f is strictly differentiable at the point X, in which case $df_X[e_r(X)] \neq 0$ (respectively, $df_X[e_l(X)]) \neq 0$), then the curve M has a right-hand (left-hand) tangent in the strong sense at the point Y = f(X). In this case the vector $a = df_X[e_r(X)]$ ($b = df_X[e_l(X)]$) is a directing vector of the right-hand (left-hand) tangent of the curve M at the point Y.*

Proof. Let us limit ourselves to the case of a right-hand tangent, since for a left-hand one all the considerations are analogous.

Let us set $a = df_X[e_r(C)]$. $\gamma = |a| > 0$. Due to the condition of the Theorem, $a \neq 0$. Since the function f is strictly differentiable at the point X, then for any X_1, X_2, such that $X_1 \in U$, $X_2 \in U$, then the following equality occurs

$$f(X_2) - f(X_1) = df_X(X_2 - X_1) + \alpha(X_1, X_2)|X_2 - X_1| \qquad (2)$$

where $\alpha(X_1, X_2) \to 0$, when $X_1 \to X$ and $X_2 \to X$.

Let us set a number $\varepsilon > 0$, such that if a unit vector e forms with $e_r(X)$ an angle less than ε, then $|df_X(e)| > \gamma/2$. Then, let us find $\delta > 0$, such that if $|X_1 - X| < \delta$ and $|X_2 - X| < \delta$, then $|\alpha(X_1, X_2)| < \gamma/2$.

Let U be a right-hand semi-neighbourhood of the point X on the curve K, lying in a ball $B(X, \delta)$ and such that for any two points $X_1, X_2 \in U$ where $X_1 < X_2$, the vector $e = \overline{X_1X_2}/|\overline{X_1X_2}|$ forms an angle less than ε with $t_r(X)$. For arbitrary points $X_1, X_2 \in U$, where $X_1 < X_2$, we have:

$$|f(X_2)| - |f(X_1)| \geqslant |df_X(X_2 - X_1)| - |\alpha(X_1, X_2)||X_2 - X_1|$$
$$|X_2 - X_1|\left(|df_X(e)| - |\alpha(X_1, X_2)|\right).$$

Since the arc U lies in the ball $B(X, \delta)$, then $|\alpha(X_1, X_2)| < \gamma/2$. The vector e forms with $t_r(X)$ an angle less than ε and, hence, $|df_X(e)| > \gamma/2$. Therefore, for the given points X_1, X_2 $|f(X_2) - f(X_1)| > 0$, so that the points $f(X_1)$ and $f(X_2)$ are different.

The mapping f, therefore, is a one-to-one mapping of the arc U onto the arc V of the curve M. Let us arbitrarily choose the points $Y_1 < Y_2$ of the curve M lying on the arc V. Let $Y_1 = f(X_1)$, $Y_2 = f(X_2)$. When the points Y_1 and Y_2 converge along the arc V from the right to the point Y, then, as a result of the considerations of continuity, X_1 tends to X and X_2 tends to X along the curve K from the right. The vector $\overline{Y_1Y_2}/|\overline{Y_1Y_2}|$ is a directing vector of the secant $l(Y_1, Y_2)$ of the curve M. We have:

$$\frac{Y_2 - Y_1}{|X_2 - X_1|} = df_X(e) + \alpha(X_1, X_2)$$

where, as above, $e = \overline{X_1X_2}/|\overline{X_1X_2}|$. At $X_1 \to X$ and $X_2 \to X$ from the right, the vector e converges to the vector $e_r(X)$, since, by the condition, the curve K has a right-hand tangent in the strong sense at the point X. At $X_1 \to X$ and $X_2 \to X$ the quantity $\alpha(X_1, X_2)$ tends to zero. Therefore, when $Y_1 \to Y$ and $Y_2 \to Y$ from the right along the curve M, $Y_1 < Y_2$ the vectors $\overline{Y_1Y_2}/|\overline{X_1X_2}|$ converge to the vector $h = df_X(e_r(X)) \neq 0$. It leads one to the conclusion that the curve M has a right-hand tangent in the strong sense at the point Y. The vector h is in this case a directing vector of the tangent discussed. The Theorem is proved.

COROLLARY. *Let U be an open set in E^n, $f:U \to E^n$ be a mapping of the class C^1. In this case if a Jacobian of the mapping f is other than zero at every point $x \in U$, then f transforms any one-sidedly smooth curve in the domain U into a one-sidedly smooth one.*

3.2. Projection Criterion of the Existence of a Tangent in the Strong Sense

3.2.1. Let us establish here a certain criterion of the existence of a tangent in the strong sense, based on the consideration of orthogonal projections of the curve considered onto a straight line.

Let us first elucidate the meaning of the condition that the curve K has a tangent at the point K in the case when K lies in one straight line. A curve lying in one straight line seems to be a somewhat paradoxical subject, but it will be demonstrated later on that it plays quite an important role.

Let us assume that the curve K lies in a certain straight line l, which is arbitrarily oriented. In this case any secant of the curve K is either coincident with l, or is oppositely directed to l. Let us suggest that K has a right-hand tangent at the point X. Then there is such a right-hand semi-neighbourhood U of the point X that any secant $l(X, Y)$, where $Y \in U$ will form with this tangent an angle less than π and, hence, it will coincide with it. This means that the vectors $\overline{XY}$, where $Y \in U$ are all equally directed, i.e. the arc U, lies in the straight line l to one side off the point X. The reverse case is also clear: if a certain right-hand semi-neighbourhood of the point X lies on l to one side off the point X, then K has a right-hand tangent at the point X. In this semi-neighbourhood there can be an infinite number of the points which coincide with X spatially, as is, for instance, the case for the curve K with the parametrization $x(t) = a + \varphi(t)e$, $-1 \leqslant t \leqslant 1$, where e is a unit vector in E^1, and φ is a real function, such that $\varphi(0) = 0$ and $\varphi(t) = t \sin^2(\pi/t)$ at $t \neq 0$.

Analogous conclusions are valid for any left-hand tangent.

If the curve K has a right-hand tangent in the strong sense at the point X, then a certain right-hand semi-neighbourhood U of the point X is a simple arc. In the case when the curve K lies in one straight line the inverse case, as can be seen, is also valid: if a certain arc $[XY]$, where $X < Y$, of the curve K is simple, then K has a right-hand tangent in the strong sense at the point X. Analogous considerations are valid for any left-hand tangent.

Let l be an arbitrarily directed straight line in E^n. Let us denote by π_l the operator of orthogonal projecting onto the straight line l, so that for an arbitrary point X $\pi_l(X)$ is an orthogonal projection of X onto l. Let k be a directed straight line in E^n, such that $(l, k) \neq \pi/2$. We assume that $\pi_l(k) = l$, if $(l \frown k) < \pi/2$ and $\pi_l(k) = -l$, if $(l \frown k) > \pi/2$ ($-l$ is the straight line derived from l by changing the orientation of l). If a is a directing vector of the straight line k, then $\pi_l(a)$ is, obviously, a directing vector of the straight line $\pi_l(k)$.

Let a number $\varepsilon > 0$ be given. The set A of the points of the sphere Ω^{n-1} is called an ε-net of the sphere if for any point $u \in \Omega^{n-1}$ there is a vector $v \in A$, such that $(u \frown v) < \varepsilon$. The set Ω^{n-1} is compact and, using the Borel theorem, we can easily show that for any $\varepsilon > 0$ there exists a finite ε-net on the sphere Ω^{n-1} (i.e., a finite set which is an ε-net).

Let us fix an arbitary point O in the space E^n. For an arbitrary vector $u \in \Omega^{n-1}$ the symbol l_u will denote a straight line passing through the point O and parallel to the vector u. The mapping of orthogonal projecting onto the straight line l_u will be further denoted by the symbol π_u.

THEOREM 3.2.1. *Let K be a non-degenerate curve in E^n, and let X be a point of the curve K. Let us assume that for any $\varepsilon > 0$ there can be found a finite ε-net A on the sphere Ω^{n-1} such that for any $u \in A$ the curve $\pi_u(K)$ has a right-hand (left-hand) tangent in the strong sense at the point $\pi_u(X)$. In this case the curve K has a right-hand (left-hand) tangent in the strong sense at the point X.*

Proof. Let us limit ourselves to the case of the right-hand tangent since for the left-hand one the considerations are analogous.

Let us arbitrarily set $\varepsilon > 0$ and construct such a finite ε-net A of the sphere Ω^{n-1}, so that for any vector $u \in A$ the curve $\pi_u(K)$ has a right-hand tangent in the strong sense at the point $\pi_u(X)$. Let $u_1, u_2, \dots, u_r$ be the elements of the set A. The straight line l_{u_i} will be further denoted by l_i, and the curve $\pi_{u_i}(K)$ and its point $\pi_{u_i}(X)$ by K_i and X_i, respectively. According to the condition, at every $i = 1, 2, \dots, r$ the curve K_i has a right-hand tangent in the strong sense at the point X_i. Therefore, one may find a point Y_i of the curve K_i, such that $X_i < Y_i$ and the arc $[X_iY_i]$ is simple. Let $[XZ_i]$ be the arc of the curve K, the orthogonal projection of which onto the straight line l_i is the arc $[X_iY_i]$, $X < Z_i$. Obviously, $<\overline{XZ_i}, u_i> \neq 0$. Let us denote by P_i the semi-space of the space V^n defined in the following way. If $<\overline{XZ_i}, u_i> < 0$, then $P_i = \{x \mid <x, u_i> \leqslant 0\}$, and if $<\overline{XZ_i}, u_i> > 0$ then $P_i = \{x \mid <x, u_i> \geqslant 0\}$. In as much as the arc $[X_iY_i]$ of the curve K_i is simple, then for any two points $X', X'' \in (XZ_i)$ such that $X' < X''$ the vector $\overline{X_i'X_i''}(X_i' = \pi_{u_i}(X'), X_i'' = \pi_{u_i}(X''))$, is either a zero one or has the same direction as the vector $\overline{X_iY_i}$. This means that the vector $\overline{X'X''}$ belongs to the semi-space P_i.

Let Z be the utmost left point of the set of points $Z_1, Z_2, \dots, Z_r$. The above proved yields that for any $X', X'' \in (XZ)$ where $X' < X''$ the vector $\overline{X'X''}$ belongs to the semi-space P_i for all $i = 1, 2, \dots, r$. Let us set

$$Q = \bigcap_{i=1}^{r} P_i.$$

The set Q, obviously, is not empty and is a closed convex cone in V^n.

Let us demonstrate that the angle between any two vectors $u, v \in Q$ is less than ε. Indeed, let $u \in Q$, $v \in Q$, $u \neq v$. Let us assume that $|u| = 1$, $|v| = 1$, so that u and v can be considered as points of the unit sphere Ω^{n-1} of the space V^n. Let us note that, first of all, $(\widehat{u, v}) = \pi$. Indeed, let us assume, on the contrary, that $(\widehat{u, v}) = \pi$. Then $u = -v$. Let us find a point $u_i \in A$, such that $(\widehat{u, u_i}) < \varepsilon_1$. Since $\varepsilon_1 < \pi/2$, then only one of the vectors u and v belongs to the semi-space P_i and, therefore, they cannot both belong to Q. The assumption that $(u, v) = \pi$ thus leads to a contradiction. Let us construct a two-dimensional plane S which passes through the vectors u and v. Let Γ be a circumference along which this plane intersects the sphere Ω^{n-1}. Let us denote by w the middle of the shortest arc of this circumference with the terminal points u and v. Let us set $w' = -w$ and let z and z' be the ends of the diameter of the circumference Γ which is perpendicular to the diameter ww', in which case the point u lies on the semi-circumference wzw'. Let us find a point $u_i \in A$, such that $(\widehat{z, u_i}) < \varepsilon_1$. The plane $P = \{x \mid (x, u_i) = 0\}$ does not contain Γ since $(\widehat{z, u_i}) < \varepsilon_1 < \pi/2$, and hence P intersects Γ at certain points a and a'. The arc uwv lies in the cone Q, which demonstrates that neither of the points a and a' can lie inside this arc. We have $a' = -a$. Since $(\widehat{z, u_i}) < \pi/2$, then the diameter aa' does not coincide with the diameter zz' of the circumference Γ and, of two points a and a', only one lies on the semi-circumference $wz' = \Gamma_1$. Let us assume that $a \in \Gamma_1$. It is obvious that a belongs to one of the arcs zu, vz' of the semi-circumference Γ_1. Let $a \in zu$. Let us construct a hyperplane Q which is orthogonal to a. It intersects the semi-circumference wzw' at a certain point b. We obviously have: $(\widehat{a, b}) = \pi/2$. This yields $(\widehat{b, z}) = (\widehat{a, w})$. The value of $(\widehat{b, z})$ is, however, equal to the distance measured over the sphere Ω^{n-1} up to the $(n-2)$-dimensional sphere $\Omega^{n-1} \cap Q$. We have: $u_i \in Q$ and, hence, $(\widehat{b, z}) \leqslant (\widehat{z, u_i}) < \varepsilon_1 \leqslant \varepsilon/2$. Therefore, $(\widehat{a, w}) < \varepsilon/2$. But $(\widehat{u, v}) = 2(\widehat{u, w}) \leqslant (\widehat{a, w}) < \varepsilon$. In the case when a lies on the arc wz' of the semi-circumference Γ, we use the same considerations with only one exception: instead of u_i we consider the point $u_i = -u_i$. Therefore, we have established that the angle between any two vectors $u, v \in Q$ is less than ε.

In particular, it results from the above proof that an angle between any two secants of the arc (XZ) of the curve K is less than ε and, since $\varepsilon > 0$ has been arbitrarily chosen, we have established that the curve K has a right-hand tangent in the strong sense at the point X. The Theorem is proved.

3.3. Characterizing One–Sidedly Smooth Curves with Contingencies

3.3.1. Let us describe here another way of introducing the class of one-sidedly smooth curves, based on the notion of contingencies.

Henceforth K will denote an arbitrary curve in the space E^n.

A right-hand (left-hand) contingency of the curve K at the point X is a totality $q_r(X)(q_l(X))$ of all directed straight lines passing through the point X, each of which is a limit of a certain sequence of the secants $(l(X, Y_m))$, $m = 1, 2, \ldots$, where the points $Y_m \to X$ from the right (left).

The definition of the notion of the tangent directly yields the notion that for the existence of a left-hand (right-hand) tangent at a point of the curve it is necessary and sufficient that the corresponding contingency consist of the only straight line (this straight line, obviously, will also be a tangent).

LEMMA 3.3.1. *Let $K = [AB]$ be an arbitrary non-degenerate curve in E^n. Let us assume that there exists a unit vector e and a not more than countable set S of the points of the curve K such that $A \in S$ and $B \in S$ and at every point X of the curve K which does not belong to S any straight line of the left-hand contingency of the curve K at the point X makes an angle less than $\pi/2$ with the vector e. In this case the curve K is a simple arc and the secant l (AB) makes an angle less than $\pi/2$ with the vector e.*

Proof. In the space E^n let us introduce an orthogonal system of Cartesian coordinates in such a way that the vector e is its first basis vector. Let $x(t)$, $0 \leqslant t \leqslant 1$, be any normal parametrization of the curve K, $x_1(t)$, $x_2(t)$, $\ldots$, $x_n(t)$ be the coordinates of the point $x(t)$ in the system of coordinates considered.

Let us show that x_1 is a strictly increasing function on the intercept $[0, 1]$. Let us first prove that x_1 is a non-decreasing function, i.e., that for any $t_1, t_2 \in [0, 1]$, such that $t_1 < t_2$ we always have $x_1(t_1) \leqslant x_1(t_2)$. Let us assume that this is not the case. Then there are values of $t_1, t_2 \in [0, 1]$ for which $t_1 < t_2$ and $x_1(t_1) > x_1(t_2)$. Let M be a set of those $t \in [0, 1]$ for which $x(t) \in S$. Since the parametrization $x(t)$ is normal, the set M is not greater than countable. Finally, let $M' = x_1(M)$. The set M' is not greater than countable and, therefore, there is a value of h such that $x_1(t_1) > h > x_1(t_2)$ and $h \notin M'$. Let t_0 be the greatest lower boundary of the values $t \in [t_1, t_2]$, for which $x_1(t) < h$. Due to the continuity of x_1, obviously, $x_1(t_0) = h$ and $x_1(t) > h$ at $t_1 \leqslant t < t_0$. The point $X_0 = x(t_0)$ does not belong to the set S. Let us arbitrarily take a sequence of the values of (t_m) such that $t_m \in [t_1, t_0)$ for all m and $t_m \to t_0$ at $m \to \infty$, and let $X_m = x(t_m)$. Each of the secants $l_m = l(X_0 X_m)$, where $m = 1, 2, \ldots$ forms with the vector e an angle greater than $\pi/2$. Choosing a converging subsequence from the sequence of the secants (l_m), $m = 1, 2, \ldots$, we find that the contingency $q_l(X_0)$ contains a straight line making an angle not less than $\pi/2$ with e. This fact, however, contradicts the condition of the Lemma. Therefore, having assumed that the function x_1 is not non-decreasing, we come to a contradiction.

Therefore we have established that x_1 is a non-decreasing function. Let us assume that x_1 is not a strictly increasing function. In this case there can be found a segment $[\alpha, \beta] \subset [0, 1]$, such that x_1 is constant within it, $x_1(t) = h$ for all $t \in [\alpha, \beta]$. In this case the arc of the curve K corresponding to the segment $[\alpha, \beta]$ lies in the plane $\{x_1 = h\}$ and, hence, the contingency at any point of this arc lies in the same plane and all the straight lines belonging to these contingencies form an angle equal to $\pi/2$ with l. And again we come to a contradiction and, hence, the function x_1 is strictly increasing.

What has been proved above shows that the curve K is a simple arc. We have:

$$<\overline{AB}, e> = x_1(1) - x_1(0) > 0$$

whence we conclude that $(l(A, \widehat{B}), e) < \pi/2$. The Lemma is proved.

COROLLARY. *Let* $K = [AB]$ *be a non-degenerate curve in* E^n. *Let us assume that there exists a unit vector* e *and a not greater than countable set* S *of the points of the curve* K *such that* $A \in S$, $B \in S$ *and at every point* $X \in K$, *which does not belong to* S, *any straight line* $t \in q_l(X)$ *forms an angle less than* α *with* e, *where* $0 < \alpha < \pi/2$. *In this case the curve* K *is a simple arc and the vector* $\overline{AB}$ *makes an angle less than* α *with* e.

Proof. The fact that K is a simple arc results directly from the Lemma, all the conditions of which are fulfilled. Moreover, the Lemma also yields $(\widehat{\overline{AB}, e}) < \pi/2$. Let u be a unit vector lying in the plane of the vectors e and $\overline{AB}$ and making with e an angle equal to $(\pi/2) - \alpha$ and such that u and $\overline{AB}$ are directed to the opposite sides of the vector e. Let $X \in K \setminus S$ and $t \in q_l(X)$. Then, due to the condition of the corollary, $(\widehat{t, e}) < \alpha$. Hence we conclude that $(\widehat{t, u}) < (\widehat{u, e}) + (\widehat{e, t}) < (\pi/2) - \alpha + \alpha = \pi/2$. Therefore, for any $t \in q_l(X)$, where $X \in K \setminus S$ $(\widehat{t, u}) < \pi/2$. On the basis of the lemma it yields $(A \widehat{\ } B) < \pi/2$. We have:

$$(\widehat{AB, e}) = (\widehat{\overline{AB}, u}) - (\widehat{e, u}) = (\widehat{\overline{AB}, u}) - \frac{\pi}{2} + \alpha.$$

This leads us to the conclusion that $(\widehat{\overline{AB}, e}) < \alpha$, which is the proof of the corollary.

As an application of Lemma 3.3.1 we get a certain criterion of the existence of a tangent in the strong sense.

Let us define the notion of the limit of contingencies at a curve point. Let there be given a curve $K = [AB]$ in the space E^n, X be a point of the curve K. Let $X < B$ $(A < X)$ and let M be a subset of the arc (XB) (the arc AX), for which X is a limiting point. Let us say that the straight line t passing through the point X is a limit of the contingencies $q_l(Y)$ when Y tends to X from the right (respectively, from the left) via the set M, if for any $Y \in U \cap M$ any straight line $t \in q_l(Y)$ forms with t an angle less than ε. If in this formulation we substitute q_l for q_r then we shall get a definition of the notion of a limit of the right-hand contingencies at Y tending to X from the right (respectively, from the left) via the set M.

THEOREM 3.3.1. *Let us assume that the straight line t is a right-hand (left-hand) tangent in the strong sense at the point X of the curve K = [AB]. In this case t is a limit for both the right-hand and left-hand contingencies at the point Y, when Y tends to X from the right (respectively, from the left) along the curve K. Inversely, if t is a limit for the contingency $q_l(Y)$ at Y → X tending to X from the right (from the left) via the set M, such that S = [AB]\M is not greater than countable, then t is a right-hand (respectively, left-hand) tangent in the strong sense at the point X of the curve K. The left-hand contingency in the latter formulation can be replaced with the right-hand one.*

Proof. For the sake of definitness let us consider the case of a right-hand tangent. Let $t = t_r(X)$ be a right-hand tangent in the strong sense at the point X of the curve K. Let us arbitrarily set $\varepsilon > 0$ and find a right-hand semi-neighbourhood U of the point X on the curve K, such that any secant $l(Y, Z)$ of the arc U forms with t an angle less than $\varepsilon/2$. Let us choose an arbitrary point $Y \in U$. Any straight line $t \in q_l(Y) \cup q_r(Y)$ is a limit for the secants $l(Y, Y_m)$ when Y_m tends to Y along the arc U. Therefore, any straight line $t \in q_l(Y) \cup q_r(Y)$ comprises with t an angle not greater than $\varepsilon/2 < \varepsilon$. Due to the arbitrariness of $\varepsilon > 0$ the first statement of the theorem is proved.

Let us now assume that t is a limit for the left-hand contingencies $q_l(Y)$ when Y tends to X from the right via the set M, such that $[XB]\M$ is not greater than countable. Let us arbitrarily choose $\varepsilon > 0$, $\varepsilon < \pi/2$ and find a semi-neighbourhood $U = (XZ)$ of the point Z, such that for any $Y \in (XZ) \cap M$ any straight line $t \in q_l(X)$ forms with t an angle less than ε. Let us arbitrarily choose a secant $l(Y_1, Y_2)$ of the arc (XZ). Applying the corollary of Lemma 3.3.1 to the arc $[Y_1Y_2]$ we see that the straight line $l(Y_1, Y_2)$ forms an angle less than ε with t. In as much as $\varepsilon > 0$ is arbitrary, it means that t is a right-hand tangent in the strong sense at the point X of the curve K. The theorem is proved.

The following two corollaries to the theorem should be noted.

COROLLARY 1. *Let K be a one-sidedly smooth curve, and let X be an arbitrary point of the curve. In this case if the points (X_m), m = 1, 2, ..., converge to the point X from the left (right) then the tangents $t_l(X_m)$ and $t_r(X_m)$ converge to the left-hand (right-hand) tangent at the point X of the curve K.*

Let K be an arbitrary curve, at every internal point of which there exist both one-sided tangents, the right-hand tangent existing at the beginning, and the left-hand one at the end. A curve is called smooth if at every point X both one-sided tangents coincide, i.e. $t_l(X) = t_r(X)$, and each time when the points X_m of the curve converge along the curve to the point X, $t(X_m) \rightarrow t(X)$.

COROLLARY 2. *For a curve to be smooth it is necessary and sufficient that it be one-sidedly smooth and have no angular points.*

THEOREM 3.3.2. *A set of angular points of any one-sidedly smooth curve is not greater than countable.*

Proof. Let S_m be a set of all points of the curve K, at which the angle between the left-hand and the right-hand tangents exceed $\pi/(m+1)$. The set S_m is finite since in the opposite case there would exist a point X_0 which is a limit of the sequence of the points (X_k), $k = 1, 2, \ldots$, of the set S_m which are located to one side of the point X_0. In line with Corollary 1 of Theorem 3.3.1, the tangents $t_l(X_k)$ and $t_r(X_k)$ converge to one and the same straight line. This, however, contradicts the fact that an angle between them is always greater than $\pi/(m+1)$. A set of all angular points of the curve K is $\bigcup_{m=1}^{\infty} S_m$, which means that it is not greater than countable. The Theorem is proved.

REMARK. As follows from the proof of Theorem 3.3.2, for any one-sidedly smooth curve a set of angular points at which an angle between the left-hand and the right-hand tangents is greater than a certain $\alpha < 0$ is finite. In particular, any one-sidedly smooth curve has a finite number of the points of return.

THEOREM 3.3.3. *Let K be a one-sidedly smooth curve and let $x(s)$ be a radius-vector of the point $X(s)$ of the curve with respect to a fixed point O in the space (the parameter s, $0 \leqslant s \leqslant l$, is the arc length read off from the beginning). In this case at every s the vector-function $x(s)$ has both left-hand and right-hand derivatives $x_l'(s)$ and $x_r'(s)$, in which case $|x_l'(s)| = |x_r'(s)| = 1$. The vector-function $x_l'(s)\,(x_r'(s))$ is continuous from the left (respectively, from the right) at all $s \in [0, l]$, and at every s $\lim_{t \to s-0} x_r(t)$ $(\lim_{t \to s+0} x_l'(t))$ exists and is equal to $x_l'(s)$ (respectively, $x_r'(t)$).*

Proof. Let $s_0 \in [0, l]$. In this case we have:

$$\frac{x(s)-x(s_0)}{s - s_0} = \frac{x(s)-x(s_0)}{|x(s)-x(s_0)|} \cdot \frac{|x(s)-x(s_0)|}{s - s_0}.$$

At $s \to s_0+0$ the unit-vector $[x(s)-s(s_0)]/|x(s)-s(s_0)|$ converges to the vector $e_r[x(s_0)]$. The quantity $|x(s)-x(s_0)|$ is the length of the chord connecting the points $x(s_0)$ and $x(s)$, and, according to Lemma 3.1.1, at $s \to s_0$

$$\frac{|x(s) - x(s_0)|}{|s - s_0|} \to 1.$$

The remaining statements of the theorem obviously result from the above proved properties of one-sidedly smooth curves.

THEOREM 3.3.4. *Let us assume that the curve K in E^n allows the parametrization $x(t)$, $a \leqslant t \leqslant b$, such that the following conditions are fulfilled:*

(1) the vector-function $x(t)$ has a left-hand derivative

$$x_l'(t) = \lim_{u \to t - 0} \frac{x(t) - x(u)}{t - u}$$

for all $t \in M$ where M is a set of all points of the length $[a, b]$, such *that $[a, b] \setminus M$ is not more than countable, in which case $x_i'(t) \neq 0$ for all such t;*

(2) *the function $x_i'(t)$ has other than zero limits over the set M from the left and from the right at every point $t_0 \in (a, b)$ on the right at the point a, on the left at the point b.*

In this case the curve K is one-sidedly smooth.

REMARK. In the formulation of the theorem the left derivative can be replaced with the right one.

Proof. Let the curve K have the parametrization which satisfies all the conditions of the theorem. Let N be a set of all the points X of the curve K which correspond to the values of $t \in M$. The set $K \setminus N$ is not greater than countable. Condition (1) yields that for any $X \in K \setminus N$ the relation

$$\frac{\overline{XY}}{|XY|}$$

has a limit at Y tending to point X from the left. Therefore, the curve K has a left-hand tangent at every point $X \in K \setminus N$. Condition (2) makes it possible to conclude that at every point X of the curve K the limits from the left and from the right of the left-hand contingency over the set $K \setminus N$ consist of a single straight line. In view of Theorem 3.3.1 it leads one to the conclusion that the curve K has a left-hand and a right-hand tangents in the strong sense at every point X and, hence, is one-sidedly smooth. The Theorem is proved.

3.4. One–Sidedly Smooth Functions

3.4.1. Let us first make some remarks referring to mathematical analysis.

In connection with the notion of a one-sidedly smooth curve it would be natural to introduce the notion of a one-sidedly smooth function of a real variable.

Let there be given a continuous function $f : [a, b] \to \mathbb{R}$. Let us choose an arbitrary point $t_0 \in [a, b]$. The function f is said to be strictly differentiable from the right (left) at the point t_0 if the relation

$$\frac{f(u) - f(t)}{u - t}$$

tends to a finite limit when $u \to t_0$, $t \to t_0$ from the right (left) to the point t_0 along the segment $[a, b]$. In this case the function f, obviously, has a right (respectively, left) derivative $f_r'(t) (f_l'(t))$ at the point t_0, this derivative coinciding with the limit given here. Let us assume that the function $f : [a, b] \to \mathbb{R}$ is one-sidedly smooth if it is strictly differentiable from the right at every point $t \in [a, b)$ and strictly differentiable from the left at all points $t \in (a, b]$. The point $t_0 \in (a, b)$ is called angular for the function f if $f_l'(t_0) \neq f_r'(t_0)$.

The following suppositions are proved by analogy with the case for curves.

THEOREM 3.4.1. *A set of angular points of a one-sidedly smooth function* $f:[a, b] \to \mathbb{R}$ *is not greater than countable.*

THEOREM 3.4.2. *Let the function* $f:[a, b] \to \mathbb{R}$ *be strictly differentiable from the right (left) at the point* $t_0 \in [a, b)$ *(respectively,* $t_0 \in (a, b]$*) and there exists a* $\delta > 0$*, such that* $t_0 + \delta \leqslant b$ *(*$a \leqslant t_0 - \delta$*) and at all the points of the interval* $(t_0, t_0 + \delta)$ *(the interval* $(t_0 - \delta, t_0)$*) there exist derivatives* $f'_l(t)$ *and* $f'_r(t)$*. In this case if* $t < b$*, then* $f'_r(t_0) = \lim_{t \to t_0 + 0} f'_l(t) = \lim_{t \to t_0 + 0} f'_r(t)$*, and if* $a < t$*, then* $f'_l(t) = \lim_{t \to t_0 - 0} f'_l(t) = \lim_{t \to t_0 - 0} f'_r(t)$*.*

THEOREM 3.4.3. *Let there be given a function* $f:[a, b] \to \mathbb{R}$ *and a point* $t_0 \in [a, b)$ *(*$t_0 \in (a, b]$*). Let us assume that there exists a neighbourhood* $U = (t_0 - \delta, t_0 + \delta) \subset [a, b]$ *of the point* t_0*, such that at every point* $t \in U$*, other than* t_0 *and lying to the right of* t_0 *(to the left of* t_0*) the function* f *has finite derivatives* $f'_l(t)$ *and* $f'_r(t)$*. In this case if the limits*

$$\lim_{t \to t_0 + 0} f'_l(t), \quad \lim_{t \to t_0 + 0} f'_r(t)$$

(respectively, the limits $\lim_{t \to t_0 - 0} f'_l(t)$*,* $\lim_{t \to t_0 - 0} f'_r(t)$*) do exist, are finite and equal one another, then* f *is strictly differentiable from the right (left) at the point* t_0*.*

Correlating the above theorems with Theorem 3.3.3 we come to the conclusion that for a rectifiable curve x to be one-sidedly smooth it is necessary and sufficient that its parametrization $x(s)$, where the parameter s is the arc length, $0 \leqslant s \leqslant l$ be a one-sidedly smooth function, i.e. each of its components $x_1(s)$, $x_2(s)$, ..., $x_n(s)$ with respect to the Cartesian orthogonal system of coordinates in E^n be a one-sidedly smooth function.

3.5. Notion of c–Correspondence. Indicatrix of Tangents of a Curve

3.5.1. Let K be a one-sidedly smooth curve in E^n, X be a point of the curve K. If the point X is smooth, then it is corresponded to by the only straight line which is a tangent of the curve at this point. Let us denote by $e(X)$ a directing unit vector of this tangent. Let us assume that X is an angular point. In this case the left-hand and the right-hand tangential unit vectors $e_l(X)$ and $e_r(X)$, and $e_l(X) \neq e_r(X)$ are defined. The vectors $e_l(X)$ and $e_r(X)$ are points on the unit sphere Ω^{n-1}. Let us connect them on Ω^{n-1} with the shortest line. If X is not a point of return, then this shortest line is the only one. If $(e_l(X) \frown e_r(X)) = \pi$, so that $e_l(X) = -e_r(X)$, the points $e_l(X)$ and $e_r(X)$ are connected on Ω^{n-1} with an infinite number of shortest lines, all of them having the length π. In this case let us arbitrarily choose one of them, denoting the arc of the circumference of a large circle so obtained, connecting the points $e_l(X)$ and $e_r(X)$ by $\tau(X)$. Any straight line passing through the point X, the directing unit vector of which is an arbitrary vector $e \in \tau(X)$, will be called an intermediate tangent of the curve K at the point X. All intermediate tangents at point X lie in one two-dimensional plane, i.e., in the plane passing through the straight lines $e_l(X)$ and $e_r(X)$. In the case of a point of return this plane is uniquely determined.

A set of tangents of the one-sidedly smoothe curve K (including the intermediate tangents at angular points) becomes ordered if we assume that in the case when $X < Y$ any tangent at the point X precedes any tangent at the point Y, and if we obtain ordered tangents at one point X in the following way. Let e_1 and e_2 be the directing unit vectors of the intermediate tangents at the point X. In this case let us assume that among of these tangents the preceding one is that for which the corresponding unit vector on the arc $\tau(X)$ is located closer to the point $e_l(X)$. There naturally arises an idea that a set of all tangents of the curve K can be considered as a curve in a certain space (namely, in a space the elements of which are directed straight lines in E^n). Let us introduce a general notion with the help of which the above supposition can attain an exact shape.

Let there be given metrizable topological spaces $\mathfrak{M}$ and $\mathfrak{N}$ and a curve K in the space $\mathfrak{M}$. Let us also assume that to every point X of the curve K there corresponds a certain curve $\gamma(X)$ in the space $\mathfrak{N}$. The curve L in the space U is called an indicatrix of the correspondence γ or, briefly, a γ-indicatrix of the curve K if to every point $X \in K$ there can be placed in correspondence a closed arc $\tilde{\gamma}(K)$ of the curve K in such a way that the following conditions be fulfilled:

(1) as an independent curve, the arc $\tilde{\gamma}(X)$ coincides with the curve $\gamma(X)$ at every X;
(2) each point of the curve L belongs to at least one of the arcs $\tilde{\gamma}(X)$;
(3) if $X_1 < X_2$, and $Y_1 \in \tilde{\gamma}(X)$, $Y_2 \in \tilde{\gamma}(X_2)$ are arbitrary points of the arcs $\tilde{\gamma}(X_1)$ and $\tilde{\gamma}(X_2)$, then $Y_1 \leqslant Y_2$.

Let us now consider the case of one-sidedly smooth curves in the space E^n. Let K be such a curve. Then we can correspond to every one of its points X a certain curve on the sphere Ω^{n-1}. Let us assume that X is an angular point. In this case we have defined the arc $\tau(X)$ of the circumference of a large circle. Let us arbitrarily set a topological mapping φ of a segment $[0, 1]$ onto the arc $\tau(X)$, such that $\varphi(0) = e_l(X)$, $\varphi(1) = e_r(X)$, and consider a curve one of the parametrizations of which is the mapping φ. This curve, obviously, is independent of the choice of the mapping φ, which satisfies the above conditions, and it will also be denoted by $\tau(X)$. If X is a smooth point, then let $\tau(X)$ be a curve degenerating into a point and coinciding with $e(X)$. Therefore, to every point X of the curve K there corresponds a certain curve $\tau(X)$ on the sphere Ω^{n-1}. If the curve L on Ω^{n-1} is an indicatrix for the correspondence τ defined in such a way, then we shall say that L is an indicatrix of the tangents of the curve K.

Let us state here some general conditions which are necessary and sufficient for the described type of correspondence $X \mapsto \gamma(X)$ to have an indicatrix. As will be further demonstrated, the correspondence τ defined here for one-sidedly smooth curves satisfies these conditions. In this way we shall establish the existence of an indicatrix of the tangent for a one-sidedly smooth curve in K.

When passing from a curve to its indicatrix of the tangents a certain loss of information occurs. It would therefore be expedient to consider another object, which will be called a complete indicatrix of the curve tangents.

Let us denote through $\hat{T}(E^n)$ the product $E^n \times \Omega^{n-1}$. The elements $\hat{T}(E^n)$ are pairs (X, u), where u is an arbitrary unit vector in E^n, $X \in E^n$. Let us also define the mappings $v:(X, e) \mapsto X$ and $h:(X, u) \mapsto u$. The former will be referred to as a vertical projection of the space $\hat{T}(E^n)$, and the latter as the horizontal one.

Let K be a one-sidedly smooth curve in E^n. For every one of its points X there is a certain curve $\tau(X)$ defined, as was described above, on the sphere Ω^{n-1}. In $\hat{T}(E^n)$ let us construct a curve $\omega(X)$ by $\tau(X)$. Namely, let $\xi(t)$, $a \leqslant t \leqslant b$ be an arbitrary parametrization of $\tau(X)$. Let us assume $\eta(t) = (X, \xi(t))$, $a \leqslant t \leqslant b$. Then η is a parametrized curve lying in the space $\hat{T}(E^n)$. It has the following property: $v[\eta(t)] = X$ for all $t \in [a, b]$. The curve in $\hat{T}(E^n)$, the parametrization of which is η, will be denoted by $\omega(X)$. Obviously, $\omega(X)$ is independent of the choice of the parametrization of the arc $\tau(X)$ and $v[\omega(X)] = X$. The correspondence ω defined here will be referred to as a tangential correspondence of the curve K. The curve M in the space $\hat{T}(E^n)$, which is an indicatrix of the tangential correspondence, is called a complete indicatrix of the tangents of the curve K. Obviously, if M is a complete indicatrix of the tangents of the curve K, then $K = v(M)$, and $h(M)$ is its common indicatrix of the tangents. Therefore the curve M is one-to-one defined by its complete indicatrix of the tangents and, hence, no loss of information on the curve occurs when passing to its complete indicatrix. Such an indicatrix of the tangents, on the other hand, gives a convenient for application description of the differentially-geometrical structure of the curve K.

Let us now find the conditions under which the correspondence $\gamma:X \rightarrow \gamma(X)$ guarantees the existence of an indicatrix of correspondence.

Let us assume that a sequence of the curves (K_m), $m = 1, 2, \ldots$, in a topological space $\mathcal{R}$ converges to a point X_0, if for any neighbourhood U of the point X_0 there exists a number m_0 such that, at $m > m_0$, K_m lies in the neighbourhood U.

Let us arbitrarily set metrizable topological spaces $\mathfrak{M}$ and $\mathfrak{N}$. Let K be an arbitrary curve in $\mathfrak{M}$. Let us assume that to every point $X \in K$ there corresponds a certain curve $\gamma(X)$ in the space $\mathfrak{N}$. We shall say that $\gamma:X \rightarrow \gamma(X)$ is a c-correspondence provided the following condition is fulfilled.

If the points (X_m), $m = 1, 2, \ldots$, of the curve K converge along the curve to the point X_0 from the right (left), then the curves $\gamma(X_m)$ converge in the space $\mathfrak{R}$ to the end (respectively, to the beginning) of the curve $\gamma(X_0)$.

Let us now show that the correspondence ω for one-sidedly smooth curves constructed above is a c-correspondence. Let (X_m), $m = 1, 2, \ldots$, be an arbitrary sequence of points of the curve K converging to a certain point X_0 from one side; for instance, from the right. According to Corollary 1 of Theorem 3.3.1, at $m \to \infty$ $e_l(X_m) \to e_r(X_0)$ and $e_r(X_m) \to e_r(X_0)$, whence we easily conclude that the curves $\omega(X_m)$ converge to the point $(X_0, e_r(X_0))$ in (E^n), which is the required proof. Because of the equality $\tau = h \circ \omega$, we see that the correspondence τ constructed above is also a c-correspondence.

LEMMA 3.5.1. *If for the correspondence $X \mapsto \gamma(X)$ defined for the curve K an indicatrix exists, then γ is a c-correspondence.*

Proof. Let $\tilde{\gamma}$ be the correspondence between the points of the curve K and the arcs of its γ-indicatrix L as has been discussed in the definition of a γ-indicatrix. Let (X_m), $m = 1, 2, \ldots$, be an arbitrary sequence of the points of the curve K converging to the point X from the left, let Y_m and Z_m be the terminal points of the arc $\gamma(X_m)$ and Y_0 be the beginning of the arc $\gamma(X)$. It is obvious that we always have $Y_m \leqslant Z_m \leqslant Y_0$ and to prove the lemma it is sufficient to demonstrate that $Y_m \to Y_0$ at $m \to \infty$.

If Y_0 is the beginning of the curve L, then $Y_m = Z_m = Y_0$ for all m and in this case there is nothing to prove. Let us then assume that Y_0 is not the beginning of L. Let us arbitrarily set a point $Y' \in L$, such that $Y' < Y_0$ and let the point X' be such that $Y' \in \tilde{\gamma}(X')$. In this case $X' \leqslant X$, since if $X < X'$ then $Y_0 \leqslant Y'$. The equality $X' = X$ is impossible since in the opposite case $Y' \in \tilde{\gamma}(X)$ and $Y_0 \in \tilde{\gamma}(X)$, which contradicts the fact that Y_0 is the origin of the arc $\tilde{\gamma}(X)$. Therefore $X' < X$. Since $X_m \to X$ from the left, then there is m_0 such that at $m \geqslant m_0$ $X' < X_m < X$. For such m we have: $Y' \leqslant Y_m \leqslant Z_m \leqslant Y$. In so far as $Y' < Y_0$ has been arbitrarily chosen, we have thus established that $Y_m \to Y$ and $Z_m \to Y$ at $m \to \infty$.

In an analogous way we can establish that if $X_m \to X$ from the right, then $\gamma(X_m) \to Z$, i.e., to the end of the arc $\gamma(X)$.

The lemma is proved.

The above lemma shows that if for the curve K there is a certain given correspondence γ, then for a γ-indicatrix to exist it is necessary that it be a c-correspondence. Theorem 3.5.1, which will be proved below, allows us to conclude that these conditions are also sufficient. Let us first investigate certain properties of c-correspondeces and γ-indicatrices.

LEMMA 3.5.2. *To every point X of the curve K let there be a corresponding curve $\gamma(X)$. In this case, if the correspondence γ is a c-correspondence, then the set of points $X \in K$ for which the curve $\gamma(X)$ does not degenerate into a point is not greater than countable.*

Proof. The curve K lies in a topological space $\mathfrak{M}$ and at every $X \in K$ $\gamma(X)$ is a curve in a metrizable topological space $\mathfrak{N}$. Let us introduce into $\mathfrak{N}$ a metric ρ in such a way that the topology defined by ρ coincides with that of the space $\mathfrak{N}$. Let E_k be a set of all points of the curve K for which the diameter of the curve $\gamma(X)$ is greater than $1/k$. Let us assume that the set E_k is infinite. In this case we can single out from it a certain subsequence (X_m), $m = 1, 2, \ldots$, converging from the left or from the right to a certain point X_0 of the curve K. Due to the definition of a c-correspondence, the curves $\gamma(X_m)$ converge to one of the terminal points of the curve $\gamma(X_0)$ and, hence, at $m \to \infty$ their diameters tend to zero. This fact, however, contradicts the condition that the diameters of all the curves $\gamma(X)$ where $X \in E_k$ are greater than $1/k$. The contradiction obtained proves that the set E_k is finite. The union of all sets E_k coincides with the set of all points X for which $\gamma(X)$ does not degenerate into a point. Therefore this set is not greater than countable, which is the required proof.

To every point X of the curve K let there correponds a curve $\gamma(X)$. Let us assume that for the correspondence γ there exists an indicatrix L. Let us denote by $\tilde{\gamma}(X)$ the arc of the curve which corresponds to the point X of the curve K by the definition of a γ-indicatrix. Let Y be a point of the curve L. Let us denote through $\delta(Y)$ a set of all points X of the curve K such that $Y \in \gamma(X)$. Therefore, to every point of the curve L there is brought into a correspondence a certain set $\delta(Y)$ of the points of the curve K.

LEMMA 3.5.3. *For any* $Y \in L$ *the set* $\delta(L)$ *is a closed arc of the curve* K *and the curve* K *is a* δ-*indicatrix of* L.

Proof. Let $Y \in L$. If $\delta(Y)$ consists of the only point, then $\delta(Y)$ is a closed arc of the curve K. Suppose that this is not the case. Arbitrarily choose the points $X' \in \delta(Y)$ and $X'' \in \delta(Y)$, such that $X' < X''$. Let $\gamma(X') = [P'Q']$, $\gamma(X'') = [P''Q'']$. In this case $Q' \leqslant P''$, $Y \in [P'Q']$ and $Y \in [P''Q'']$. This is only possible when $Q' = Y = P''$. For any X such that $X' < X < X''$ all the points of the arc $\gamma(X)$ must be located between Q' and P'', i.e., for all such X, $\gamma(X) = Y$. In particular, we come to the conclusion that all the points of the arc $[X'X'']$ belong to $\delta(Y)$.

It follows from what has been proved above that $\delta(Y)$ is a certain arc. Let R and T be, respectively, its beginning and its end, $R < T$. Let us arbitrarily set a sequence (X_m) of the points of the arc (RT) such that $X_m \to R$ at $m \to \infty$. Let $[Y_m Z_m] = \gamma(X_m)$, and let Z_0 be an end of the arc $\gamma(R)$. At $m \to \infty$ the arcs $\gamma(X_m)$ converge to Z_0. For every m $Y \in [Y_m Z_m]$. This gives $Y = Z_0$, i.e., $Y \in \gamma(R)$ and, hence, $R \in \delta(Y)$. In an analogous way we establish that $T \in \delta(Y)$ and we, hence, find that $\delta(Y) = [RT]$. Therefore, we have proved that $\delta(Y)$ is a closed arc of the curve K at any $Y \in L$.

Let us now demonstrate that K is a δ-indicatrix of the curve L. Conditions (1) and (2) of the definition of an indicatrix are automatically fulfilled here.

Let us prove that the condition (3) is also fulfilled. Let $Y_1 < Y_2$ and let $X_1 \in \delta(Y_1)$, $X_2 \in \delta(Y_2)$. Let us assume that $X_2 < X_1$. In as much as $Y_1 \in \gamma(X_1)$, $Y_2 \in \gamma(X_2)$, then $Y_2 \leqslant Y_1$, which contradicts the condition. Therefore $X_1 \leqslant X_2$ and the lemma is proved.

LEMMA 3.5.4. *Let us assume that to each point X of the curve K there corresponds a certain curve $\gamma(X)$. In this case if the correspondence γ has an indicatrix, then it is the only indicatrix.*

Proof. Let L_1 and L_2 be γ-indicatrices of the curve K. The task is to prove that $L_1 = L_2$.

For $X \in K$ let $\gamma_1(X)$ and $\gamma_2(X)$ be the arcs of the curves L_1 and L_2 corresponding to the point X in line with the definition of a γ-indicatrix, and let $\gamma_1(X)$ and $\gamma_2(X)$, as independent curves, coincide with $\gamma(X)$.

Let us denote by $L_1 \times L_2$ a set of all pairs (Y_1, Y_2), where Y_1 is a point of the curve L_1, and Y_2 is a point of the curve L_2. In the set $L_1 \times L_2$ let us choose a certain subset Δ. Namely, let us assume that a pair $(Y_1, Y_2) \in \Delta$ if and only if there exists a point X of the curve K such that $Y_1 \in \gamma_1(X)$, $Y_2 \in \gamma_2(X)$ and the points Y_1 and Y_2 coincide as the points of the curve $\gamma(X) = \gamma_1(X) = \gamma_2(X)$.

For every point $Y \in L_1$ there can be found a pair $(Y_1, Y_2) \in \Delta$, such that $Y = Y_1$. Indeed, let X be such that $Y \in \gamma_1(X)$. Let us set $Y_1 = Y$ and let Y_2 be a point of the arc $\gamma_2(X)$, which coincides with Y_1 as a point of the curve $\gamma(X)$. The obtained pair (Y_1, Y_2) is the required one. In the same way we can conclude that for any $Y \in L_2$ there is a pair $(Y_1, Y_2) \in \Delta$, such that $Y_2 = Y$.

Let the pairs $(Y_1, Y_2) \in \Delta$ and $(Z_1, Z_2) \in \Delta$ be such that $Y_1 < Z_1$. Let us prove that in this case $Y_2 < Z_2$ as well. Indeed, let $X_1 \in K$ and $X_2 \in K$ be such that Y_1 and Y_2 are the corresponding points of the arcs $\gamma_1(X_1)$ and $\gamma_2(X_1)$, and Z_1 and Z_2 are the corresponding points of the arcs $\gamma_1(X_2)$ and $\gamma_2(X_2)$. In this case $X_1 \leqslant X_2$ since, if it were $X_2 < X_1$, then the relation $Z_1 \leqslant Y_1$ would be fulfilled, which contradicts the condition. If $X_1 = X_2 = X$, then Y_1 and Z_1 lie on the arc $\gamma_1(X)$ and, as the points of the curve $\gamma(X)$, they coincide with Y_2 and Z_2, respectively. As $Y_1 < Z_1$, then the relation $Y_2 < Z_2$ is also fulfilled. Let us assume that $X_1 < X_2$. Then $Y_2 \leqslant Z_2$. Our aim, however, is to prove that $Y_2 < Z_2$, i.e., the case when $Y_2 = Z_2$ should be excluded. Indeed, let us assume that, on the contrary, $Y_2 = Z_2$. For any X such that $X_1 < X < X_2$ all the points of the arc $\gamma_2(X)$ must be located to the right of the point $Y_2 \in \gamma_2(X_1)$ or coincide with it, and to the left of the point $Z_2 \in \gamma_2(X_2)$ or coincide with it. In as much as Y_2 and Z_2 are one and the same point of the curve L_2, it is possible only if $\gamma_2(X)$ coincides with $Y_2 = Z_2$. Therefore, for all X such that $X_1 < X < X_2$, the curve $\gamma_2(X)$ consists of the only point $Y = Y_2 = Z_2$. Hence, for all X such that $X_1 < X < X_2$, the curve $\gamma(X)$ consists of the only point. For any $Y \in \gamma_2(X)$ we have: $Y = Z_2 = Y_2$, whence we conclude that Y_2 is the end of the arc $\gamma_2(X_1)$. In the same way we conclude that Z_2 is the beginning of the arc $\gamma_2(X_2)$. Hence, Y_1 is the end of the arc $\gamma_1(X_1)$, and Z_1 is the beginning of the arc $\gamma_1(X_2)$. The points Y_1 and Z_1 are spatially coincident. As far as $Y_1 <$ Z_1, then there can be found a point $T \in L$, such that $Y_1 < T < Z_1$ and T coincides spatially neither with Y_1 nor with Z_1. This fact, however, contradicts the condition that at $X_1 < X < X_2$ the curve $\gamma(X)$ consists of the only point which coincides spatially with both Y_1 and Z_1. The point T also cannot belong to the arc $\gamma_1(X_1)$ since Y_1 is the end of this arc and, by an analogous reason, T cannot belong to the arc $\gamma_1(X_2)$ as well. Therefore, having assumed that $Y_2 = Z_2$ we have obtained a contradiction.

So long as the curves L_1 and L_2 can be considered as equal partners, the above proof reveals that if the pairs $(Y_1, Y_2) \in \Delta$ and $(Z_1, Z_2) \in \Delta$ are such that $Y_2 < Z_2$, then $Y_1 < Z_1$ as well.

The above proof also shows that for any point $Y \in L_1$ the point $Z \in L_2$ is such that $(Y, Z) \in \Delta$ is the only one. Indeed, if there existed two different points $Z_1, Z_2 \in L_2$, such that $(Y, Z_1) \in \Delta$ and $(Y, Z_2) \in \Delta$, then we would get $Y < Y$, which is absurd. Corresponding to $Y \in L_1$ the point $\varphi(Y) \in L_2$, such that $(Y, \varphi(Y)) \in \Delta$, and we therefore get a mapping φ L_1 onto L_2. This mapping is a mapping 'onto'. At $Y_1 < Y_2$ we have: $\varphi(Y_1) < \varphi(Y_2)$, i.e., φ is a similarity of the curves L_1 and L_2 as ordered sets. It should be remarked that the points Y and $\varphi(Y)$ always coincide spatially. It follows from what has been proved above that the curves L_1 and L_2 coincide.

To every point X of the curve K let there correspond a curve $\gamma(X)$. Let us assume that there exists a γ-indicatrix L and that $\tilde{\gamma}$ is the correspondence between the points of the curve K and the arcs of the indicatrix, as was explained in the definition of a γ-indicatrix. The parametrizations $x(t)$ and $y(t)$, $a \leqslant t \leqslant b$, of the curves K and L are called γ-consistent, if for every $t \in [a, b]$ $y(t)$ is a point of the arc $\gamma[x(t)]$. The parametrization $x:[a, b] \to$ $\mathfrak{M}$ of the curve K is called γ-consistent if there exists a parametrization y of the curve L, such that the parametrizations x and y are γ-consistent.

3.5.2. Let us now prove the basic results of this section.

THEOREM 3.5.1. *To every point X of the curve K let there correspond a curve $\gamma(X)$. In this case if the correspondence γ is a c-correspondence, then the curve K has a γ-indicatrix. The curve and its γ-indicatrix allow γ-consistent parametrizations, and for any parametrization $x:[a, b] \to \mathfrak{M}$ of the curve K and for any $\varepsilon > 0$ we can find a γ-consistent parametrization $x_1:[a, b] \to \mathfrak{M}$, such that*

$$\forall t \in [a, b] \, \rho[x_1(t), x(t)] < \varepsilon.$$

(*Here ρ is a metric in $\mathfrak{M}$, such that the topology generated by it coincides with that given in $\mathfrak{M}$*).

Proof. The points X of the curve K for which $\gamma(X)$ does not degenerate into a point will be called special. A set of points, by Lemma 3.5.2, is not greater than countable.

Let $x:[a, b] \to \mathfrak{M}$ be such a parametrization of the curve K that to every special point there corresponds a segment $\lambda(X) = [\alpha, \beta]$, $\alpha < \beta$, wherein $x(t)$ is constant and coincides with the given special point X of the curve. The problem of constructing such a parametrization will be discussed below. Let us construct a certain function $y(t)$ by $x(t)$ in the following way.

Let $X_1, X_2, \ldots$ be all special points of the curve K and let α_k and β_k be the least and the largest values of the parameter $t \in [a, b]$, such that $x(t) = X_k$ at all $t \in [\alpha_k, \beta_k]$. For every k let us define a parametrization of the curve $\gamma(X_k)$ in such a way that the domain of the changes of the parameter t is the segment $[\alpha_k, \beta_k]$. Let $y_k(t)$, $\alpha_k \leqslant t \leqslant \beta_k$, be this parametrization.

If t does not belong to one of the segments $[\alpha_k, \beta_k]$, then the curve $\gamma[x(t)]$ degenerates into a point and in this case we set $y(t) = \gamma[x(t)]$. At $t \in [\alpha_k, \beta_k]$ we set $y(t) = y_k(t)$.

The function $y(t)$ is defined at all $t \in [a, b]$. Since γ is a c-correspondence, then y is continuous. The function $y(t)$ defines the curve M which, as is clear from the construction of $y(t)$, is the γ-indicatrix of the curve K.

It follows from the above considerations that the parametrization $x : [a, b] \rightarrow \mathfrak{M}$ is γ-consistent if to every special point there corresponds a segment of t values of a non-zero length wherein x is constant. Let us prove the existence of the parametrizations of this kind. Let $x_0 : [a, b] \rightarrow \mathfrak{M}$ be an arbitrary normal parametrization of the curve K, and let (t_m), $m = 1, 2, \ldots$, be a sequence of the parameter t values, such that the sequence $(x(t_m))$ contains all special points of the curve. Let us set

$$\lambda(t) = \begin{cases} \sum_{t_m \leqslant t} \dfrac{1}{2^m} & \text{at} \quad a < t \leqslant b, \\ 0 & \text{at} \quad t = a \end{cases} \tag{3}$$

At $t = t_m$ the function λ is considered to be multi-valued and to assume all the values from the segment $[\lambda(t-0), \lambda(t+0)]$. Now let

$$\psi_h(t) = h\lambda(t) + kt + l$$

where $0 < h < (b-a)/2$, and the numbers k and l are defined from the condition: $\psi_h(a) = a$, $\psi_h(b) = b$. The function is strictly increasing. It is continuous everywhere except at the points t_m. At each of these points it is multi-valued, its values filling in a certain closed segment of a non-zero length. Let now φ_h be a function inverse to ψ_h, $\varphi_h[\psi_h(t)] \equiv t$; the function φ_h being non-decreasing and continuous. To every point t_m there corresponds a segment wherein φ_h is continuous and equal to t_m. Assuming that $x_h(t) = x[\varphi_h(t)]$, we obtain the required parametrization of the curve K, as can easily be seen.

The function ψ_h is defined in the segment $[a, b]$. As is seen, the estimate $|\psi_h(t) - t| < Ch$ is valid, where C is a constant. Let $t \in [a, b]$ and let u be such that $\psi_h(u) = t$ (here the equality is understood in the sense that one of the values of the function $\psi_h(u)$ at the point u is equal to t). Then

$$u - Ch \leqslant \psi_h(u-0) \leqslant t \leqslant \psi_h(u+0) \leqslant u + Ch,$$

$$\varphi_h(u-Ch) \leqslant \varphi_h(t) = u \leqslant \varphi_h(u+Ch)$$

which yields at $h \to 0$ $\varphi_h(t) \to t$ uniformly in the segment $[a, b]$.

The function $x_h(t)$, $a \leqslant t \leqslant b$, is a parametrization of the curve K obeying all the required conditions. At $h \to 0$ $x_h(t) \to x(t)$ uniformly in the segment $[a, b]$, which means that for any $\varepsilon > 0$ there is an h_0 such that, at $0 < h < h_0$,

$$\rho[x_h(t), x(t)] < \varepsilon.$$

The theorem is proved.

Applying the theorem to the case when $\mathfrak{M} = E^n$, $\mathfrak{M} = (E^n)$ and γ is a tangential correspondence, we find that any one-sidedly smooth curve in E^n has a complete indicatrix of the tangents, which fact means that an ordinary indicatrix of the tangents also exists.

3.6. One–Sidedly Smooth Curves in Differentiable Manifolds

3.6.1. Let us introduce the following notion. Let $x:[a, b] \to \mathbb{R}^n$ be one-sidedly smooth if at every point $t \in (a, b]$ the function $x(t)$ has a left derivative $x_l'(t) = \lim_{h \to -0}(1/h)[x(t+h) - x(t)]$, and at every point $t \in [a, b)$ there exists a right derivative $x_r'(t) = \lim_{h \to +0}(1/h)[x(t+h) - x(t)]$, in which case the following conditions are fulfilled: (1) $x_l'(t) \neq 0$, $x_r'(t) \neq 0$ for any t, for which at least one of their derivatives is defined. (2) For any $t \in (a, b]$

$$x_l'(t) = \lim_{\tau \to t-0} x_l'(\tau) = \lim_{\tau \to t-0} x_r'(\tau)$$

and for any $t \in [a, b)$

$$x_r'(t) = \lim_{\tau \to t+0} x_l'(\tau) = \lim_{\tau \to t+0} x_r'(\tau).$$

According to Theorem 3.5.1, any one-sidedly smooth curve has a parametrization which is a one-sidedly smooth path, and vice versa: if the curve K has such a parametrization, it is one-sidedly smooth. Let U be an open set in $\mathbb{R}^n$, $f:U \to \mathbb{R}^n$ be a mapping of the class C^1, such that at every point $x \in U$ a Jacobian of the mapping f is other than zero. Let $x:[a, b] \to U$ be a one-sidedly smooth path, lying in U. In this case the path $y(t) = f(x(t))$ is evidently also one-sidedly smooth. Therefore, in particular, an image of any one-sidedly smooth curve $K \subset U$ at the mapping f is a one-sidedly smooth curve in $\mathbb{R}^n$.

Let M be an n-dimensional differential manifold of the class C^r, $r \geqslant 1$, K be an arbitrary curve in M. Let us say that the curve K is one-sidedly smooth if it can be subdivided into a finite number of arcs $L_1 = [X_{i-1}X_i]$, where $X_0 = A < X_1 < \cdots < X_{m-1} < X_m = B$ (A is the beginning, B is the end), for each of which there exists an admissible chart $\varphi_i:U_i \to \mathbb{R}^n$ of the manifold M, such that $L_i \subset U_i$ and $\varphi_i(L_i)$ is a one-sidedly smooth curve in R^n.

Making use of the above remark concerning an image of a one-sidedly smooth curve at a mapping of the class C^1 with a non-zero Jacobian, we can easily demonstrate that if K is a one-sidedly smooth curve in the manifold M, then for any admissible chart $\varphi:U \longrightarrow \mathbb{R}^n$ of the manifold M and any arc L of the curve K, which is contained in the domain of the definition of this chart, the curve $\varphi(L)$ in the space $\mathbb{R}^n$ is one-sidedly smooth.

The path $p:[a, b] \longrightarrow M$ in an n-dimensional manifold M of the class C^r, $r \geqslant 1$, is called one-sidedly smooth if for any admissible chart $\varphi:U \longrightarrow \mathbb{R}^n$ of the manifold M and any segment $[\alpha, \beta] \subset [a, b]$, such that $p([\alpha, \beta]) \subset U$, the path $x(t) = \varphi(p(t))$ in the space $\mathbb{R}^n$ is one-sidedly smooth. For the path $p:[a, b] \longrightarrow M$ to be one-sidedly smooth, it is evidently sufficient that there could be found such a subdivision of the segment $[a, b]$ into segments by the points $\alpha_0 = a < \alpha_1 < \cdots < \alpha_m = b$ and the admissible charts $\varphi_i:U_i \longrightarrow \mathbb{R}^n$, such that $p([\alpha_{i-1}, \alpha_i]) \subset U_i$ at every $i = 1, 2, \ldots, n$ and $x_i(t) = \varphi_i(p(t))$, $t \in [\alpha_{i-1}, \alpha_i]$ be a one-sidedly smooth path in $\mathbb{R}^n$.

Any one-sidedly smooth curve K in the manifold M of the class C^r, $r \geqslant 1$ allows the parametrization $p:[a, b] \longrightarrow M$, which is a one-sidedly smooth path in M.

For any one-sidedly smooth curve on a differential manifold the notion of a complete indicatrix can be defined, which can serve as a basis for constructing a certain classification of the curves on differential manifolds, which will be discussed below. Let us first consider certain general problems of the geometry of differentiable manifolds (in Sections 3.6.2 and 3.6.3).

3.6.2. Let M be an n-dimensional differential manifold of the class C^r, $r \geqslant 1$. Using it, we can construct a certain manifold $T(M)$ of the dimension $2n$ and class C^{r-1}, which will be called a tangential manifold for M. Since it is common practice to consider only the case $r = \infty$, then let us cite here the necessary definitions.

All the further considerations will refer to a certain fixed n-dimensional manifold M of the class C^r, where $r \geqslant 1$.

Let G be an open set in M. Let us assume that $f:G \longrightarrow \mathbb{R}$ is a function of the class C^s, where $1 \leqslant s \leqslant r$, if for any admissible chart $\varphi:U \longrightarrow \mathbb{R}^n$, where $U \subset G$, of the manifold M the function $\tilde{f} = f \circ \varphi^{-1}$ belongs to the class $C^s(V)$, where $V = \varphi(U)$ is an open set in $\mathbb{R}$. Let $p \in U$, $q = \varphi(p)$. The derivatives $\partial f(q)/\partial t_i$, $i = 1, 2, \ldots, n$, will henceforth be denoted by the symbol $\partial f(p)/\partial t_i$. For every point $p \in U$ there are n real numbers defined $t_i(p)$, $i = 1, 2, \ldots, n$ which are components of the point $\varphi(p)$ in $\mathbb{R}^n$. The functions $t_i:U \longrightarrow \mathbb{R}$ are called the coordinate functions corresponding to the chart φ. It is evident that they belong to the class C^r. Let $\varphi:U \longrightarrow \mathbb{R}^n$ and $\psi:U \longrightarrow \mathbb{R}^n$ be two admissible charts in M, let $t_1, t_2, \ldots, t_n$ be the coordinate functions of the former chart, and let $u_1, u_2, \ldots, u_n$ be those of the latter chart. For the function f of the class C^s, $1 \leqslant s \leqslant r$, which is defined in U, let $\partial f(p)/dt_i$ be its derivatives with respect to the coordinates t_i, while $\partial f(p)/\partial u_j$ be the derivatives of the same function with respect to the coordinates u_j. Formally:

$$\frac{\partial f}{\partial t_i}(p) = \frac{\partial(f \circ \varphi^{-1})}{\partial t_i} [\varphi(p)], \quad \frac{\partial f}{\partial u_j}(p) = \frac{\partial(f \circ \psi^{-1})}{\partial u_j} [\psi(p)].$$

In conformity with the rule of differentiating a superposition, known from the analysis, we have:

$$\frac{\partial f}{\partial u_j}(p) = \sum_{k=1}^{n} \frac{\partial f}{\partial t_k}(p) \frac{\partial t_k}{\partial u_j}(p).$$

In accordance with the well-known relation in differential geometry, let us assume that a summation is carried out with respect to the index repeated twice from 1 to n, so that the latter equality can be simplified to the form:

$$\frac{\partial f}{\partial u_j}(p) = \frac{\partial f}{\partial t_k}(p) \frac{\partial t_k}{\partial u_j}(p) \tag{4}$$

Let p be an arbitrary point of the manifold M. Let us denote by $C'(p)$ the totality of all real functions of the class C^1, each of which is defined in the neighbourhood of the point p (its own for any function $f \in C^1(p)$). Let us call a vector at the point p of the manifold M any functional $z : C^1(p) \to \mathbb{R}$, such that the following condition is fulfilled. There exists an admissible chart $\varphi : U \to \mathbb{R}^n$ and numbers $z_1, z_2, \ldots, z_n \in \mathbb{R}$, such that for any function $f \in C^1(p)$ the following equality holds:

$$z(f) = \sum_{i=1}^{n} \frac{\partial f}{\partial X_i}(p) z_i \tag{5}$$

where $x_1, x_2, \ldots, x_n$ are the coordinates with respect to the given chart φ.

Let us recall the properties of the vector which immediately result from the definition.

Let $\psi : V \to \mathbb{R}^n$ be another admissible chart defined in the neighbourhood of the point p. Let $y_1, y_2, \ldots, y_n$ be the coordinates with respect to the chart ψ. We have: $\partial f(p)/\partial x_k = \partial f(p)/\partial y_i \, \partial y_i(p)/\partial x_k$. Substituting this expression into (5) we get:

$$z(f) = \sum_{i=1}^{n} \frac{\partial f}{\partial y_i}(p) z_i \tag{6}$$

where

$$z_i' = \sum_{k=1}^{n} \frac{\partial y_i}{\partial x_k}(p) z_k \tag{7}$$

Therefore, the quantity $z(f)$ is expressed in terms of the derivatives of the function f by formula (5) in any admissible system of coordinates which is defined in the neighbourhood of p.

The numbers $z_1, z_2, \ldots, z_n$ are defined by the vector z in a one-to-one way. Indeed, let us set $f \equiv x_i$. We have: $\partial f/\partial x_i = 1$, $\partial f/\partial x_j = 0$ at $j \neq i$ and, hence, $z(f) = z_i$, i.e., $z_i = z(x_i)$. Let us say that $z_1, z_2, \ldots, z_n$ are the coordinates of the vector z with respect to the given chart $\varphi : U \to \mathbb{R}^n$. Equality (7) shows in which way one can find the vector coordinates with respect to one chart provided its coordinates in another chart are known.

A totality of all vectors at the point p of the manifold M is expressed by the symbol $T_p(M)$ and is called a tangential space of the manifold M at the point p. Let z and h be two arbitrary elements of the space $T_p(M)$. Let us arbitrarily select an admissible chart φ in the neighbourhood of the point p. In the case the quantities $z(f)$ and $h(f)$, where $f \in C^1(p)$ are expressed through the derivatives of f with respect to the coordinates in this chart and through the coordinates of the vectors z and h by the formulas:

$$z(f) = \frac{\partial f}{\partial t_k}(p)z_k; \quad h(f) = \frac{\partial f}{\partial t_k}(p)h_k.$$

We therefore find that for $\alpha, \beta \in \mathbb{R}$

$$(\alpha z + \beta h)(f) = \frac{\partial f}{\partial t_k}(p)\,(\alpha z_k + \beta h_k).$$

The functional $\alpha z + \beta h$ is, hence, also a vector at the point p. In this case the mapping, which brings the point $(z_1, z_2, \ldots, z_n) \in \mathbb{R}^n$ into correspondence with the vector z, where z_i are the coordinates of z with respect to the chart φ, is linear. This mapping, which will be henceforth denoted as $\varphi'(p)$, is evidently a one-to-one mapping of $T_p(M)$ onto $\mathbb{R}^n$. This fact, in particular, affords that $T_p(M)$ is an n-dimensional vector space.

Let $f \in C^1(p)$. Bringing a number $z(f)$ into correspondence to an arbitrary $z \in T_p(M)$, we get a linear functional on the vector space $T_p(M)$, which will be denoted through the symbol $f'(p)$. We have: $z(f) = f'(p)(z)$.

Let us consider an example that is important for further analysis. Let $q : [a, b] \to M$ be a one-sidedly smooth path in the manifold M, $p = q(t_0)$, where $a \leqslant t_0 < b$. Let f be a function of the class $C^1(p)$. In this case for a certain $\delta > 0$ the function $f[p(t)]$ is defined in the segment $(t_0, t_0 + \delta)$. Let us prove that it has a derivative $(D_r f)(p(t_o)) = \lim_{t \to +0}(f[p(t_0 + h)] - f[p(t_o)])/h$. Indeed, let φ be an arbitrary admissible chart defined in the neighbourhood of the point p, and let $x_1, x_2, \ldots, x_n$ be the coordinates with respect to this chart. Let us set $x(t) = \varphi[p(t)]$, $\tilde{f}(x) = f[\varphi^{-1}(x)]$. Let x be defined and be one-sidedly smooth in a certain segment $[t_0, t_0 + \delta]$. We have: $f[q(t)] = \tilde{f}[x(t)]$. The function $\tilde{f}$ belongs to the class C^1. In conformity with the known results of the calculus, this immediately implies the existence of a derivative

$$D_r f[q(t_0)] = \lim_{h \to +0} \frac{f[q(t_0 + h)] - f[q(t_0)]}{h}$$

$$= \lim_{h \to +0} \frac{\tilde{f}[x(t_0+h)] - \tilde{f}[x(t_0)]}{h} = \sum_{k=1}^{n} \frac{\partial f}{\partial X_k}(p) \; x'_{k,r}(t_0).$$

The functional $f \mapsto D_r f[q(t_0)]$ defined in such a way, is a vector at the point $p = q(t_0)$. Let us call it a right derivative of the path q at the point t_0 and denote it by the symbol $q'_r(t_0)$. In an analogous way a left derivative $q'_l(t_0)$ is defined, where $a < t_0 \leqslant b$, as a functional on $C^1(q(t_0))$, such that, for $f \in C^1[q(t_0)]$ $q'_r(t_0)(f) = \lim_{h \to -0}(f[q(t_0+h)] - f[q(t_0)])/h$.

Let M be an n-dimensional manifold of the class C^r, where $r \geqslant 1$. Let us set

$$T(M) = \bigcup_{p \in M} T_p(M).$$

For an arbitrary $h \in T(M)$ let $\pi(h) = p$, where p is such that $h \in T_p(M)$. In the manifold $T(M)$ the structure of the $2n$-dimensional differentiable manifold of the class C^{r-1} will be defined in the following way.

Let $\varphi : U \to M$ be an arbitrary admissible chart in M. Let us set $\tilde{U} = \pi^{-1}(U)$ and define the mapping $\tilde{\varphi} : \tilde{U} \to \mathbb{R}^{2n}$, setting for $z \in U$ $\tilde{\varphi}(z) = (\varphi(\pi(z)), \varphi'(\pi(z))z) \in \mathbb{R}^{2n}$. We have: $\tilde{\varphi}(z) = (t_1, t_2, \ldots, t_n, z_1, z_2, \ldots, z_n)$, where $t_1, t_2, \ldots t_n$ are the coordinates of the point $p = \pi(z)$, and $z_1, z_2, \ldots, z_n$ are those of the vector z with respect to the chart φ. Let us set $\varphi(U) = G$. It is obvious that $\tilde{\varphi}$ is a one-to-one mapping of $\tilde{U}$ onto the open set $G \times \mathbb{R}^n$ of the space $\mathbb{R}^{2n}$. Let $\psi : V \to \mathbb{R}^n$ be another arbitrary chart in M. Let us define the mapping ψ in an analogous way: $\tilde{\psi} : z \in V \mapsto (\psi[\pi(z)], \varphi'[\pi(z)](z)) \in \mathbb{R}^{2n}$. Let us assume that $U \cap V$ is not empty and see what the function $\tilde{\psi} \circ \tilde{\varphi}^{-1}$ is. Let $p \in U \cap V$, $z \in T_p(M)$, $t = \varphi(p)$, $\xi = \varphi'(p)(z)$, $u = \psi(p)$, $\eta = \psi'(p)(z)$. If we set $\psi \circ \varphi^{-1} = \theta$, then we have: $u = \theta(t)$. In line with (7) we have $\eta_i = \Sigma(\partial u_i/\partial t_k)(p)\xi_k$. It should be noted that $\partial u_i(p)/\partial t_k = \partial \theta_i(t)/\partial t_k$, and the expression of η in terms of ξ can be written in the following way: $\eta = \theta'(t)\xi$. Therefore, the transition function $\tilde{\psi} \circ \tilde{\varphi}^{-1}$ has the form: $(t, \xi) \mapsto (\theta(t), \theta'(t)\xi)$. In particular, we come to the conclusion that $\tilde{\psi} \circ \tilde{\varphi}^{-1}$ is a function of the class C^{r-1}. A totality of all the mappings $\tilde{\varphi} : \pi^{-1}(U) \to \mathbb{R}^{2n}$ corresponding to different admissible charts in M forms, as follows from what has been proved above, a certain atlas of the class C^{r-1} in the set $T(M)$, and, therefore, the structure of the differentiable manifold of the class C^{r-1} is defined in $T(M)$.

3.6.3. For an arbitrary manifold M of the class C^r let us now define another one which will be called a reduced tangential manifold for M.

Let R be an n-dimensional vector space. Let us call any directed straight line in R, passing through the point 0, an axis in R. Let λ be an arbitrary axis in R, and u be any directed vector along it. A ray formed by the vectors x of the type $x = \alpha u$, where $\alpha \geqslant 0$, will be denoted by the symbol λ^+ and will be called a positive semi-axis of the axis λ. If v is a non-zero vector from R, then the axis for which v is a directed vector will be denoted as $\lambda(v)$, and its positive semi-axis. $\lambda^+(v)$, respectively. A set of all axes of the vector space $\mathbb{R}$ will be denoted by $\hat{R}$.

Let M be an n-dimensional differentiable manifold of the class C^k. For any point $p \in M$ let the vector space $T_p(M)$ and the set $\hat{T}_p(M)$ of the axes in $T_p(M)$ be defined. Any axis in $T_p(M)$ will also be called a one-dimensional direction of the manifold M at the point p. Let us set

$$\hat{T}(M) = \bigcup_{p \in M} \hat{T}_p(M).$$

For arbitrary $\lambda \in \hat{T}(M)$ let us denote by $\hat{\pi}(\lambda)$ a point p of the manifold M, such that $\lambda \in \hat{T}_p(M)$.

Let $T^*(M)$ be a totality of all non-zero vectors $v \in T(M)$. The set $T^*(M)$ is open in $T(M)$. Let $v \in T^*(M)$, $p = \pi(v)$, and let $\lambda(v)$ be an axis in $T_p(M)$ with the directed vector v. We obtain a mapping $\lambda : v \mapsto \lambda(v)$ of the set $T^*(M)$ onto $\hat{T}(M)$.

Let us now introduce into $\hat{T}(M)$ the structure of a differentiable manifold of the class C^r. Let us first make the following remark. Let G be an open set in $\mathbb{R}^n$, and let Ω^{n-1} be a unit sphere in $\mathbb{R}^n$, $\Omega^{n-1} = \{x \in \mathbb{R}^n \mid |x| = 1\}$. The set $G \times \Omega^{n-1}$ is a $(2n-1)$-dimensional manifold of the class C^∞ (or even of the class C^ω).

Let $\varphi : U \to \mathbb{R}^n$ be an arbitrary admissible chart in $\mathbb{R}^n$. Let us set $\hat{u} = \hat{\pi}^{-1}(u)$, and let $G = \varphi(u)$ be an open set in $\mathbb{R}^n$. Let $\lambda \in \hat{T}p(M)$, where $p \in u$, v is a directing vector of λ. Let us set:

$$\hat{\varphi}(\lambda) = \left[\varphi(p), \frac{\varphi'(p)(v)}{|\varphi'(p)(v)|}\right]. \tag{8}$$

As far as $v \neq 0$ and $\varphi'(p)$ is a bijective mapping of $T_p(M)$ onto $\mathbb{R}^n$, then $\varphi'(p)(v) \neq 0$. The ratio $\varphi'(p)(v)/|\varphi'(p)(v)|$ is independent of the choice of v, and, hence, the right-hand part of equality (8) depends only on the choice of $\lambda \in \hat{U}$. Let us set $\varphi'(p)(v)/|\varphi'(p)(v)| = \tau_\varphi(\lambda)$. The point $\hat{\varphi}(\lambda)$ of the space $\mathbb{R}^{2n}$ belongs to the $(2n-1)$-dimensional manifold $G \times \Omega^{n-1}$. The mapping $\hat{\varphi} : \hat{U} \to G \times \Omega^{n-1}$ defined in such a way is bijective. Let $\varphi : U \to \mathbb{R}^n$ and $\psi : V \to \mathbb{R}^n$ be arbitrary admissible charts of the manifold M, such that $U \cap V = W$ is not empty. Let $H_1 = \varphi(W)$, $H_2 = \psi(W)$, M_1 and M_2 be open sets in $\mathbb{R}^n$. In this case defined are the mappings $\hat{\varphi} : \hat{U} \to \mathbb{R}^{2n}$ and $\hat{\psi} : \hat{U} \to \mathbb{R}^{2n}$, and the mapping $\hat{\psi} \circ \hat{\varphi}^{-1}$ of the manifold $H_1 \times \Omega^{n-1}$ onto the manifold $H_2 \times \Omega^{n-1}$. Let us consider the latter in more detail. Let $\lambda \in \hat{T}_p(M)$, $(t, \xi) = \hat{\varphi}(\lambda)$, $(u, \eta) = \hat{\psi}(\lambda)$, $p \in W$. We have: $t = \varphi(p)$, $u = \psi(p)$. Let $h \neq 0$ be a directed vector from λ. In this case $\varphi'(p)(h) = \alpha\xi$, $\psi'(p)(h) = \beta\eta$, where $\alpha < 0$, $\beta > 0$. If we set $\theta = \varphi \circ \varphi^{-1}$, then $\beta\eta = \theta'(t)(\alpha\xi)$. The vector η is a unit vector, and, hence, $\eta = \theta'(t)(\xi)/|\theta'(t)(\xi)|$. The mapping $\hat{\theta} = \hat{\psi} \circ \hat{\varphi}^{-1}$ is consequently expressed in the following way:

$$\hat{\theta} : (t, \xi) \in H_1 \times \Omega^{n-1} \mapsto (\theta(t), A_\theta(t)(\xi)) \in H_2 \times \Omega^{n-1}$$

where $A_\theta(t)(\xi) = \theta'(t)(\xi)/|\theta'(t)(\xi)|$. This shows that θ is a mapping of the class C^{r-1}.

On the set $\hat{T}(M)$ there exists a structure, and only one, of a differentiable manifold of the class C^{r-1}, such that for any admissible chart $\psi:U \to \mathbb{R}^n$ of the manifold M the mapping $\hat{\varphi}:U \to \mathbb{R}^{2n}$ constructed above is a diffeomorphism of the class C^{r-1} of the set $\hat{u}$ onto the manifold $\varphi(u) \times \Omega^{n-1}$. We are not going to dwell upon this statement in detail here, since it will require somewhat cumbersome, though, in essence, quite trivial considerations.

3.6.4. Let M, as before, denote an n-dimensional differentiable manifold of the class C^r, where $r \geqslant 1$.

Let K be a one-sidedly smooth curve in M. Let us arbitrarily choose a point $X \in K$. Let $\varphi:U \to \mathbb{R}^n$ any admissible chart of the manifol M defined in the neighbourhood of the point X. Let us find an arc $L = [YZ]$ of the curve K, which is contained in the domain of the definition of this chart and such that $X \in L$, in which case if X is an internal point of the curve K, then X is also an internal point of the arc L. With a mapping φ the arc L is transformed into a certain one-sidedly smooth curve $L_1 = \varphi(L)$. Let $X_1 = \varphi(X)$, e_l, e_r are the left-hand and the right-hand unit vectors of the curve L_1 at the point X_1. Let us denote through u_l and u_r the vectors from $T_X(M)$, such that $\varphi'(X)(u_l) = e_l$, $\varphi'(X)(u_r) = e_r$ and let $t_l(X)$ and $t_r(X)$ be the axes in $T_X(M)$, with the directed vectors u_l and u_r, respectively. Let us call $t_l(X)$ a left-hand, and $t_r(X)$ a right-hand tangents at the point X of the curve K. If X is a terminal point of the curve K, then only one of the straight lines $t_l(X)$ and $t_r(X)$ is defined.

In the above definition there is one troublesome peculiarity associated with the fact that, when defining the straight lines $t_l(X)$ and $t_r(X)$, use was made of a certain admissible chart, given in a certain neighbourhood of the point X. In order to prove the independence of the straight lines $t_l(X)$ and $t_r(X)$ of the choice of the chart φ, let us arbitrarily set a one-sidedly smooth parametrization $p:[a, b]$ of the curve K. Let $X = p(t_0)$ and $[\alpha, \beta]$ be a set of the values of the parameter t corresponding to the arc $L = [YZ]$. Let $x(t) = \varphi[p(t)]$. The path $x(t)$, $\alpha \leqslant t \leqslant \beta$, in the space $\mathbb{R}^n$ is one-sidedly smooth. At the point t_0 the vector $u = p'_l(t_0)$ and $v = p'_r(t_0)$ are defined, their definition being independent of the choice of a chart in the neighbourhood of the point φ. We have: $\varphi'(X)(u) = x'_l(t_0)$, $\varphi'(X)(v) = x'_r(t_0)$. The vectors $x'_l(t_0)$ and $x'_r(t_0)$ are directed for the tangents $t_l(X_1)$ and $t_r(X_1)$ at the point X_1 of the curve $L_1 = \varphi(L)$. Therefore, $p'_l(t_0)$ and $p'_r(t_0)$ are directed vectors for the axes l and r in the space $T_X(M)$. Therefore, l and r are independent of the choice of the chart $\varphi:U \to \mathbb{R}^n$. Besides, we come to the conclusion that if $p:[a, b] \to M$ is a one-sidedly smooth parametrization of the curve K, then the vectors $p'_l(t_0)$ and $p'_r(t_0)$ are directed ones for the left-hand and right-hand tangents at the point $X = p(t_0)$ of the curve K.

Let K be one-sidedly smooth curve in the manifold M. Let us define a certain correspondence $X \mapsto \hat{\tau}(X)$, where X is an arbitrary point of the curve K and $\hat{\tau}(X)$ is a curve in $\hat{T}(M)$. If X is a smooth point of the curve K, then $\hat{\tau}(X)$ consists of a single point of the axis $t(X) = t_l(X) = t_r(X)$. Let us assume that X is an angular point. Let us denote by λ and ρ the positive semi-axis of the tangents $t_l(X)$ and $t_r(X)$, $\lambda = t_l(X)$, $\rho = t_r(X)$. In this case $\hat{\tau}(X)$ is a curve obtained when the axis t in the space $T(M)$ rotates around the point o, covering an angle limited by the semi-axes λ and ρ. In the case when $t_l(X)$ and $t_r(X)$ differ only in the direction, so that X is a point of return of the curve K, let us assume that such an angle is an arbitrarily chosen two-dimensional semi-plane, at the boundary of which the straight lines $t_l(X)$ and $t_r(X)$ lie. In a formal way the curve $\hat{\tau}(X)$ in $\hat{T}_X(M)$ is defined in the following way. Let us assume that $t_l(X)$ and $t_r(X)$ are not colinear. Let u and v be directed vectors of the axes $t_l(X)$ and $t_r(X)$. Let us denote by $\lambda(t)$ an axis in $\hat{T}_X(M)$ with the directed vector $u(t) = (1-t)u + tv$, where $0 \leqslant t \leqslant 1$. We have: $\lambda(0) = t_l(X)$, $\lambda(1) = t_r(X)$. We thus obtain the path $\lambda(t)$, $0 \leqslant t \leqslant 1$, in the space $\hat{T}_X(M)$, which defines the curve in $\hat{T}_X(M)$ which is assumed to be $\hat{\tau}(X)$. The curve $\hat{\tau}(X)$ can easily be proved to be independent of the choice of the vectors u and v. In the case when the straight lines $t_l(X)$ and $t_r(X)$ are colinear and differ only in the direction, the curve $\hat{\tau}(X)$ is obtained in the following way. We choose in $T_X(M)$ an arbitrary two-dimensional semi-plane P, the boundaries of which are the straight lines $t_l(X)$ and $t_r(X)$. Then we arbitrarily choose a directed vector u of the tangent $t_l(X)$ and the vector v directed to the inside of p. Then we set $\lambda(t) = \cos \pi t \cdot u + \sin \pi t \cdot v$, $0 \leqslant t \leqslant 1$. We have: $\lambda(0) = u$, $\lambda(1) = -u$, i.e., $\lambda(1)$ is a directed vector of the axis $t_r(X)$. As $\hat{\tau}(X)$ we choose a curve with the parametrization $\lambda(t)$, $0 \leqslant t \leqslant 1$. It can be easily proved that the curve $\hat{\tau}(X)$ is independent of the choice of the vectors u and v.

Let us now demonstrate that the correspondence $X \mapsto \hat{\tau}(X)$ is a c-correspondence. Let $\varphi : U \to \mathbb{R}^n$ be an arbitrary admissible chart of the manifold M, and let $L = [Z_0 Z_1]$ be an arc of the curve K, such that $U \supset L$, $X \in L$, in which case if X is an internal point of the curve, then X is also an internal point of the arc L. Let $R = \varphi(L)$, $Y = \varphi(X)$. The curve R in the space $\mathbb{R}^n$ is one-sidedly smooth. Let $G = \varphi(u)$. In the space $\hat{T}(M)$ defined are the open set $\hat{U}$ and a topological mapping $\hat{\varphi}$ of the set $\hat{U}$ onto $G \times \Omega^{n-1}$. Let us construct a complete indicatrix of the tangents of the curve K_1. It will be a curve lying in the set $G \times \Omega^{n-1}$. Let us choose an arbitrary point $Z \in L$. The mapping $\varphi'(Z)$ is linear and the directed vectors of the left-hand and right-hand tangents at the point Z of the curve L are transformed into the directed vectors of the corresponding tangents at the point $P = \varphi(Z)$ of the curve R. From this we can see that with the mapping $\hat{\varphi}$ the arc $\hat{\tau}(Z)$ is transformed into a shortest line on the sphere Ω^{n-1}, which connects the points $e_l(P)$ and $e_r(P)$, ($e_l(P)$ and $e_r(P)$ are the left-hand and the right-hand tangential unit vectors of the curve R at the point P). We thus come to the conclusion that $\hat{\varphi}(\hat{\tau}(Z))$ is an arc $\sigma(X)$ of the complete indicatrix of the tangents of the curve R. Therefore, $\hat{\tau}$ is a c-correspondence. Indeed, let (X_m), $m = 1, 2, \ldots$, be an arbitrary sequence of the points of the curve K converging to the point X either from the left or from the right, $Y_m = \varphi(X_m)$. As $m \to \infty$ the points Y_m converge to the point Y from one side, and, hence, the curves $\sigma(Y_m)$ converge to the left end of the curve $\sigma(Y)$, if $Y_m \to Y$ from the left, and to the right end, if $Y_m \to X$ from the right. Taking into account the fact that for all m for which $Z_0 < X_m < Z_1$ $\hat{\varphi}(\hat{\tau}(X_m)) = \sigma(Y_m)$, we see that $X \mapsto \hat{\tau}(X)$ is a c-correspondence.

Any manifold is a metrizable topological space. In conformity with the results obtained in Section 3.5 (Theorem 3.5.1), we come to the conclusion that for any one-sidedly smooth curve K in the manifold M of the class C^r, $r \geqslant 1$, the correspondence $X \mapsto \hat{\tau}(X)$ has an indicatrix, which is called a complete indicatrix of the tangents of the curve K.

It follows from the constructions carried out above that if $\varphi : U \to \mathbb{R}^n$ is an admissible chart, such that K lies in the domain of the definition of U of the chart φ, then the mapping $\hat{\varphi}$ transforms the curve $\hat{\tau}(K)$ into a curve $\sigma(L)$, where $L = \varphi(K)$ is a complete indicatrix of the tangents of the curve K.

The notion of a complete indicatrix of the tangents can be considered as a kind of analogue on the notion of a derivative for a function of one variable. On its basis one can construct a certain classification of the curves on an arbitrary differentiable manifold.

For any Lipshitz manifold the class of rectifiable curves is determined. Let M be a manifold of the class C^r, $r \geqslant 1$. Let us construct a manifold $\hat{T}(M)$, which belongs to the class C^{r-1}. Let us introduce the following notations. Let the totality of all one-sidedly smooth curves on the manifold M be denoted through $D_1(M)$, and the totality of all rectifiable curves in M be denoted through $R_1(M)$. Let K be a one-sidedly smooth curve in M, and let $\tau(K)$ be its complete indicatrix in the manifold $\hat{T}(M)$. If $r \geqslant 2$, then $\hat{T}(M) \in C^1$, and hence $\hat{T}(M)$ is a Lipshitz manifold. Therefore the class of rectifiable curves and the class of one-sidedly smooth curves in $\hat{T}(M)$ are defined. We shall say that K is a curve of a finite turn if $\hat{\tau}(K) \in R_1[\hat{T}(M)]$. The curve K is called one-sidedly smooth in the second order if $t(K)$ is a one-sidedly smooth curve in the manifold $\hat{T}(M)$. In the former case we will write $K \in R_2(M)$, and in the latter $K \in D_2(M)$. We thus have the classes of the curves $R_2(M)$ and $D_2(M)$. Evidently, $R_1(M) \supset D_1(M) \supset R_2(M) \supset D_2(M)$. This construction can be further extended. Let us assume that for any manifold of the class C^r, $r \geqslant 1$, the classes $R_r(M)$ and $D_r(M)$ are determined and let M be a manifold of the class C^r. Let us say that the curve $K \in D_1(M)$ belongs to the class $R_{r+1}(M)$ if its complete indicatrix of the tangents $\tau(K)$ belongs to the class $R_r(M)$ and $K \in D_{r+1}(M)$ if $\tau(K) \in D_r(M)$. By way of induction we can easily see that at any r the following inclusions occur:

$$R_r(M) \supset D_r(M) \supset R_{r+1}(M) \supset D_{r+1}(M).$$

In this book we are not going to consider the classes of the curves $R_r(M)$ and $D_r(M)$ introduced here at an arbitrary r.

By way of conclusion, let us consider a certain general supposition on the transformation of one-sidedly smooth curves with respect to differentiable mappings of manifolds. Let us first make some remarks of a general character on the mappings of differentiable manifolds. Let M_1 and M_2 be differentiable manifolds of the class C^k, $k \geqslant 1$, of the dimension n and m, respectively, $f : M_1 \rightarrow M_2$ be a mapping of the class C^k. In this case for any point $p \in M_1$ defined is a linear mapping $f'(p)$ of the vector space $T_p(M_1)$ onto the vector space $T_q(M_2)$, where $q = f(p)$, which is called a derivative (or, more frequently, a differential) of the mapping f at the point p. We say that the mapping f is non-degenerate at the point p if the rank of the linear mapping $f'(p)$, i.e., the dimensionality of the vector space $f'(p)[T_p(M_1)]$ is equal to the lower of the numbers n and m. The mapping f defines a certain mapping f^* of the space $T(M_1)$ in $T(M_2)$ Namely, let $u \in T(M_1)$, $p = \pi(u)$, $q = \varphi(u)$. Then $f^*(u)$ is an element $f'(p)(u)$ of the layer $T_q(M_2)$ of the manifold M_2. Let $\varphi : U \rightarrow \mathbb{R}^n$, $\psi : V \rightarrow \mathbb{R}^m$ be the charts in the manifolds M_1 and M_2, respectively. Let us assume that $f(U) \subset V$. In this case the mapping $f_1 = \psi \circ f \circ \varphi^{-1}$, which is called a coordinate representation of f through the charts φ and ψ, is defined. Therefore the chart $\varphi^* : \pi^{-1}(U) \rightarrow \mathbb{R}^{2n}$ is defined in $T(M_1)$ and $\psi^* : \pi^{-1}(V) \rightarrow \mathbb{R}^{2m}$ in $T(M_2)$. Let $f_1^* = \psi^* \circ f^* \circ \varphi^{*-1}$ be a coordinate representation of f^* through the charts φ^* and ψ^*. The mappings f_1 and f_1^* are interrelated in the following way: for any pair $(x, z) \in \varphi(u) \times \mathbb{R}^n$ $f_1^*(x, z) = (f_1(x), f_1'(x)z)$. In this case the following relation is valid interconnecting $f_1'(x)$ and the mapping $f'(p)$. Let $p \in U$, $q = f(p) \in V$, $x = \varphi(p)$. Then $f_1'(x) = \psi'(q) \circ f'(p) \circ [\varphi'(p)]^{-1}$.

We also have the mappings $\hat{T}(M_1)$ and $\hat{T}(M_2)$. Generally speaking, to the mapping f there corresponds no mapping $\hat{T}(M_1)$ onto $\hat{T}(M_2)$, since $f'(p)$ can transform some straight lines in $T_p(M_1)$ into one-point sets. Let us assume that $n \leqslant m$ and let U be a set of those $p \in M_1$ for which the linear mapping $f'(p) : T_p(M_1) \rightarrow T_p(M_2)$ is non-degenerate, i.e., the rank of $f'(p)$ is equal to n. The set U is open. Let us consider the set $\hat{U} = \hat{\pi}^{-1}(U)$. Let $\lambda \in \hat{U}$, $p = \hat{\pi}(\lambda)$. The linear mapping $f'(p)$ is non-degenerate as far as $p \in U$, and, hence, it transforms the axis λ into a certain axis μ, in the space $T_q(M_2)$, $q = f(p)$. Therefore, the mapping $\hat{f} : \hat{T}_p(M_1) \rightarrow \hat{T}_q(M_2)$ is defined. Since $p \in U$ is arbitrary, the mapping $\hat{f} : \hat{T}_u(M_1) \rightarrow \hat{T}_u(M_2)$ is thus defined.

THEOREM 3.6.1. *Let M_1 and M_2 be differentiable manifolds of the class C^k, of the dimensions n and m, respectively, $f : M_1 \rightarrow M_2$ be a mapping of the class C^k, K be a one-sidedly smooth curve in M_1. In this case, if for every point X of the curve K the linear mapping $f'(X)$ does not turn to zero on the straight lines $t_l(X)$ and $t_r(X)$, then the curve $Q = f(K)$ is one-sidedly smooth. In this case $f'(X)[t_l(X)]$ is a left-hand tangent of the curve Q at the point $Y = f(X)$, and $f'(X)[t_r(X)]$ is a right-hand one.*

If $n \leqslant m$ and the mapping f is non-degenerate at every point X of the curve K, then it will be non-degenerate in a certain open U, containing the curve K and the mapping $\hat{f} : \hat{T}_U(M_1) \rightarrow \hat{T}(M_2)$ transforms the complete indicatrix of the tangents of K into a complete indicatrix of the tangents of $f(K)$.

Proof. Let us first prove a statement concerning the image of a one-sidedly smooth curve. Let $f:M_1 \to M_2$ be a mapping of the class C^k, $k \geqslant 1$, and let K be a one-sidedly smooth curve in M_1. Let us assume that for any point $X \in K$ the mapping $f'(X)$ does not turn identically to zero on every of the straight lines $t_l(K)$ and $t_r(K)$. Let $\varphi:U \to \mathbb{R}^n$ and $\psi:V \to \mathbb{R}^m$ be admissible charts of the manifolds M_1 and M_2, respectively, such that $f(U) \subset V$, $f_1 = \psi \circ f \circ \varphi^{-1}$ and let L be a closed arc of the curve K lying in U. Using the Borel theorem we can easily arrive at the conclusion that the curve K can be divided into a finite number of arcs L in such a way that for any of them we can construct the charts φ and ψ obeying all the above conditions. Let us set $L' = \varphi(L)$, $S = f(L)$, $S' = \psi(S)$, $R = f(K)$. The definition of the notion of a one-sidedly smooth curve on a manifold yields that L' is a one-sidedly smooth curve in $\mathbb{R}^n$. Let us arbitrarily choose a point X' of the curve L'. Let $X = \varphi^{-1}(Y)$, Y be a point $f(X)$ of the curve R, and let $Y' = \psi(Y)$ be a point of the curve S. Let t_l', t_r' be a left-hand and a right-hand tangents of the curve L at the point X'; let t_l, t_r be a left-hand and a right-hand tangents of the curve K at the point X. In line with the condition of the theorem, the mapping $f'(X)$ does not turn identically to zero on the straight lines t_l and t_r and, consequently, it transforms them into certain straight lines $\tilde{t}_l$ and $\tilde{t}_r$ in the space $T_Y(M_2)$. The mappings $\varphi'(X)$ and $\psi'(Y)$ are one-to-one, and, hence, the linear mapping $f_1'(X') = \psi'(Y) \circ f'(X) \circ [\varphi'(X)]^{-1}$ does not turn to zero on the straight lines t_l and t_r. We thus come to the conclusion that the mapping $f_1 = \psi \circ f \circ \varphi^{-1}$, belonging to the class C^k, is such that for any point X' of the curve L' its differential does not turn to zero on the straight lines t_l and t_r which are the left-hand and the right-hand tangents on the curve L' at this point. Therefore, in line with Theorem 3.1.3, the curve $S' = f_1(L') = \psi(S)$ is one-sidedly smooth and, hence, the arc S of the curve $R = f(L)$ is one-sidedly smooth. We come to the conclusion that the curve R can be divided into a finite number of one-sidedly smooth arcs and, hence, that it is one-sidedly smooth. The first statement of the theorem is thus proved.

Let us assume that $n \leqslant m$ and the mapping $f:M_1 \to M_2$ is non-degenerate at every point X of the curve K, i.e., the rank of the linear mapping $f'(X)$ equals n at all the points of the curve. From the considerations of continuity it follows that U is an open subset of the manifold M_1. For any $X \in U$ the mapping $f'(X)$ is one-to-one and, hence, it maps any axis of the space $T_X(M_1)$ onto any axis of the space $T_X(M_2)$, i.e., it defines the mapping $\hat{f}_X:\hat{T}_X(M_1) \to \hat{T}_X(M_2)$, $\hat{f}_X(X) = f'(X)(\lambda)$ for any axis $\lambda \in \hat{T}_X(M_1)$. The mapping f, therefore, defines a certain mapping $\hat{f}:\hat{T}_U(M_1) \to \hat{T}(M_2)$, such that for any $\lambda \in \hat{T}_U(M_1)$, $\hat{\pi}[\hat{f}(\lambda)] = f[\hat{\pi}(\lambda)]$. In other words, the layer $\hat{T}_X(M_1)$ of the manifold $\hat{T}(M_1)$ corresponding to the point $X \in M_1$ is transformed into $\hat{T}_Y(M_2)$ which is the layer of the manifold $\hat{T}(M_2)$, corresponding to the point $Y = f(X)$.

The curve $R = f(K)$, as a consequence of what has been proved above, is one-sidedly smooth. Let $x(t)$ and $\lambda(t)$, $a \leqslant t \leqslant b$, be corresponding parametrizations of the curve K and its complete indicatrix of the tangents $\tau(K)$ respectively. The curve $\tau(K)$ lies in $\hat{T}_u(M_1)$, and, hence, the curve $S = \hat{f}[\tau(K)]$ is defined. The function $M(t) = \hat{f}[\lambda(t)]$, $a \leqslant t \leqslant b$, is a parametrization of the curve S, and $y(t) = f[x(t)]$ is a parametrization of R. At every t $\lambda(t)$ belongs to the layer $\hat{T}_{x(t)}(M_1)$ of the manifold $\hat{T}(M_1)$. In line with the above listed properties of the mapping $\hat{f}$, the point $\mu(t) = \hat{f}[\lambda(t)]$ belongs to $\hat{T}_{y(t)}(M_2)$, so that $\hat{\pi}[\mu(t)] = y(t)$. Our aim is to prove that S is a complete indicatrix of the tangents of the curve R, $S = \hat{\tau}(R)$, in which case $y(t)$, $\mu(t)$, $a \leqslant t \leqslant b$, are the corresponded parametrizations of the curve R and its indicatrix $\tau(R)$. It, however, directly results from the fact that the arc $\hat{\tau}_K(X)$ is transformed, with the mapping $\hat{\varphi}$, into the arc $\hat{\tau}_R(\varphi(X))$ of the curve R.

The theorem is proved.

Some Facts of Integral Geometry

4.1. Manifold G_k^n of k-Dimensional Directions in V^n

4.1.1. Integral geometry is a field of studies that arose about 50 years ago. Some ideas pertaining to it were mentioned earlier in the works by Poincaré and Crofton, but it was in the papers by W. Blaschke and co-workers, published in the 'thirties, that integral geometry was formulated as an independent branch of research. This field of geometry proved to be quite viable and is still attracting the attention of scholars. For a description of the history of the problem and a survey of the data obtained, the reader is referred to the monograph [33].

Let us try to present some basic ideas of integral geometry in its classical form. Integral geometry studies different sets of geometrical images: straight lines, planes or convex surfaces in the Euclidean space or, in the general case, in an arbitrary homogeneous space, i.e. in a space where a group of transformations operates. In this case one considers sets which are invariant with respect to the action of the group of transformations that defines the geometry of the space. The first basic problem of integral geometry is the construction of a measure, which is invariant with respect to the group of transformations of the considered space, on a given set of geometrical images. The second problem is somewhat indefinite and can be formulated as follows.

Let us suppose that in a certain space there is a certain object of a geometrical or analytical nature, let us denote it through $\mathcal{F}$, and a certain set S of geometrical images which is invariant with respect to the action of the group of space transformations. Let us then assume that a certain number $\varphi(x, \mathcal{F})$ corresponds to each $x \in S$. This can be, for instance, the number of the points of intersection of the surface $\mathcal{F}$ and the straight line x (in the case when S is a set of straight lines), the integral along the straight line x from the square of the derivative of the function $\mathcal{F}$ in the direction of the straight line x, and so on. The problem is to get all possible information on $\mathcal{F}$ while studying the function $\varphi(x, \mathcal{F})$. The simplest problem arising in this case is to establish what the integral from the function $\varphi(x, \mathcal{F})$ with respect to the invariant measure on S is equal to?

Let, for instance, $\mathcal{F}$ be a curve on a plane, and $\varphi(p, \mathcal{F})$ be a number of points of the curve $\mathcal{F}$ lying on the straight line p. On the set S of all straight lines on the plane one can define a measure which is invariant with respect to the plane motion. Let us suppose that there is a Cartesian orthogonal coordinate system fixed on the plane. Let us then denote by $h(p)$ the distance from the origin of the coordinates to the straight line p, and let $\varphi(p)$ be the angle formed by the ray outgoing from O orthogonally to p and the positive semi-axis Ox, $0 \leqslant \varphi(p) \leqslant 2\pi$. In the coordinates $h = h(p)$, $\varphi = \varphi(p)$ the invariant measure on the set of straight lines on the plane is expressed by the integral

$$\tfrac{1}{2} \iint dh d\varphi \tag{1}$$

Let K be an arbitrary plane curve and let $\varphi(p,K)$ be a number of the points of the curve lying on the straight line p. In this case an integral of the function $\varphi(p,K)$ with respect to the measure expressed by equality (1) equals the length of the curve K. This is one of the classical results of integral geometry.

Another example: let $f(x)$ be an arbitrary function on the plane. Let us denote by $\tilde{f}(p)$ an integral of the function f along the straight line p. The function $\tilde{f}$, defined in such a way on the set of all straight lines on the plane, is termed a Radon transform of the function f. The problem of restoring the function through its Radon transform is important in an applied sense, since it serves as the basis for a new field of medical diagnostics - tomography.

In their papers, W. Blaschke and his followers mainly considered the case when the number $\varphi(x, \mathcal{F})$ was a certain characteristic of the intersection of $\mathcal{F}$ and x (in this case $\mathcal{F}$ is, therefore, either a curve or a surface, a convex body, etc.). One can, however, also consider characteristics of some other nature. The present book mainly concentrates on the integral geometrical relations pertaining to the characteristics which are not functions of an intersection. It should be noted that there is a certain connection between the relations considered in this book and the integral geometrical formulas of Blaschke type, which will be shown below.

4.1.2. Here E^n, as earlier, denotes an n-dimensional Euclidean space, V^n is a vector Euclidean space whose elements are vectors of the space E^n.

Every k-dimensional subspace of V^n will be termed a k-dimensional direction of E^n, where $1 \leqslant k \leqslant n - 1$. Let P be an arbitrary k-dimensional plane in E^n; H be any k-dimensional direction. Then we shall say that the plane P is parellel to H and write $P||H$ if for any two points $X, Y \in P$ the vector $u = XY$ belongs to H. A set of all k-dimensional directions of the space E^n will be denoted through G_k^n.

Let us further suppose that a Cartesian orthogonal system of coordinates is given in V^n. Let $e_1, e_2, \ldots, e_n$ be the basis of this coordinate system. In this case each vector $u \in V^n$ is corresponded to by a certain vector $(x_1, x_2, \ldots, x_n)$ of the arithmetic n-dimensional Euclidean space $\mathbb{R}^n$, such that $u =$

$x_1 e_1 + x_2 e_2 + \ldots + x_n e_n$, the numbers $x_1, x_2, \ldots, x_n$ being the coordinates of the vector u in the coordinate system considered. For simplicity, let us write 'the vector $u = (x_1, x_2, \ldots x_n)$ is given' instead of 'the vector u with the coordinates $x_1, x_2, \ldots, x_n$ is given'.

Hereafter it would be in some cases convenient to assume that there is a fixed point O, called the origin, in E^n. Setting the vector $x = OX$ to the point $X \in E^n$, we get a one-to-one mapping of the space E^n onto V^n. To every k-dimensional direction H there corresponds the only k-dimensional plane passing through O, which is parallel to the direction H. This plane will also be denoted by H. The set G_k^n, accordingly, will be identified with the set of the k-dimensional planes in E^n, passing through the point O.

The symbol $\mathbb{O}^n$ will denote the totality of all linear orthogonal transformations of the space V^n, $\mathbb{D}^n$ will denote the set of all motions of E^n. If there is an origin given in E^n, then any transformation $f \in \mathbb{D}^n$ can be expressed in the formula

$$y = f_L(x) + a,$$

where x is the radius-vector with respect to O of an arbitrary point $x \in E^n$; $x = OX$; $y = OY$ is the radius-vector of the point $Y = f(X)$, $a \in V^n$, and f_L is an orthogonal transformation. It should be noted that f_L is independent of the choice of the origin O, and is called a linear part of the motion f.

Any transformation $\varphi \in \mathbb{O}^n$ transfers the arbitrary k-dimensional subspace H of the space V^n into some other k-dimensional subspace $\varphi(H)$ of V^n, and thus defines the action of the group $\mathbb{O}^n$ on G_k^n.

Let H be a k-dimensional direction in E^n, and let f be an arbitrary motion of the space E^n. Any k-dimensional plane P, which is parallel to H, is transformed by the motion f into the plane $Q = f(P)$, which, as is easily seen, is parallel to the k-dimensional direction $f_L(H)$.

Let $H \in G_k^n$. Also let $\nu(H)$ denote the totality of all vectors $u \in V^n$ which are orthogonal to the plane H. The set $\nu(H)$ is an $(n-k)$-dimensional subspace of V^n, and thus one obtains a certain mapping ν of the set G_k^n onto G_{n-k}^n. It can be easily proved that this mapping is bijective and that $\nu \circ \nu$ is the identity mapping. The mapping ν is invariant with respect to the action of the group $\mathbb{O}^n$ in the following sense: for any $H \in G_k^n$ and any $\varphi \in \mathbb{O}^n$ we have $\varphi[\nu(H)] = \nu[\varphi(H)]$. Hereafter ν will be referred to as a canonical correpondense of the sets G_k^n and G_{n-k}^n.

4.1.3. On the set G_k^n one can introduce the structure of a differentiable manifold of the class C^∞ (or even of the class C^ω). Let us demonstrate how to do it.

Let $P, Q \in G_k^n$. P is considered to be orthogonal to Q provided P contains at least one non-zero vector u, which is orthogonal to Q.

Let P_0 be an arbitrary element of the set G_k^n. Let us denote by $U(P_0)$ the set of all k-dimensional subspaces of V^n, none of which is orthogonal to P_0. Let us also introduce in V^n a Cartesian orthogonal system of coordinates in such a way that the subspace P_0 be determined in it by a system of equations $x_{k+1} = x_{k+2} = \cdots = x_n = 0$. For an arbitrary $u \in V^n$ the vector $x = (x_1, x_2, \ldots, x_n)$ in $\mathbb{R}^n$, which is formed by the coordinates of vector u, will be viewed as a pair $x = (y, z)$, where $y = (x_1, x_2, \ldots, x_k) \in \mathbb{R}^k$, $z = (x_{k+1}, x_{k+2}, \ldots, x_n) \in \mathbb{R}^{n-k}$. The plane Q is given through a system of $(n - k)$ linear equations which can be written as

$$Ay + Bz = 0 \tag{2}$$

where A is a matrix consisting of $n - k$ rows and k colums, and B is a square matrix of order $n - k$. Let us show that $\det B \neq 0$. Indeed, if we assume that, on the contrary, $\det B = 0$, then there will be a vector $z_0 \neq 0$ such that $Bz_0 = 0$. Let u_0 be a vector in V^n, with the coordinate vector $x_0 = (0, z_0)$. The vector x_0 obeys the system (2) and, hence, $u_0 \in Q$. The vector u_0 is evidently orthogonal to P_0, which contradicts the fact that the plane Q is not orthogonal to P_0. This contradiction proves that $\det B \neq 0$. Solving the equality (2) with respect to z, one discovers that the equation of the plane Q can be presented in the following form:

$$z = H_Q y \tag{3}$$

where $H_Q = -B^{-1}A$ is a matrix consisting of $n - k$ rows and k columns. The matrix H_Q is uniquely defined by the plane Q. Indeed, let us assume that the matrices H_1 and H_2 are such that each of the equations $z = H_1 y$ and $z = H_2 y$ defines the plane Q. Let us take an arbitrary $y \in \mathbb{R}^k$. Let $z_1 = H_1 y$, $z_2 = H_2 y$. The vectors in V^n, which have the coordinate vectors $x_1 = (y, z_1)$ and $x_2 = (y, z_2)$, belong to Q and, hence, their difference, i.e. the vector v, the coordinate vector for which is $x_1 - x_2 = (0, z_1 - z_2)$, belongs to Q and is orthogonal to P_0. Since Q is not orthogonal to P_0, then $v = 0$ and, hence, $x_1 - x_2 = 0$, i.e. $z_1 = z_2$. Thus for any $y \in \mathbb{R}^k$ $H_1 y = H_2 y$, and hence $H_1 = H_2$. The matrix H_Q will hereafter be referred to as the coordinate matrix of the plane Q. Let us interpret H_Q as a point of the space $\mathbb{R}^m$, where m is the number of the elements of matrix H_Q, i.e. $m = (n - \kappa)k$. The correspondence $\theta P_0 : Q \mapsto H_Q$ between the set $U(P_0)$ and the space $\mathbb{R}^m$ is bijective. The mapping $\theta P_0 : U(P_0) \mapsto \mathbb{R}^m$ will be called a canonical system of coordinates in G_k^n which corresponds to the plane P_0. Let $\theta P_1 : U(P_1) \mapsto \mathbb{R}^m$ and $\theta P_2 : U(P_2) \mapsto \mathbb{R}^m$ be two canonical systems of coordinates in G_k^n. One can easily prove that the transition functions $\theta P_2 \circ \theta P_1^{-1}$ and θP_2^{-1} are defined on some open sets of the space $\mathbb{R}^m$ and are analytical (and even rational) functions.

Since $P_0 \in U(P_0)$ and $P_0 \in G_k^n$ are arbitrary, then the sets $U(P)$ in their totality cover G_k^n. The system of sets $U(P)$ and mappings $\theta_P : U(P) \mapsto \mathbb{R}^m$ forms, due to what has been said above, a certain atlas of the class C^ω on the set G_k^n, and thus the structure of the analytical manifold of the dimension $m = (n - \kappa)k$ is defined in G_k^n.

4.1.4. Let $P \in G_k^n$. Let us consider a canonical system of coordinates $\theta_P : U(P)$

$\mapsto \mathbb{R}^m$ and let $\varphi \in O^n$ be an orthogonal transform which maps the plane P into itself. Let $Q \in U(P)$. Then $Q' = \varphi(Q) \in U(P)$ as well. Let us clarify the interconnections of the coordinate matrices H_Q and $H_{Q'}$ of the planes Q and Q'. In the space V^n we have a Cartesian orthogonal system of equations in which the plane P is defined by the system of equations $x_{k+1} = x_{k+2} = \cdots = x_n = 0$. Let $e_1, \ldots, e_k, e_{k+1}, \ldots, e_n$ be the basis vectors of the coordinate system considered. If $\varphi(P) = P$, then $\varphi[\nu(P)] = \nu(P)$ as well. $\nu(P)$ is a plane stretched onto the vectors $e_{k+1}, \ldots, e_n$. If $x = (x_1, \ldots, x_k, x_{k+1}, \ldots, x_n)$, then, as above, we shall write $x = (y, z)$, where $y = (x_1, \ldots, x_k) \in \mathbb{R}^k$, $z = (x_{k+1}, \ldots, x_n) \in \mathbb{R}^{n-k}$. Let $u' = \varphi(u)$. We have: $u' = (y', z')$, where $y' = Ay$, $z' = Bz$ and A and B are orthogonal matrices of the orders k and $n - k$, respectively. The point (y, z) belongs to Q' if and only if $(A^{-1}y, B^{-1}z)$ belongs to Q, i.e., if $A^{-1}y = H_Q B^{-1}z$. This means that Q' is defined by the system of equations $y = AH_Q B^{-1}z$ and, hence

$$H_{Q'} = AH_Q B^{-1}.$$

This expression is the one required for $H_{Q'}$. Here A and B are orthogonal matrices of orders k and $n - k$, respectively.

4.1.5. Let us demonstrate that the canonical correspondence ν of the sets G_k^n and G_{n-k}^n is a mapping of the class C^ω. Let us set in V^n an arbitrary orthonormal basis $e_1, \ldots, e_k, e_{k+1}, \ldots, e_n$. Let P_0 be a k-dimensional plane which is stretched onto the vectors $e_1, \ldots, e_k$, and let Q_0 be an $(n - k)$-dimensional plane stretched onto the vectors $e_{k+1}, \ldots, e_n$. It is obvious that $Q_0 = \nu(P_0)$. Let $P \in U(P_0)$. The plane P is given by the equation $z = Hy$. Let $\xi = (\eta, \zeta) \in \nu(P)$. Then

$$\langle \eta, y \rangle + \langle \zeta, Hy \rangle = 0 \tag{4}$$

for any vector $y \in \mathbb{R}^k$. We have: $\langle \eta, y \rangle + \langle \zeta, Hy \rangle = \langle \eta, y \rangle + \langle H^*\zeta, y \rangle = \langle \eta + H^*\zeta, y \rangle$. The condition (4), in view of the arbitrariness of $y \in \mathbb{R}^k$, yields that $\eta + H^*\zeta = 0$. Thus we find that the plane $\nu(P)$ is defined by the system of equations

$$y = -H^*z. \tag{5}$$

It follows from the last equality that $\nu(P)$ belongs to the canonical neighbourhood $U(Q_0)$. Thus, ν maps $U(P_0)$ into $U(Q_0)$. In this case, if the plane P has a coordinate matrix H, then the matrix $-H^*$ will be a coordinate matrix of the plane $\nu(P)$. From this we immediately conclude that ν is the mapping of the class C^ω of the set G_k^n onto G_{n-k}^n.

4.1.6. Let us prove that the manifold G_k^n is compact. For this purpose it is sufficient to show that G_k^n is covered over with a finite number of compact sets. Let us set in V^n a Cartesian orthogonal system of coordinates with the basis $u_1, u_2, \ldots, u_n$. The plane stretched onto the vectors $u_{i_1}, u_{i_2}, \ldots, u_{i_k}$, where $1 \leqslant i \leqslant i_2 < \cdots < i_k \leqslant n$, will be denoted by $P_{i_1 i_2 \ldots i_k}$. Let us assume that $P \in G_k^n$ and let $Q = \nu(P)$, $a_1, a_2, \ldots, a_{n-k}$ be the vectors forming an orthogonal basis of the subspace Q. Then the plane P will be defined by the system of equations

$$<u,a_1> = 0, \quad i = 1, 2, \ldots, n - k.$$

The same system in coordinate form is as follows:

$$Ax = 0 \tag{6}$$

where A is the $(n - k) \times n$ matrix, with the elements of its ith row being the coordinates of vector a_1. Since the system of the vectors a_i is orthonormal, the sum of the squared minors of the order $n - k$ of the matrix A equals 1. Let us find the largest among them by the absolute value minor. Let it be the minor formed by the columns numbered $j_1 < j_2 < \cdots < j_{n-k}$. Let $i_1 < i_2 < \cdots < i_k$ be the numbers of the remaining columns of the matrix A. In this case the system (6) can be written as $By + Cz = 0$, where $y = (x_{i_1}, x_{i_2}, \ldots, x_{i_k})$, $z = (x_{j_1}, x_{j_2}, \ldots, x_{j_{n-k}})$. The number of the minors of the order $n - k$ of the matrix A is $C_n^{n-k} = C_n^k$, and, since the sum of their squares equals 1, the largest of these squares cannot be less than $1/C_n^k$, i.e. $|\det C| \geqslant 1/\sqrt{C_n^k}$. Solving the equation $By + Cz = 0$ with respect to z, one discovers that the plane P is given by the system of equations

$$z = Hy$$

where $H = -C^{-1}B$. It can easily be proved that in view of the condition $|\det C| \geqslant 1/\sqrt{C_n^k}$. Let us denote through $\tilde{U}_{i_1 i_2 \ldots i_k}$ the set of all the planes $P \in U(P_{i_1 i_2 \ldots i_k})$, the elements of the coordinate matrix H_P of which are less than L. The set $\tilde{U}_{i_1 i_2 \ldots i_k}$ is compact. The plane P is chosen arbitrarily. It proves that any plane $P \in G_k^n$ belongs, at least, to one of the compact sets $\tilde{U}_{i_1 i_2 \ldots i_k}$ and thus, the compactness of G_k^n has been demonstrated.

4.2. Imbedding of G_k^n into a Euclidean Space

4.2.1. Here we are going to describe a construction which enables one to consider a compact manifold G_k^n as a closed surface of a Euclidean vector space of the dimension $(n^2+n)/2$. It can be used to solve the problem of constructing on G_k^n a measure which is invariant with respect to orthogonal transformations. The imbedding of G_k^n into a Euclidean space can be achieved by way of corresponding a certain linear mapping to an arbitrary k-dimensional subspace, namely, the mapping which is an orthogonal projection of V^n onto a given subspace, and of considering the set of all mappings of the space V^n onto itself as a vector Euclidean space.

In literature one can usually find a more complicated construction of G_k^n into a Euclidean space of a fairly high dimension which is based on Grassman's algebra. In this case one gets an embedding of a set of oriented k-dimensional planes but not of what is understood here as G_k^n. For the purpose in view, the use of a set of oriented k-dimensional planes instead of G_k^n would be an unnecessary complication.

By way of introducing, let us make some remarks concerning linear mappings.

Let L_n^n be a set of all linear mappings of the space V^n onto itself. In V^n let us set a Cartesian orthogonal system of coordinates in an arbitrary way. Any mapping $Z \in L_n^n$ is uniquely defined by its matrix in this coordinate system and any square matrix of the order n defines the mapping $Z \in L_n^n$. It follows from here that L_n^n is an n^2-dimensional vector space. Let $X \in L_n^n$. Let us denote by X^* the mapping $Y \in L_n^n$, which is conjugate to X, i.e. such that for any vectors u, $v \in V^n$ the equality

$$<Xu, v> \ = \ <u, Yv>$$

is valid. In a Cartesian orthogonal system of coordinates the matrix of the mapping X^* is a transposed matrix of X.

For any $X \in L_n^n$ we have: $(X^*)^* = X$. For any pair of mappings $X, Y \in L_n^n$ $(XY)^* = Y^*X^*$. Let us set, for arbitrary $Z, T \in L_n$: $<Z,T>$ = trace (Z^*T). If $(z_{ik})_{i,k=1,2,...,n}$, $(t_{ik})_{i,k=1,2,...,n}$ are matrices of the mappings Z and T in a Cartesian orthogonal coordinate system, then

$$<Z,T> \ = \ \sum_{i=1}^{n} \sum_{k=1}^{n} z_{ik}t_{ik}.$$

Therefore, $<Z, T>$ = trace $(T^*Z) = <T, Z>$. One can also easily see that $<Z, T>$ = trace (ZT^*) =trace (TZ^*). The quantity $<Z,T>$ will be called a scalar product of the mappings Z and T. The set L_n^n, with the scalar product $<Z, T>$ allowed for, is an n^2-dimensional Euclidean vector space.

The mapping $Z \in L_n^n$ is called symmetrical if $Z^* = Z$. This is equivalent to the fact that for any vectors u, $v \in V^n$ $<Zu, v> = <u, Zv>$. The mapping $Z \in L_n^n$ is symmetrical if and only if the matrix of Z is symmetrical in any Cartesian orthogonal system of coordinates in V^n. The dimension of the Euclidean space S_n^n of symmetrical linear mappings of the space V^n is, therefore, equal to $(n^2 + n)/2$.

The mapping $Z \in L_n^n$ belongs to the group $\mathbb{O}^n$ if and only if the equality

$$Z^*Z = I$$

holds, where I is the identical mapping. It yields, in particular, that for any $Z \in \mathbb{O}^n$ $Z^{-1} = Z^*$.

If $T \in S_n^n$, then for any $Z \in L_n^n$ we also have $U = ZTZ^* \in S_n^n$. Indeed for any A, $B, C \in L_n^n$ we have: $(ABC)^* = C^*B^*A^*$. From this one can conclude that $(ZTZ^*)^* = (Z^*)^*T^*Z^* = ZT^*Z^*$ and, hence, if $T^* = T$, then $U^* = U$ as well, i.e. if $T \in S_n^n$, then $U \in S_n^n$ too, which is the required proof.

Let P be a k-dimensional subspace of V^n, and let $\eta(P)$ be mapping which corresponds to every point $x \in V^n$ its orthogonal projection onto the plane P. Let us prove that the mapping $\eta(P)$ is linear and symmetrical. Let $a_1, a_2, ..., a_k$ be an arbitrary system of vectors forming an orthonormal basis of the subspace P. Let $x \in V^n$, $y = \eta(P)(x)$. The vector $y \in P$ and, hence, y has the form $y = \sum_{i=1}^{k} \eta_i a_i$. The vector $x - y$ is orthogonal to the plane P and, therefore, we have: $<x - y, a_i> = 0$ for any $i = 1, 2, ..., k$. We have $<a_i, a_j> = 0$ when $i \neq j$ and $<a_i, a_i> = 1$ for any $i = 1, 2, ..., k$. Therefore, $\eta_i = <y, a_i> = <x, a_i>$

and, hence,

$$y = \eta(P)x = \sum_{i=1}^{k} <x, a_i> a_i. \tag{7}$$

It proves that $\eta(P)$ is linear. Let x and y be arbitrary vectors in V^n. In view of (7) we get

$$<\eta(P)x, y> = <\sum_{i=1}^{k} <x, a_i> <a_i, y> = \sum_{i=1}^{k} <x, a_i> <y, a_i>.$$

The right-hand part of this equality is symmetrical with respect to x and y, which results in $<\eta(P)x, y> = <x, \eta(P)y>$ and, hence, because of the arbitrariness of x and y, $(\eta(P))^* = \eta(P)$, i.e. $\eta(P) \in S_n^n$.

Let us see the way the mapping $\eta(P)$ changes with orthogonal transformations of V^n. Let $\varphi \in O^n$, $Q = \varphi(P)$. If we take an arbitrary point $x \in V^n$, then $y = \eta(Q)x$ can be found in the following way. Let us first find the point $x' = \varphi^{-1}x$. Let $y' = \eta(P)x' = \eta(P)(\varphi^{-1}x)$. The transformation φ transfers the point x' into the point x and the plane P into the plane Q, while y' is transferred into the point of the plane Q nearest to x, i.e. $\varphi y' = \eta(Q)x$, which results in $\varphi(\eta(P)(\varphi^{-1}x)) = \eta(Q)x$ and, hence, $\eta(Q) = \varphi\eta(P)\varphi^{-1}$. As far as $\varphi \in O^n$, then $\varphi^{-1} = \varphi^*$ and, therefore, if φ is an orthogonal transformation transferring the plane P into the plane Q, then

$$\eta(Q) = \varphi\eta(P)\varphi^* \tag{8}$$

Let P_0 be a plane given in a Cartesian orthogonal coordinate system by the equations $x_{k+1} = x_{k+2} \cdots x_n = 0$, and let P be an arbitrary k-dimensional plane, with φ an orthogonal transform, such that $P = \varphi(P_0)$. In this case, in line with (8), $\eta(P) = \varphi\eta(P_0)\varphi^*$. The matrix of the mapping $\eta(P_0)$ has the form $\begin{bmatrix} E_k, & 0 \\ 0, & 0 \end{bmatrix}$, where E_k is a unit matrix of the order k. Therefore, we see that for any plane P the matrix of linear mapping $\eta(P)$ can be presented as

$$\eta(P) = \varphi \begin{bmatrix} E_k, & 0 \\ 0, & 0 \end{bmatrix} \varphi^* \tag{9}$$

where $\varphi \in O^n$. It also shows the symmetry of $\nu(P)$.

By corresponding the mapping $\eta(P)$ to the subspace P, we get a certain mapping η of the set G_k^n in S_n^n. For any $P \in G_k^n$ we evidently have: $\eta(P)(V^n) = P$. If, therefore, $P_1 \neq P_2$, then $\eta(P_1) \neq \eta(P_2)$, and hence the mapping

$$\eta : P \in G_k^n \mapsto \eta(P) \in S_n^n$$

is one-to-one.

Let us show that the mapping η belongs to the class C^∞, with its rank in every point of the set G_k^n being equal to $(n-k)k$. Let us set an arbitrary $P_0 \in G_k^n$ and introduce in V^n a Cartesian coordinate system in which the plane P_0 is defined by the system $x_{k+1} = x_{k+2} = \cdots x_n = 0$. Let $U(P_0)$ be a canonical neighbourhood of the plane P_0. Any plane $P \in U(P_0)$ can be presented by the equation:

$$z = H_P y$$

where H_P is a matrix made of k columns and $n - k$ rows. The elements of the matrix H_P are the coordinates of $P \in G_k^n$ in the canonical coordinate system in G_k^n corresponding to P_0. Let us take an arbitrary point $x = (y, z) \in V^n$. Let $u = \eta(P)x = (v, w)$, where $v \in \mathbb{R}^k$, $w = H_P v \in \mathbb{R}^{n-k}$. The vector $x - u$ is orthogonal to any vector $p = (q, r)$ which lies in the plane P. If $p = (q, r) \in P$, then $r = H_P q$, and the condition of the orthogonality of the vector $x - u$ and the plane P yields that for any $q \in \mathbb{R}^k$

$$<y - v, q> + <z - H_P v, H_P q> = 0.$$

This equality can be rewritten in the following way:

$$<y - v + H_P^* z - H_P^* H_P v, q> = 0.$$

Since q is an arbitrary vector from $\mathbb{R}^k$, we have the equality:

$$y + H_P^* z = (E_k + H_P^* H_P)v. \tag{10}$$

Here E_k denotes a unit matrix of the order k. The determinant of the matrix $\Theta_P = E_k + H_P^* H$ is other than zero because this matrix is symmetrical, and the quadratic form $<\Theta_P v, v> = |v|^2 + |H_P v|^2$ is positively determined, with all its eigenvalues not less than 1. The equality (10) yields

$$v = \Theta_P^{-1} y + \Theta_P^{-1} H_P^* z.$$

Taking into account that $w = H_P z$, we can write the matrix of the mapping $\eta(P)$ in the following way:

$$\begin{bmatrix} \Theta_P^{-1}, & \Theta_P^{-1} H_P^* \\ H_P \Theta_P^{-1}, & H_P \Theta_P^{-1} H_P^* \end{bmatrix} \tag{11}$$

From this equality one can see that the elements of the matrix $\eta(P)$ are infinitely differentiable (and even rational) functions of the elements of the matrix H_P. But it is the elements of the matrix H_P that are the coordinates of the plane P in the given canonical chart of the set G_k^n. Therefore, the mapping η of the set G_k^n in L_n^n belongs to the class C^∞. Due to the fact that $\eta(P)$ is symmetric, the matrix (11) is also symmetrical, which is, incidentally, evident by itself.

The mapping

$$\Phi : H \longmapsto \begin{bmatrix} (\Theta(H))^{-1}, & (\Theta(H))^{-1} H^* \\ H(\Theta(H))^{-1}, & H(\Theta(H))^{-1} H^* \end{bmatrix} \tag{12}$$

where H is an $(n - k) \times k$-matrix, $\Theta(H) = E_k + H^* H$, is a coordinate representation of the mapping $\eta(P)$ with respect to the given canonical system of coordinates in G_k^n. To prove that the rank of the mapping $\eta : G_k^n \to L_n^n$ is equal to $(n-k)k$, due to the arbitrariness of P_0, it is sufficient to show that the rank of the mapping Φ at the point $H = 0$ is equal to $(n-k)k$. To this end it is sufficient to find the differential of the mapping Φ at the point $H = 0$. We have:

$$\Phi(0) = \begin{pmatrix} E_k, & 0 \\ 0, & 0 \end{pmatrix}.$$

If we neglect the terms of order with respect to H exceeding 1 in the expression $\Phi(H) - \Phi(0)$, then we get

$$\Phi'(0)(H) = \begin{pmatrix} 0, & H^* \\ H, & 0 \end{pmatrix}.$$

From this we can see that $\Phi'(0)$ is an injective mapping of the space of the $(n - k) \times k$-matrices into the space of the square matrices of the order n and, hence, the rank of $\Phi'(0)$ is equal to $(n-k)k$. This proves that the rank of the mapping $\eta : G_k^n \to L_n^n$ equals $(n-k)k = \dim G_k^n$.

Let us then assume $\eta(G_k^n) = \Pi_k^n$. The set Π_k^n, due to what has been proved above, is an $(n-k)k$-dimensional submanifold of the class C^ω in the vector space S_n^n. The set Π_k^n is diffeomorphic to G_k^n and the mapping η is a certain canonical diffeomorphical mapping of G_k^n onto Π_k^n. Since G_k^n is compact, then Π_k^n is a compact set of S_n^n.

The equality (9) yields that for any $P \in G_k^n$ the equality $<\eta(P), \eta(P)> =$ trace $(\eta(P))^2 = $ trace $\begin{pmatrix} E_k, 0 \\ 0, 0 \end{pmatrix} = k$ is valid. It means that the set $\Pi_k^n = \eta(G_k^n)$ lies on a sphere of the radius $\sqrt{k}$ in the Euclidean space S_n^n.

4.2.2. Any orthogonal transformation φ of the space V^n defines a certain transformation of the manifold G_k^n, which sets to the arbitrary plane P the plane $\varphi(P)$ which is an image of P with respect to the mapping φ. It would be thus natural to denote this transformation of G_k^n also by the symbol φ. Any mapping $f : G_k^n \to G_k^n$ defines a certain mapping $\hat{f} = \eta \circ f \circ \eta^{-1}$ of the manifold Π_k^n in Π_k^n. Let us consider the interrelations of the mappings φ and $\hat{\varphi}$ in case when $\varphi \in \mathcal{O}^n$. Let $X \in \Pi_k^n, P = \eta^{-1}(X)$. Then $X = \eta(P)$ and $\hat{\varphi}(X) = \eta[\varphi(P)]$. In view of the equality (8) we have $\eta[\varphi(P)] = \varphi \eta(P) \varphi^*$, which means

$$\hat{\varphi}(X) = \varphi X \varphi^*. \tag{13}$$

The right-hand part of the equality (13) should be interpreted as a superposition of the linear mappings φ, X and φ^* of the space V^n.

Let us demonstrate that for $\varphi \in \mathcal{O}^n$ the transformation $\hat{\varphi}$ of the space L_n^n which is defined by the equality $\hat{\varphi}(X) = \varphi X \varphi^*$ is orthogonal with respect to the scalar product in L_n^n, defined by the formula $<X, Y> = $ trace (X^*Y). Indeed, let $X' = \varphi X \varphi^*$, $Y' = \varphi Y \varphi^*$. Then $X'^* = \varphi X^* \varphi^*$ and hence $X'^* Y = \varphi X^* \varphi^* \varphi Y \varphi^* = \varphi X^* Y \varphi^* = (X\varphi^*)^* Y \varphi^*$. We have: trace $(X'^* Y') = $ trace (XY^*), i.e. $<X'Y'> = <X, Y>$ and thus the orthogonality of $\hat{\varphi}$ has been proved.

To any transformation φ of the manifold G_k^n which results from the orthogonal transformation of the space V^n, there corresponds the transformation $\hat{\varphi}$ of the manifold Π_k^n, defined by the formula $\hat{\varphi} = \eta \circ \varphi \circ \eta^{-1}$. According to (13), $\hat{\varphi}(X) = \varphi X \varphi^*$ for any $X \in \Pi_k^n$. The interrelation of the mappings φ and $\hat{\varphi}$ operating on G_k^n and Π_k^n, respectively, can be described in a formal way as follows. The mapping diagram

$$\begin{array}{ccc} G_k^n & \xrightarrow{\varphi} & G_k^n \\ \eta \downarrow & & \eta \downarrow \\ \Pi_k^n & \xrightarrow{\hat{\varphi}} & \Pi_k^n \end{array}$$

is commutative, i.e. $\eta \circ \varphi = \hat{\varphi} \circ \eta$,

The manifold G_1^{n+1} is also called an n-dimensional projective space and is denoted by the symbol $\mathbb{P}^n$. Its elements are the straight lines of the vector space V^{n+1}, passing through the point O. Let us consider the sphere $\Omega^n = \{x \in V^{n+1} \mid |x| = 1\}$. Each straight line $l \in G_1^{n+1}$ crosses Ω^n in two diametrically opposing points. We, therefore, say that $\mathbb{P}^n$ is obtained through identification any two diametrically opposing points of the sphere Ω^n. By way of setting to the point $x \in \Omega^n$ the straight line $\pi(x) \in G_1^{n+1}$ passing through it, we get the mapping $\pi : \Omega^n \to \mathbb{P}^n$ which is called a canonical covering of $\mathbb{P}^n$. For any $x \in \Omega^n$ we evidently have $\pi(-x) = \pi(x)$. The construction described here gives an embedding of the space $\mathbb{P}^n$ into a Euclidean sphere of the dimension $\frac{1}{2}(n+2)(n+1) - 1$.

Due to the known general results obtained in topology, any compact n-dimensional manifold of the class C^∞ allows an embedding of the dimension $2n + 1$. The general result given makes it possible to estimate the degree of 'excessiveness' of the embedding of the manifold G_k^n into a sphere constructed here. It should be noted that, setting $n = 3$, $k = 1$, we obtain an embedding of the projective plane $\mathbb{P}^2$ into a 5-dimensional sphere. It is known that there exists no embedding into a 4-dimensional sphere for the manifold $\mathbb{P}^2$, and, hence, in the particular case considered, the construction described here gives a $\mathbb{P}^2$ embedding into a sphere of the smallest dimensionality possible.

4.3. Existence of Invariant Measure in G_k^n

4.3.1. Let M be an arbitrary differentiable manifold. Let us denote by $\mathcal{B}(M)$ the totality of all Borel sets of the manifold M, which is the least family of the M subsets containing all open sets and such that the difference of any two and the union of a not more than countable number of sets, belonging to the family, also belongs to it.

The non-negative function $\mu : \mathcal{B}(M) \to \bar{\mathbb{R}}$ is referred to as the measure on the manifold M provided it obeys the following conditions:

(1) $\mu(\varnothing) = 0$;
(2) For any sequence (A_m), $m = 1, 2, \ldots$, of mutually non-intersecting sets from $\mathcal{B}(M)$ the equality

$$\mu \left(\bigcup_{m=1}^{\infty} A_m \right) = \sum_{m=1}^{\infty} \mu(A_m)$$

holds.

The measure μ on the manifold M is called non-trivial if there exists a set $A \in \mathcal{B}(M)$, such that $0 < \mu(A) < \infty$.

Let $\varphi : M \to M$ be an arbitrary topological mapping of the manifold M onto itself, μ be the measure on the manifold M. Let us say that the measure μ is invariant with respect to φ, if for any $A \in \mathcal{B}(M)$

$$\mu[\varphi(A)] = \mu(A).$$

Let S be an arbitrary set of the topological mappings of the manifold onto itself. The measure μ is called invariant with respect to S if it is invariant with respect to any mapping $\varphi \in S$.

We have already defined the manifold G_k^n of the k-dimensional directions of the n-dimensional Euclidean space E^n. Any $\varphi \in \mathbb{O}^n$ defines the mapping $P \mapsto \varphi(P)$ of the set G_k^n onto itself, which is denoted through φ as well. Orthogonal transformations of the space V^n therefore generate a certain set of transformations of the manifold G_k^n. Our nearest aim is to prove that on G_k^n there exists a non-trivial measure which is invariant with respect to this set of transformations. For this purpose let us make use of the embedding of G_k^n into a Euclidean vector space S_n^n constructed as in Section 4.2. In S_n^n there is a compact manifold Π_k^n of dimension $(n-k)k$. In this case the topological mapping $\eta : G_k^n \to \Pi_k^n$ of the manifold G_k^n onto the surface Π_k^n is defined. The mappings η and η^{-1} belong to the class C^∞. To any transformation $\varphi \in \mathbb{O}^n$ there corresponds a certain orthogonal transformation $\hat{\varphi}$ of the space S_n^n which is defined by the formula $\hat{\varphi}(X) = \varphi X \varphi^*$. Let $\tilde{\mu}$ be a measure on the manifold Π_k^n. For an arbitrary Borel set $A \subset G_k^n$ let us set $\mu(A) = \tilde{\mu}[\eta(A)]$. The set function μ obtained in such a way is a measure on the manifold G_k^n. Let $\varphi \in \mathbb{O}^n$, A be an arbitrary Borel set in G_k^n. We have $\mu[\varphi(A)] = \tilde{\mu}[\eta(\varphi[A])]$. In view of the relations between the mappings φ and η, the equality $\eta[\varphi(A)] = \hat{\varphi}[\eta(A)]$ holds and, hence,

$$\mu[\varphi(A)] = \tilde{\mu}[\hat{\varphi}(\eta[A])] \tag{14}$$

for any Borel set A in the manifold G_k^n. The equality (14), yields, in particular, that if the measure $\tilde{\mu}$ in Π_k^n is invariant with respect to the transformations $\hat{\varphi}$, where $\varphi \in \mathbb{O}^n$, then the measure μ will be invariant with respect to the transformations of the manifold G_k^n resulting from orthogonal transformations in V^n. The inverse is also true: if the measure μ in G_k^n is invariant with respect to the action of the group $\mathbb{O}^n$ on G_k^n, then the measure $\tilde{\mu}$ on Π_k^n is invariant with respect to the transformations $\hat{\varphi}$, where $\varphi \in \mathbb{O}^n$.

The problem of constructing an invariant measure on G_k^n is thus reduced to an analogous problem for the surface Π_k^n in the Euclidean space S_n^n. This second problem is simply solved, i.e. the area in the sense of differential geometry is the required invariant measure on Π_k^n.

4.3.2. Let us cite some fundamentals pertaining to the notion of a surface area. Let $\mathcal{F}$ be an r-dimensional surface of class C^∞ in an m-dimensional Euclidean space E^m, i.e. $\mathcal{F}$ be an r-dimensional submanifold of the class C^∞ of the space E^m. Let $\varphi : V \to \mathbb{R}^r$ be an arbitrary admissible local system of coordinates of the manifold $\mathcal{F}$. Then $U = \varphi(V)$ is an open set in $\mathbb{R}^r$, the inverse mapping $f = \varphi^{-1}:U \to V$ belongs to the class C^∞ and the vectors

$$\partial_i f(t) = \frac{\partial f}{\partial t_i}(t), \quad i = 1, 2, \ldots, r,$$

are linearly independent. The vector-function $f = \varphi^{-1}$ is called a local parametrization of the surface $\mathcal{F}$.

Let $f:U \to \mathcal{F}$ be an arbitrary local parametrization of the surface $\mathcal{F}$. Let us assume $g_{ij}(t) = \langle \partial_i f(t), \partial_j f(t) \rangle$ and let

$$g(t) = \det \|g_{ij}(t)\|, \quad i, j = 1, 2, \ldots, r.$$

The quadratic form $\sum_{i, j=1}^r g_{ij}(t) \xi_i \xi_j$ is called a metric form of the surface $\mathcal{F}$ corresponding to the given parametrization f. Let A be an arbitrary Borel set lying on the surface $\mathcal{F}$. If we assume that $A \subset V = f(U)$, then the number

$$\mu(A) = \int_{f^{-1}(A)} \sqrt{g(t)}\, dt$$

is determined. The quantity $\mu(A)$ is called the area of the set A. Let us show that it is independent of the choice of the local parametrization f. Let $f_1:U_1 \to \mathcal{F}$ be another arbitrary local parametrization of the surface $\mathcal{F}$, such that $A \subset V_1 = f_1(U_1)$. Let us assume that $g_{ij}^{(1)}(t) = \langle \partial_i f_1(t), \partial_j f_1(t) \rangle$, $i, j = 1, 2, \ldots, r$. Applying, then, the classical formula of changing the variables in a multiple integral, we get

$$\int_{f_1^{-1}(A)} \sqrt{g_1(t)}\, dt = \int_{f^{-1}(A)} \sqrt{g(t)}\, dt$$

then proving the $\mu(A)$-independence of the choice of parametrization.

Let us suppose that it is impossible to choose a local parametrization $f:U \to \mathcal{F}$ of the surface $\mathcal{F}$ for the set $A \in \mathcal{B}(\mathcal{F})$, such that $A \subset f(U)$. In this case A can be presented as a union $\bigcup A_\nu$ of a not more than countable number of mutually non-crossing sets from $\mathcal{B}(\mathcal{F})$, for each of which the local parametrization does exist. Let us set

$$\mu(A) = \sum_\nu \mu(A_\nu).$$

One can easily prove that the sum in the right-hand part is independent of the choice of representation $A = \bigcup A_\nu$. Therefore, a certain function of the set μ is defined on the totality of all Borel subsets of the surface $\mathcal{F}$. We can readily prove that μ is a measure. For any point $p \in \mathcal{F}$, for any neighbourhood U in E^m $\mu(\mathcal{F} \cap U) > 0$, which proves that the measure μ is non-trivial. If the surface $\mathcal{F}$ is compact, the quantity $\mu(\mathcal{F})$ is finite. Indeed, for any point $p \in \mathcal{F}$ there evidently exists a neighbourhood U such that $\mu(\mathcal{F} \cap U) < \infty$. Due to its compactness, $\mathcal{F}$ is covered with a finite number of such neighbourhoods, and thus $\mu(\mathcal{F}) < \infty$.

The area is invariant with respect to the motions of the space E^m in the following sense. Let $h:E^m \to E^m$ be an arbitrary motion of the space E^m, $\mathcal{F}$ be any r-dimensional surface in E^m, $\mathcal{F}' = h(\mathcal{F})$ be a surface into which $\mathcal{F}$ is transformed through the motion h. In this case for any Borel set $A \subset \mathcal{F}$ the surface measures of the sets A and $h(A)$ coincide. Indeed, let us assume that there exists such a local parametrization $f : U \to \mathcal{F}$ of the surface $\mathcal{F}$ that $A \subset f(U)$. If we set $f_1 = h \circ f$, the function f_1 is then a parametrization of the surface $\mathcal{F}'$. In this case $h(A) \subset f_1(U)$ and $f_1^{-1}[h(A)] = f^{-1}(A)$. Since the scalar products are invariant with respect to orthogonal transformations, we have

$$g_{ij}^{(1)}(t) = <\partial_i f_1(t), \ \partial_j f_1(t)> = <\partial_i f(t), \ \partial_j f(t)> = g_{ij}(t)$$

and, hence

$$g_1(t) = \det \|g_{ij}(t)\| = g(t).$$

This tells us that the surface measures of the sets A and $h(A)$ are equal to the same integral and, hence, that they coincide. The case when one cannot find the parametrization $f : U \to \mathcal{F}$ for the set A such that $A \subset f(U)$ is evidently reduced to the case considered above. We come, in particular, to the conclusion that if the motion h transforms the surface $\mathcal{F}$ into itself, then the area of the set is a measure on $\mathcal{F}$, invariant with respect to h.

Applying what has been written above to the particular case of the surface Π_k^n in the Euclidean space S_n^n, we find that on Π_k^n there exists a non-trivial measure, invariant with respect to the transformations $\hat{\varphi}$, where $\varphi \in \mathbb{O}^n$, i.e. it is the area of the set on Π_k^n that is the measure. If the measure $\tilde{\mu}$ is the area on the surface Π_k^n, then, due to the compactness of Π_k^n, the quantity $\omega_{n,k} = \tilde{\mu}(\Pi_k^n)$ is finite. The function of the set $\tilde{\mu}_{n,k} = 1/\omega_{n,k} \tilde{\mu}$ is also an invariant measure on Π_k^n. We have $\tilde{\mu}_{n,k}(\Pi_k^n) = 1$. The measure corresponding to the measure $\tilde{\mu}_{n,k}$ on the manifold G_k^n will be denoted through $\mu_{n,k}$. The set function $\mu_{n,k}$, defined on $\mathcal{B}(G_k^n)$ by the method described above, will be further referred to as a normalized invariant measure of G_k^n.

4.4. Invariant Measure in G_k^n and Integral. Uniqueness of an Invariant Measure

4.4.1. Let M be a compact differentiable manifold, and let $C(M)$ be the totality of all continuous real functions defined in M, and μ be the measure of the manifold M, i.e. a non-negative and completely additive function given on the totality of all Borel subsets M. In this case for any function $f \in C(M)$ the number

$$I_\mu(f) = <f, \mu> = \int_M f(x) \ d\mu(x) \tag{15}$$

is determined, which means that a linear functional is defined on $C(M)$. This functional is non-negative, i.e. if $f(x) \geqslant 0$ in M, $f \in C(M)$, then $I_\mu(f) \geqslant 0$. In view of the known results of functional analysis, if $l : C(M) \to \mathbb{R}$ is a non-negative linear functional, then there exists a measure μ, such that $l(f) = <f, \mu>$ for any function $f \in C(M)$. This measure μ is unique, which, in particular, implies that if the measures μ_1 and μ_2 on the manifold M are such that for any function $f \in C(M)$ $<f, \mu_1> = <f, \mu_2>$, then the given measures coincide.

The symbol $\bar{\mathbb{R}}$ will denote a set of real numbers with two ideal elements added, i.e., the points $-\infty$ and ∞. The function $f:M \rightarrow \bar{\mathbb{R}}$ is called measurable in the Borel sense if its Lebesgue sets (i.e. the sets $f^{-1}(S)$, where S is an arbitrary interval of the set $\bar{\mathbb{R}}$) are of the Borel type. Let us denote by $\mathfrak{M}(M)$ the totality of all measurable in the Borel sense functions $f:M \rightarrow \bar{\mathbb{R}}$. Let $\mathfrak{M}_+(M)$ be the totality of all non-negative functions of $\mathfrak{M}(M)$. The set of all functions $f \in \mathfrak{M}(M)$, integrable with respect to the given measure μ in the manifold M, will be denoted through $\mathcal{L}(M, \mu)$. The quantity

$$\int_M f(x)\,d\mu(x) \tag{16}$$

is determined for any function $f \in \mathcal{L}(M, \mu)$, as well as for all $f \in \mathfrak{M}_+(M)$. (In the latter case the value ∞ is allowed for the integral).

Let us recall some known facts from the theory of integrals.

Let $A \subset M$. An indicator of the set A is the function $\chi_A:M \rightarrow \mathbb{R}$, defined by the condition: $\chi_A(x) = 1$, if $x \in A$, $\chi_A(x) = 0$, if $x \notin A$. The function $f:M \rightarrow \mathbb{R}$ is called a step-function if it allows the representation

$$f = a_1\chi_{A_1} + a_2\chi_{A_2} + \cdots + a_m\chi_{A_m} \tag{17}$$

where $a_1, a_2, \ldots, a_m$ are real numbers and $A_1, A_2, \ldots, A_m$ are Borel sets. For any function f of type (17)

$$\int_M f(s)\,d\mu(x) = \sum_{i=1}^{m} a_i\mu(A_i).$$

For any function $f \in \mathfrak{M}^+(M)$ there exists a sequence of step-functions (f_m), $m = 1, 2, \ldots$, such that at any $x \in M$ the sequence $(f_m(x))$, $m = 1, 2, \ldots$, is increasing and $f_m(x) \rightarrow f(x)$ at $m \rightarrow \infty$. For any such sequence of step-functions

$$\lim_{m\to\infty} \int_M f_m(x)\,d\mu(x) = \int_M f(x)\,d\mu(x).$$

Further on in the book we shall prove various kinds of integral identities. In many cases their proof can be carried out by the following scheme. First we establish the validity of the identity for the case when the function is an indicator of a certain set (which is generally the main part of the proof). The next step is to establish whether the indentity considered is valid for all step-functions and then, by way of a limiting transition, we prove that it is valid for any non-negative measurable function, which readily yields the validity of the given identity for all integrable functions. Considerations related to the transition from the case when the function is an indicator of the set to the general case are based on the properties of the integral cited above. In most cases we are not going to give all considerations referring to the given part of the proof.

4.2.2. Let us cite a general formula of the change of variables in an integral with respect to the measure. Let μ be the measure in the manifold M, and let $\varphi: M \to M$ be a topological mapping of M onto itself. For an arbitrary $A \in \mathcal{B}(M)$ we set $\mu_{\varphi-1}(A) = \mu[\varphi^{-1}(A)]$. The function of the set $\mu_{\varphi-1}$, as can easily be verified, is also a measure in M. If f belongs to one of the classes $\mathfrak{M}_+(M)$ or $\mathcal{L}(M, \mu_{\varphi-1})$, then $f \circ \varphi$ is the function of the class $\mathfrak{M}_+(M)$, respectively, of the class $\mathcal{L}(M, \mu)$, in which case

$$\int_M f[\varphi(x)] \, d\mu(x) = \int_M f(x) \, d\mu_{\varphi-1}(x). \tag{18}$$

To prove the equality (18) it is sufficient to consider the case when f is a step-function. This latter case is evidently reduced to one when $f = \chi_A$, and where $A \in \mathcal{B}(M)$. We have: $\chi_A[\varphi(x)] = 1$, if $\varphi(x) \in A$, i.e., if $x \in \varphi^{-1}(A)$, $\chi_A[\varphi(x)] = 0$, if $\varphi(x) \notin A$, i.e., if $x \notin \varphi^{-1}(A)$. This means that $\chi_A \circ \varphi$ is an indicator of the set $\varphi^{-1}(A)$, $\chi_A \circ \varphi = \chi_{\varphi^{-1}(A)}$. We therefore conclude that

$$\int_M \chi_A[\varphi(x)] \, d\mu(x) = \mu[\varphi^{-1}(A)] = \mu_{\varphi-1}(A) = \int_A \chi_A(x) \, d\mu_{\varphi-1}(x)$$

and for the case in question the validity of equality (18) is thus established, from which we infer that it is valid for the general case, too.

Let us assume that there is a set S of the transformations of the manifold M, and let S be a group, i.e. for any $\varphi \in S$ we also have $\varphi^{-1} \in S$, then $\varphi \circ \psi \in S$. Let μ be a measure on M, which is invariant with respect to the action of the group S. In this case the functional I_μ, defined on S by equality (15) possesses the following property of invariance. For any $\varphi \in S$ and for any function $f \in C(M)$

$$I_\mu(f \circ \varphi) = I_\mu(f). \tag{19}$$

Indeed, according to (18), $I_\mu(f \circ \varphi) = \langle f \circ \varphi, \mu \rangle = \langle f, \mu_{\varphi-1} \rangle = I_\mu(f)$, as far as $\mu_{\varphi-1} \equiv \mu$. It should be noted that the inverse case also holds, i.e. if the non-negative linear functional l on $C(M)$ is invariant with respect to the group of transformations S in the sense that for any function $f \in C(M)$ and, for any $\varphi \in S$, $l(f \circ \varphi) = l(f)$, then the measure μ, resulting from the functional l, is invariant with respect to S. Indeed, in this case for any $\varphi \in S$ we have

$$\langle f \circ \varphi^{-1}, \mu \rangle = l(f \circ \varphi^{-1}) = l(f) = \langle f, \mu \rangle.$$

Due to (19), $\langle f \circ \varphi^{-1}, \mu \rangle = \langle f, \mu_\varphi \rangle$ and thus we get

$$\forall f \in C(M) \quad \langle f, \mu \rangle = \langle f, \mu_\varphi \rangle.$$

This demonstrates the coincidence of the measures μ and μ_φ, i.e. that $\mu(A) = \mu[\varphi(A)]$ for any $A \in \mathcal{B}(M)$. Since $\varphi \in S$ was chosen arbitrarily, the invariancy of the measure μ is thus proved.

THEOREM 4.4.1. *If the measure μ on the manifold G_k^n is invariant with respect to the action of the group O^n on G_k^n, then $\mu = C\mu_{n,k}$, where $C \geqslant 0$ is a constant.*

Proof. For arbitrary $P, Q \in G_k^n$ we set $\rho(P, Q) = |\eta(P) - \eta(Q)|$. (Here for $X \in S_n^n$ we set $|S| = \sqrt{\langle X, \overline{X} \rangle} = (\text{trace } (X^*X))^{\frac{1}{2}})$. The function $\rho(P, Q)$ is a metric on G_k^n, which is invariant with respect to the action of the group O^n, i.e. for any $\varphi \in O^n$ $\rho(P, Q) = \rho[\varphi(P), \varphi(Q)]$. Indeed, we have $\rho[\varphi(P), \varphi(Q)] = |\hat\varphi\eta(P) - \hat\varphi\eta(Q)| = \rho(P, Q)$.

Let us first prove the following additional supposition. let $\theta : [0, \infty) \to \mathbb{R}$ be a continuous function, and let μ be an invariant measure on G_k^n. In this case

$$\gamma(P) = \int_{G_k^n} \theta[\rho(P, Q)] \, d\mu(Q)$$

is also a constant. Indeed, let us arbitrarily set $P_0 \in G_k^n$ and let $\varphi \in O^n$ transform P_0 into P. We have $\rho[\varphi(P_0), Q] = \rho[P_0, \varphi^{-1}(Q)]$ and from this conclude that

$$\int_{G_k^n} \theta[\rho(P, Q)] \, d\mu(Q) = \int_{G_k^n} \theta[\rho(P_0, \varphi^{-1}(Q))] \, d\mu(Q)$$

$$= \int_{G_k^n} \theta[\rho(P_0, Q)] \, d\mu(Q), \tag{20}$$

i.e. $\gamma(P) = \gamma(P_0)$. (The last of the equalities (20) results from the invariance of the measure μ. As far as $P \in G_k^n$ is taken arbitrarily, we thus prove that $\varphi(P) \equiv$ const.

Let us arbitrarily set a continuous function $\omega(t)$ of the real variable $t \in [0, \infty)$, such that $\omega(t)$ at $0 \leqslant t < 1$ and $\omega(t) = 0$ if $t \geqslant 1$. Let $h > 0$. If we set

$$\delta(h) = \int_{G_k^n} \omega[\tfrac{1}{h} \rho(P, Q)] \, d\mu_{n,k}(Q)$$

then, obviously, $\delta(h) > 0$ at any $h > 0$. Let $f \in C(G_k^n)$. Let us assume

$$f_h(P) = \frac{1}{\delta(h)} \int_{G_k^n} \omega[\tfrac{1}{h} \rho(P, Q)] f(Q) \, d\mu_{n,k}(Q).$$

The function f_h is continuous. Let us show that at $h \to 0$ $f_h \to f$ uniformly on G_k^n. Let $\lambda(\eta)$ be the module of continuity f on G_k^n, i.e.

$$\lambda(\eta) = \sup_{\rho(P, Q) \leqslant \eta} |f(P) - f(Q)|.$$

For any $P \in G_k^n$ we have

$$|f_h(P) - f(P)| \leqslant \int_{G_k^n} \frac{1}{\delta(h)} \, \omega[\tfrac{1}{h} \rho(P, Q)] \, |f(Q) - f(P)| \, d\mu_{n,k}(Q)$$

$$= \int_{\rho[P,Q] \leqslant h} \frac{1}{\delta(h)} \, \omega[\tfrac{1}{h} \rho(P, Q)] \, |f(Q) - f(P)| \, d\mu_{n,k}(Q)$$

$$\leqslant \lambda(h) \int_{\rho[P,Q] \leqslant h} \frac{1}{\delta(h)} \, \omega[\tfrac{1}{h} \rho(P, Q)] \, d\mu_{n,k}(Q) = \lambda(h).$$

Since, at $h \to 0$, $\lambda(h) \to 0$, then at $h \to 0$ $f_h \to f$ uniformly.

Let μ be an arbitrary invariant measure on G_k^n. Let us calculate the integral

$$\int_{G_k^n} f_h(P) \, d\mu(P).$$

Applying the Fubini theorem, we get

$$\int_{G_k^n} f_h(P) \, d\mu(P) = \int_{G_k^n} \left\{ \int_{G_k^n} \frac{1}{\delta(h)} \, \omega[\tfrac{1}{h} \, \rho(P, Q)] \, d\mu(P) \right\} f(Q) \, d\mu_{n,k}(Q). \quad (21)$$

In view of the remark made at the beginning of the proof, the internal integral in the right-hand part of (21) is independent of P and, hence, we get

$$\int_{G_k^n} f_h(P) \, d\mu(P) = C(h) \int_{G_k^n} f(Q) \, d\mu_{n,k}(Q).$$

If in this equality we set $f(P) \equiv 1$, then, obviously, $f_h(P) \equiv 1$ as well, which yields

$$C(h) = \mu(G_k^n),$$

so that $C(h)$ is h-independent, $C(h) = C = $ const and thus

$$\int_{G_k^n} f_h(P) \, d\mu(P) = C(h) \int_{G_k^n} f(P) \, d\mu_{n,k}(P).$$

If in this equality we go over to the limit at $h \to 0$, we get

$$\int_{G_k^n} f(P) \, d\mu(P) = C \int_{G_k^n} f(P) \, d\mu_{n,k}(P).$$

As the function $f \in C(G_k^n)$ is chosen arbitrarily, then the measures μ and $C\mu_{n,k}$ coincide. The theorem is proved.

REMARK. The theorem in question is a consequence of certain general results. The proof of the latter requires, however, additional data and these general results will not be the subject of discussion in this book.

4.4.3. Let us consider the case when $k = 1$. Let Ω^{n-1} be a unit sphere in the space V^n, i.e. a set of the vectors $x \in V^n$, such that $|x| = 1$. The sphere Ω^{n-1} is a hypersurface in V^n and, hence, on Ω^{n-1} there is a certain measure, i.e. an area, determined. This measure will be denoted by the symbol μ_{n-1} or, when ambiguity is impossible, by μ. In line with the known results of the analysis we have:

$$\omega_{n-1} = \mu_{n-1}(\Omega^{n-1}) = \frac{2\pi^{n/2}}{\Gamma(n/2)}. \quad (22)$$

For $u \in \Omega^{n-1}$ let $\pi(u)$ be a straight line which is colinear to the vector u. Obviously, $\pi(u) = \pi(-u)$. The straight line $\pi(u)$ intersects Ω^{n-1} in two points u and $-u$. The mapping $\pi : u \in \Omega^{n-1} \to \pi(u) \in G_1^n$ is continuous. Moreover, being a mapping of the differentiable manifold Ω^{n-1} on G_1^n, it belongs to the class C^ω. The mapping π is locally topological. Let $A \subset G_1^n$ be a Borel set. Due to the continuity of π, the set $\pi^{-1}(A)$ is also a Borel set. The function of the set $A \mapsto \mu_{n-1}[\pi^{-1}(A)]$ is a measure of G_1^n. This measure can easily be verified as being invariant with respect to the action of the group $\mathbb{O}^n$ on G_1^n. Therefore it is proportional to the measure $\mu_{n,1}$. The proportionality coefficient can be found by accounting for the fact that $\mu_{n,1}(G_1^n) = 1$. We have: $\pi^{-1}(G_1^n) = \Omega^{n-1}$. Hence, obviously, $\mu_{n-1}[\pi^{-1}(A)] = \omega_{n-1}\mu_{n,1}(A)$. Let us assume that on G_1^n there is a measurable function f given. If we set $\tilde{f}(u) = f[\pi(u)]$, then the equality

$$\int_{G_k^n} f(P) \; d\mu_{n,1}(P) = \frac{1}{\omega_{n-1}} \int_{\Omega^{n-1}} \tilde{f}(u) \; d\mu_{n-1}(u) \tag{23}$$

holds. For the case when $f = \chi_A$, where A is a Borel set, the validity of this equality results from the fact that $\chi_A[\pi(u)] = \chi_{\pi^{-1}(A)}(u)$ and, hence,

$$\int_{\Omega^{n-1}} \chi_A[\pi(u)] \; d\mu_{n-1}(u) = \mu_{n-1}[\pi^{-1}(A)]$$
$$= \omega_{n-1} \cdot \mu_{n,1}(A) = \omega_{n-1} \int_{G_1^n} \chi_A(P) \; d\mu_{n,1}(P).$$

This relation yields the validity of the equality (23) for the case when f is a step-function. Then, using a limiting transition, one can easily state the validity of the equality (23) for any function f for which at least one of the integrals of (23) is determined (in which case another integral is also determined.

4.5. Some Relations for Integrals Relative to the Invariant Measure in G_k^n

4.5.1. Let k, m and n be integer numbers such that $0 < k < m < n$. Let us arbitrarily select an m-dimensional subspace Q of the space V^n. Let $G_k(Q)$ denote the totality of all k-dimensional subspaces contained in Q. The subspace Q is an m-dimensional Euclidean vector space and, hence, due to the results obtained in Sections 4.3 and 4.4, the set $G_k(Q)$ is a compact differentiable manifold of the class C^ω, and on $G_k(Q)$ there exists a measure μ invariant with respect to the orthogonal transformations of the space Q and such that $\mu(G_k(Q)) = 1$. The measure μ obeying all these conditions is unique. It will be denoted by the symbol $\mu_{Q,k}$.

Let f be a continuous function on the set G_k^n, and let Q be an m-dimensional subspace V^n, $k < m < n$. The set $G_k(Q)$ is a closed subset (or even a submanifold) of the set G_k^n. An integral of the function f over the set $G_k(Q)$ with respect to the measure $\mu_{Q,k}$ will be denoted by $(\tau f)(Q)$, i.e.

$$(\tau f)(Q) = \int_{G_k(Q)} f(P) \; d\mu_{Q,k} \; (P). \tag{24}$$

We get a certain function τf, determined on the set G_m^n of the m-dimensional subspaced V^n.

Let us show that the operator τf, defined by the equality (24), possesses the following property of invariance. For any $\varphi \in \mathcal{O}^n$ $(\tau f) \circ \varphi = \tau(f \circ \varphi)$ or, in a detailed form:

$$(\tau f)[\varphi(Q)] = \int_{G_k(Q)} f[\varphi(P)] \; d\mu_{Q,k}(P). \tag{25}$$

Indeed, let us fix Q and set $T = \varphi(Q)$. Let g be an arbitrary function from

$C[G_k(T)]$. Let us set

$$l_1(g) = \int_{G_k(T)} g(P) \, d\mu_{T,k}(P) = (\tau g)(T),$$

$$l_2(g) = \int_{G_k(Q)} g[\varphi(P)] \, d\mu_{Q,k}(P) = (\tau(g \circ \varphi))(Q).$$

Since $g \in C(G_k(T))$ is arbitrarily chosen, the data of the equality determine certain linear functionals l_1 and l_2 on $C[G_k(T)]$. If g is non-negative, then, obviously, $l_1(g) \geqslant 0$ and $l_2(g) \geqslant 0$. Finally, if $g \equiv 1$, then $l_1(g) = l_2(g) = 1$. In accordance with the property of invariance of the integral, for any orthogonal transformation ψ of the subspace T onto itself $l_1(g \circ \psi) = l_1(g)$. Let us prove that the functional l_2 also possesses the same property of invariance. Let ψ be an orthogonal transformation of the plane T into itself. If we set $\psi_1 = \varphi^{-1} \circ \psi \circ \varphi$, then ψ_1 is an orthogonal transformation of Q. We have: $g[\psi(\varphi[P])] = g(\varphi[\psi_1(P)])$ and hence

$$l_2(g \circ \psi) = \int_{G_k(Q)} g(\psi[\varphi(P)] \, d\mu_{Q,k}(P)$$

$$\int_{G_k(Q)} g(\varphi[\psi_1(P)]) \, d\mu_{Q,k}(P) = \int_{G_k(Q)} g[\varphi(P)] \, d\mu_{Q,k}(P).$$

The latter equality is a consequence of the invariance of the integral over the measure $\mu_{Q,k}$ defined in the manifold $G_k(Q)$. Therefore l_1 and l_2 are invariant non-negative linear functionals on $C[G_k(T)]$. They define some invariant measures μ_1 and μ_2 on $C[G_k(T)]$. Since $l_1(1) = l_2(1) = 1$, we have: $\mu_1[G_k(T)] = \mu_2[G_k(T)] = 1$, and hence the measures μ_1 and μ_2 coincide. In this case, naturally, the functionals l_1 and l_2 coincide too; i.e., $l_1(g) = l_2(g)$. Setting $g = f|_{G_k(T)}$, we get the equality (25).

If $f \in C(G_k(T))$, then the function τf, defined by f with equality (24), is continuous. Indeed, let (Q_ν), $\varphi = 1, 2, \ldots$, be an arbitrary sequence of the planes from G_m^n, converging at $\nu \to \infty$ to a certain plane Q_0. In this case there is a sequence of the transformations $\varphi_\nu \in \mathbb{O}^n$, such that $\varphi_\nu(Q_0) = Q_\nu$ and at $\nu \to \infty$ φ_ν tends to the identical mapping. In line with (25), at every ν,

$$(\tau f)(Q_\nu) = (\tau f)[\varphi_\nu(Q_0)] = \int_{G_k(Q_0)} f[\varphi_\nu(P)] \, d\mu_{Q_0,k}(P).$$

At $\nu \to \infty$ $f[\varphi_\nu(P)] \to f(P)$ for all $P \in G_k^n$. In this case convergence, as is seen, is uniform. This allows one to conclude that $(\tau f)(Q_\nu) \to (\tau f)(Q_0)$ at $\nu \to \infty$ and, thus, the continuity of τf is proved.

THEOREM 4.5.1. *Let $0 < k < m < n$, where k, m, n are integer numbers and let $f : G_k^n \to \bar{\mathbb{R}}$ be integrable over the measure $\mu_{n,k}$ (non-negative measurable) function. In this case, for almost all (in the sense of the measure $\mu_{n,m}$) planes $Q \in G_m^n$, the integral*

$$(\tau f)(Q) = \int_{G_k(Q)} f(P) \, d\mu_{Q,k}(P)$$

is defined, the function $Q \mapsto (\tau f)(Q)$ is integrable with respect to the measure $\mu_{n,m}$ (being, accordingly, non-negative, measurable), and the following *equality holds:*

$$\int_{G_m^n} \left\{ \int_{G_k(Q)} f(P) \, d\mu_{Q,k}(P) \right\} \, d\mu_{n,m}(Q) = \int_{G_k^n} f(P) \, d\mu_{n,k}(P).$$

Proof. The basic case is when f is a continuous function on G_k^n. Then, as has been shown above, the function τf is continuous on G_k^n. For any $\varphi \in O^n$ we have: $\tau(f \circ \varphi) = (\tau f) \circ \varphi$ and, hence

$$<\tau(f \circ \varphi), \mu_{n,m}> = <(\tau f) \circ \varphi, \mu_{n,m}> = (\tau f, \mu_{n,m}>.$$

The functional l_0 on $C(G_k^n)$, defined by the equality $l_0(f) = <\tau f, \mu_{n,m}>$, is thus invariant, and if $f \equiv 1$, then $l_0(f) = 1$. Therefore, $l_0(f) = <f, \mu_{n,m}>$, i.e. we get $<\tau f, \mu_{n,m}> = <f, \mu_{n,k}>$ for any $f \in C(G_k^n)$. The theorem is thus proved for the case $f \in C(G_k^n)$. A transition to the general case is achieved by applying common considerations of the theory of integrals, which the reader can do for himself. The theorem is proved.

COROLLARY. *Let $0 < k < m < n$, where k, m, n are integer numbers, $A \subset G_k^n$ be a measurable with respect to the measure $\mu_{n,k}$ set. For arbitrary $P \in G_m^n$ let $A_P = A \cap G_k(P)$. In this case for almost all $P \in G_m^n$ the set A_P is measurable and the equality*

$$\mu_{n,k}(A) = \int_{G_m^n} \mu_{P,k}(A_P) \, d\mu_{n,m}(P)$$

is valid.

The above statement can be obtained from the theorem if as f one chooses the indicator of the set A, i.e. function which equals zero at $Q \notin A$ and equals 1, if $Q \in A$.

At every integer k, such that $0 < k < n$ the diffeomorphism $\nu : G_k^n \rightarrow G_{n-k}^n$ is determined where, for $P \in G_k^n$, $\nu(P)$ is an $(n-k)$-dimensional subspace V^n, which is quite orthogonal to P. For an arbitrary Borel set $A \subset G_k^n$ let us set $(\nu^* \mu)(A) = \mu_{n,n-k}[\nu(A)]$. The measure $\nu^* \mu$ defined in such a way is invariant and normalized, which yields $\nu^* \mu = \mu_{n,k}$ and we thus find that for any Borel set $A \subset G_k^n$

$$\mu_{n,k}(A) = \mu_{n,n-k}[\nu(A)]. \tag{26}$$

As a supplement to Theorem 4.5.1 let us prove turning into zero of the measure of some sets in G_k^n.

LEMMA 4.5.1. *Let Q be an $(n-1)$-dimensional subspace in V^n, and let E be a set of all $P \in G_k^n$ contained in Q. In this case E is the set of zero measure in G_k^n.*

Proof. The set E is closed and hence its indicator χ_E

$$(\chi_E(P) = 1 \text{ at } P \in E, \quad \chi_E(P) = 0 \text{ at } P \notin E)$$

is a measurable function. The task is to prove that

$$\int_{G_k^n} \chi_E(P) \, d\mu_{n,k}(P) = \mu_{n,k}(E) = 0.$$

Let us calculate this integral using the result of Theorem 4.5.1. It is convenient to formulate our considerations by way of induction over the numbers $l = n - k$. Let $l = 1$. Then E consists of the only point and, hence, $\mu_{n,k}E) = \mu_{n,n-k}(E) = 0$. Let us assume that for certain l the lemma statement is true and let $n - k = l + 1$. let H be an arbitrary $(k+1)$-dimensional subspace V^n. Then in line with the theorem

$$\int_{G_k^n} \chi_E(P)\, d\mu_{n,k}(P) = \int_{G_{k+1}^n} \left\{ \int_{P \subset H} \chi_E(P)\, d\mu_{H,k}(P) \right\} d\mu_{n,\,k+1}(H)$$

$$= \int_{G_{k+1}^n} \mu_{H,k}[E \cap G_k(H)]\, d\mu_{n,\,k+1}(H). \tag{27}$$

Let E' be a set of the $H \in G_{k+1}^n$ which are contained in Q. In accordance with the induction assumption, E' is the set of zero measure in G_{k+1}^n. At $H \notin E'$ the intersection $E \cap G_k(H)$ consists of not more that one point, and, hence, in this case $\mu_{H,k}[E \cap G_k(H)] = 0$. Therefore, the subintegral expression in (27) equals zero almost everywhere in G_{k+1}^n, which yields the required result.

COROLLARY 1. *Let there be a straight line l in V^n and let E be a set of all $P \in G_k^n$ which are orthogonal to l. In this case E is a set of zero measure in G_k^n.*

Indeed, if P is orthogonal to l, then $P \subset Q = \nu(l)$, which yields the required result.

COROLLARY 2. *Let there be a straight line l in V^n passing through the point 0, and let E be a set of all $P \in G_k^n$, containing l. Then $\mu_{n,k}(E) = 0$.*

Indeed, if $P \supset l$, then $\nu(P)$ is orthogonal to l. Corollary 1, yields the fact that $\nu(E)$ is a set of zero measure in G_{n-k}^n. In view of the equality $\mu_{n,k}(E) = \mu_{n,n-k}[\nu(E)]$, the corollary is proved.

COROLLARY 3. *Let Q be an arbitrary m-dimensional subspace V^n, where $1 \leqslant m \leqslant k$, let E be a set of all $P \in G_k^n$, such that $Q \subset P$. In this case E is a set of the zero measure in G_k^n.*

Indeed, let l be an arbitrary straight line, lying in Q. If $P \supset Q$, then, obviously, $P \supset l$ and, hence, $E \subset E'$, where E' is a totality of all $P \in G_k^n$, containing l. In line with Corollary 2, $\mu_{n,k}(E') = 0$, which yields that $\mu_{n,k}(E) = 0$ as well.

4.6. Some Special Subsets of G_k^n

4.6.1. Here we are going to establish the measures of turning to zero of some special subsets of G_k^n. First let us consider the sets of k-dimensional directions associated with the arbitrary curve K in E^n.

Let K be a non-degenerate curve in E^n. The arc $[XY]$ of the curve K will be called k-dimensional, where $1 \leqslant k \leqslant n$, if this arc is contained in a certain k-dimensional plane and is not contained in any $(k-1)$-dimensional plane.

LEMMA 4.6.1. *Let K be a non-degenerate curve in E^n. Then at any k such that $1 \leqslant k \leqslant n$, the set of k-dimensional planes, which are the plane of the k-dimensional arc of the curve K, is not greater than countable.*

Proof. Let (L_m), $m = 1, 2, \ldots$, be an arbitrary sequence of the polygonal lines inscribed into the curve K and such that $\delta(L_m) \to 0$ at $m \to \infty$. Let $[XY]$ be an arbitrary k-dimensional arc of the curve K, where $1 \leqslant k \leqslant n$. Let $[X_m Y_m]$ be an arc of the polygonal line L_m inscribed into the arc $[XY]$. It is obvious that $[X_m Y_m]$ is a polygonal line and at $m \to \infty$ $[X_m Y_m] \to [XY]$ and, hence, at sufficiently great m the arc $[X_m Y_m]$ is also k-dimensional. In this case the plane of the arc $[X_m Y_m]$ at such m coincides with that of the arc $[XY]$. Since the arc $[X_m Y_m]$ does not lie in any $(k-1)$-dimensional plane, there is a sequence from $k + 1$ of the vertices $A_{i_0}, A_{i_1}, \ldots, A_{i_k}$, where $i_0 < i_1 < \cdots < i_k$ of the arc $[X_m Y_m]$ of the polygonal line L_m, which are in general position. The plane of the arc $[X_m Y_m]$ coincides with the k-dimensional plane, passing through the points $A_{i_0}, A_{i_1}, \ldots, A_{i_k}$. Therefore, if P is a plane of a certain k-dimensional arc of the curve K, then P is originated by the vertices of one of the polygonal lines L_m of the given sequence. Since the set of the k-dimensional planes, which can be defined by the vertices of the polygonal line L_m for every m, is finite, then, consequently, the set of all k-dimensional planes, each of which is a plane of a certain k-dimensional arc of the curve K, is not greater than countable and the lemma is thus proved.

COROLLARY 1. *Let K be a non-degenerate curve in E^m, S be a set of all $P \in G_k^n$, for which there exists a k-dimensional plane in E^n, parallel to P and containing an arc of the curve K. The measure of the set S equals zero.*

Proof. Let $1 \leqslant l \leqslant k$. Let us denote by S_l the set of all $P \in G_k^n$ for which there exists a k-dimensional plane, parallel to P and containing the l-dimensional arc of the curve K. In this case, obviously, $S = \bigcup_{l=1}^k S_l$. Let E_l be a set of all l-dimensional planes, each of which contains the l-dimensional arc of the curve K. Let $E_l = \{Q_1, Q_2, \ldots\}$. Let us denote through $S_{l,m}$ the set of all $P \in G_k^n$ which contain the l-dimensional subspace, parallel to Q_m. According to Corollary 3 of Lemma 4.5.1, $S_{l,m}$ is a set of the zero measure. Since $S_l = \bigcup_m S_{l,m}$, then S_l is also a set of the zero measure, which was the aim of the proof.

4.6.2. Let us first write some notes concerning the structure of the convex cone in the n-dimensional vector Euclidean space V^n.

The non-empty set $T \subset V^n$ is referred to as a convex cone if T is closed and for any vectors $u, v \in T$ and for any numbers $\lambda \geqslant 0$, $\mu \geqslant 0$ it is true that $\lambda u + \mu v \in T$. The whole space V^n is a convex cone. Any subspace V^n is a convex cone.

Any convex cone, obviously, contains the point 0.

Let us introduce the following notation. Let $a \in V^n$ and A be a subset in V^n. In this case $a + A$ denotes a totality of all vectors u of the type: $u = a + v$, where $v \in A$.

If T is a convex cone in V^n, then, as follows from the definition of a convex cone, for any $x \in T$ the set $x + T \subset T$.

The vector $h \in V^n$ is called a support vector of the convex cone T, if for all $x \in T$ the inequality $<h, x> \leqslant 0$ holds. The vector $h \in V^n$ is called a sup-port vector in a strict sense if for any vector $x \in T$, which is other than zero, $<h, x> < 0$ (a strict inequality).

The totality of all support vectors of the convex cone T in the space V^n will be denoted through T^*. The set T^* can be easily verified to be a convex cone in V^n as well. The cone T^* is called dual to the cone T.

LEMMA 4.6.2. *For any convex cone T in V^n, the cone which is dual to T^*, coincides with the initial cone T.*

Proof. The definition itself yields that any vector $u \in T$ is a support vec-tor for the cone T^*. Let us demonstrate that no vector $u \notin T$ is a support vec-tor for T^*. Indeed, let $u \notin T$. The set T is closed and, hence, $V^n \backslash T$ is an open set. This means that there is $\delta > 0$, such that a sphere of the radius δ and centered at u is contained in $V^n \backslash T$ and, hence, $|u - x| \geqslant \delta$ for all $x \in T$. Let x_0 be the point nearest to u of the closed set T. Let us set $h = u - x_0$. For any $\lambda \geqslant 0$ the point $\lambda x_o \in T$. The expression $|u - \lambda x_0|^2$ assumes its least value at $\lambda = 1$. We have:

$$|u - \lambda x_0|^2 = <u, u> - 2\lambda<u, x_0> + \lambda^2<x_0, x_0>.$$

Differentiating this expression with respect to λ and equating the derivative at the point $\lambda = 1$ to zero, we get:

$$-2<u, x_0> + 2<x_0, x_0> = 0$$

which yields $0 = <u - x_0, x_0> = <h, x_0>$. We have: $<h, u> = <h, x_0 + h> = <h, x_0> + <h, h> = |h|^2 \geqslant \delta^2 > 0$.

Let us prove that the vector $h \in T^*$. Indeed, if we choose an arbitrary vec-tor $x \in T$, then for any $\lambda \geqslant 0$ $x_0 + \lambda x \in T$ and, hence, $|u - x_0 - \lambda x| \geqslant |u - x_0|$, i.e. $|h - \lambda x| \geqslant |h|$ for all $\lambda \geqslant 0$. Therefore, $|h - \lambda x|^2 \geqslant |h|^2$ and, hence,

$$<h, h> - 2\lambda<h, x> + \lambda^2<x, x> \geqslant <h, h>$$

which yields $\lambda^2<x, x> \geqslant 2\lambda<h, x>$ and, at last, $\lambda/2<x, x> \geqslant <h, x>$ for any $\lambda \geqslant 0$. If we tend λ to zero, then we get $0 \geqslant <h, x>$ and, since the vector $x \in T$ is arbitrarily chosen, it means that h is a support vector for T. Thus, we have found the vector $h \in T^*$, such that $<h, u> > 0$. Therefore, the vector u is not a support vector of the cone T^*.

Consequently, if $u \in T$, then u is a support vector for T^*, and if $u \notin T$, then u is not a support vector for T^*. Therefore, the totality of all support vectors of the cone T^* coincides with the cone T, which is the proof required.

It follows from the lemma that if the cone T does not coincide with the whole of the space, then T has at least one support vector $h \neq 0$. Indeed, if it is not the case, i.e. $T^* = \{0\}$, then the cone $(T^*)^*$, which is dual to T^*, coincides with V^n. But, in line with the lemma, $(T^*)^* = T$ and, hence, if $T^* = \{0\}$, then $T = V^n$.

4.6.3. LEMMA 4.6.3. *Let T be a convex cone in V^n. The totality of all vectors $h \in V^n$, which are support vectors for the cone T in a strict sense, coincides* *with the set of all internal points of the cone T^*.*

Proof. Let h be a vector which is a support vector, in a strict sense, for the cone T. Let us prove that h is an internal point of T^*. The intersection of the cone T with the unit sphere $\Omega^{n-1} = \{x \in V^n \mid |x| = 1\}$ will be denoted through T_1. The set T_1 is compact. For any $u \in T_1$ there is an $\varepsilon > 0$, such that if $|h' - h| < \varepsilon$ and $|u' - u| < \varepsilon$, then $<h', u'> < 0$. Let us denote by B_u a sphere of radius ε and centre $u \in T_1$. According to the Borel lemma, there is a finite system of the points $u_1, u_2, \ldots, u_m$, such that the spheres $B_{u_1}, B_{u_2}, \ldots, B_{u_m}$ cover the set T_1. Let $\varepsilon_1, \varepsilon_2, \ldots, \varepsilon_m$ be the radii of these spheres. Let us denote by ε the least of the numbers $\varepsilon_1, \varepsilon_2, \ldots \varepsilon_m$. Let h' be such that $|h' - h| < \varepsilon$. If we arbitrarily choose a vector $u \in T$, $u \neq 0$, then $v = u/|u| \in T_1$. Let us find an i such that $v \in B_{u_i}$. In this case $|v - u_i| < \varepsilon_i$, $|h' - h| < \varepsilon \leqslant \varepsilon_i$ and, hence, $<h', v> < 0$. Therefore: $<h', u> = |u| <h', v> < 0$. We thus find that any vector h' such that $|h' - h| < \varepsilon$ is a support vector for T and, hence, h is an internal point for T^*.

Let h be an internal point for T^*. Let us find $\varepsilon > 0$, such that any vector h', for which $|h' - h| < \varepsilon$, is contained in T^*. Let us choose a vector $u_0 \in T$, such that $u_0 \neq 0$. Let us assume that $<h, u_0> = 0$. In this case, obviously, we can write $|u_0| = 1$. Let $h' = h + (\varepsilon/2)u_0$. Then $|h' - h| = \varepsilon/2 < \varepsilon$ and, hence, $h' \in T^*$. We have: $<h', u_0> = <h, u_0> + \varepsilon/2 = \varepsilon/2 > 0$. This contradicts the fact that $<h', u> \leqslant 0$ for any $u \in T^*$. Consequently, the equality $<h, u_0> = 0$ is impossible, and, hence, $<h, u> < 0$ for any vector $u \neq 0$ belonging to T.

The lemma is thus proved.

Any non-zero vector $h \in T^*$ defines the hyperplane $P(h) = \{x \mid <h, x> = 0\}$ which is called the support hyperplane of the cone T. The vector h in this case is referred to as the vector of the external normal of the hyperplane $P(h)$. If the vector h is a support vector, in a strict sense, of the cone T, then the hyperplane $P(h)$ shares the only general point with the cone T, the point 0. We shall say that the support hyperplane P touches the cone T provided it contains the points of the cone T, which are other than the point 0. In view of Lemma 4.6.3, the support hyperplane P will touch the cone T if and only if its external normal h is a boundary point for the cone T^*.

4.6.4. LEMMA 4.6.4. *The boundary of any convex cone in V^n is a set of zero measure.*

Proof. If the cone T coincides with V^n, then the boundary T is an empty set and, hence, is a set of the zero measure. If T is contained in a certain hyperplane, then the measure of the cone T itself and, hence, the measure of its boundary also, are equal to zero. If the cone T is not contained in any proper subspace V^n, then it has internal points. Indeed, in this case there are n linearly independent vectors $u_1, u_2, \ldots, u_n$ belonging to T. Let us set $v = u_1 + u_2 + \cdots + u_n$. For any $h \in T^*$, $h \neq 0$ $<h, v> = \sum_{i=1}^{n} <h, u_i> \leqslant 0$, and $<h, u_1> \leqslant 0$ for all $i = 1, 2, \ldots, n$. Therefore, if $<h, v> = 0$, then $<h, u_i> = 0$ for all $i = 1, 2, \ldots, n$. This, however, is impossible since the vectors $u_1, u_2, \ldots, u_n$ are linearly independent. Therefore, for any $h \in T^* <h, v> < 0$ (the strict inequality) and, hence, v is an internal point of the cone T.

So, let T be a convex cone having some internal points. Let a be an arbitrary internal point of T. Let us introduce a Cartesian orthogonal coordinate system into the space V^n in such a way that the vector a lies on the positive semi-axis Ox_n. The required result will be derived from the fact that in this coordinate system the boundary of the cone T is defined by the equation $x_n = f(x_1, x_2, \ldots, x_{n-1})$, where f is a continuous function. Let us introduce the following notations: we set $(x_1, x_2, \ldots, x_{n-1}) = y$. The point $y = (x_1, x_2, \ldots, x_{n-1}) \in \mathbb{R}^{n-1}$ will be identified with the point $(x_1, \ldots, x_{n-1}, 0) \in \mathbb{R}^n$.

Since a is an internal point of T, there is a $\delta > 0$ such that the closed sphere $\bar{B}(a, \delta) \subset T$. Let us consider a straight cyclic cone Q, defined by the inequality $x_n \geqslant (|a|/\delta) |y|$. The cone Q is contained in T. Indeed, let $x = (y, x_n) \in Q$. Let us set $u = \lambda x$, where $\lambda = |a|/x_n$. In this case we have: $u = (\lambda y, |a|)$. The point a has the coordinates $(0, |a|)$ and $|u-a| = \lambda |y| \leqslant \lambda x_n/|a| = \delta$, so that $u \in B(a, \delta)$ and, hence, $u \in T$. Hence, we conclude that $x = (1/\lambda)u \in T$ and the inclusion $Q \subset T$ is thus proved.

Let us consider an arbitrary point $y \in \mathbb{R}^{n-1}$. We can show that a straight line, passing through y parallel to the axis Ox_n, intersects the boundary T at the only point. Indeed, let h be an arbitrary support vector of the cone T. In this case $<a, h> < 0$, since a is an internal point of T. Therefore, there is such t_1 that the quantity $<y+ta, h> = <y, h> + t<a, h>$ is positive at $t \leqslant t_1$. The coordinate x_n of the vector $y + tq$ equals $t|a|$ and at $t \geqslant t_2 = (1/\delta)|y|$, obviously, $t|a| \geqslant (|a|/\delta)|y|$, i.e. at $t \geqslant t_2$ the point $y + ta$ belongs to the cone Q, and, hence, to T. Therefore, we have found such values of t_1 and t_2 that at $t \leqslant t_1$ $y + ta \notin T$, and at $t \geqslant t_2$ $y + ta \in T$. This means that there is such a t that $t_1 < t < t_2$ and the point $y + ta$ lies on the boundary of T. Let t_0 be the least of such t, and let $z = y + t_0 a$. The straight line $l : t \mapsto y + ta$ intersects the boundary of T at the only point. Indeed, the cone $z + T$ is contained in T and the ray of the straight line l with the origin at z and directed into the semi-space $x_n \geqslant 0$ is obviously contained in $z + T$, all the points of this ray, except for z, being the internal points of $z + T$, and hence of T. Therefore, no point of l, lying on l above the point z, is a boundary point of T, so l really intersects the boundary of T at the only point. Since $y \in \mathbb{R}^{n-1}$ is chosen arbitrarily, the boundary of T has thus been established to be set by the equation $x_n = f(y)$. Let us prove the continuity of f. Let us choose arbitrary points $y_1 \in \mathbb{R}^{n-1}$ and $y_2 \in \mathbb{R}^{n-1}$. Let $z_1 = (y_1, f(y_1))$, $z_2 = (y_2, f(y_2))$. We have: $z_1 + Q \subset T$, which yields $f(y_2) \leqslant (|a|/\delta) |y_2 - y_1| + f(y_1)$, i.e. $f(y_2) - f(y_1) \leqslant (|a|/\delta) |y_2 - y_1|$. Exchanging the roles of y_1 and y_2 we get after all that $|f(y_2) - f(y_1)| \leqslant (|a|/\delta) |y_2 - y_1|$. The function thus obeys the Lipshitz condition and is, consequently, continuous. The lemma is proved.

The set $A \subset V^n$ will be called a cone if for any point $x \in A$ for any $\lambda \geqslant 0$ the point $\lambda x \in A$. In other words, A is a cone if A is the union of a certain set of rays with their general origin at the point 0. It is obvious that the boundary of a convex cone is a cone. Let B^n the ball $\{x \in V^n \mid |x| < 1\}$, Ω^{n-1} be its boundary sphere and let μ_{n-1} denote the $(n-1)$-dimensional Lebesgue measure, i.e. an area on the sphere Ω^{n-1}. Let A be an arbitrary cone in V^n. In line with the known results from the Lebesgue theory of the integral, A is measurable with respect to the Lebesgue measure μ_n in the space V^n if and only if the intersection $A \cap \Omega^{n-1}$ is a set measurable with respect to the measure μ_{n-1}. In this case the equality $\mu_n(A \cap B^n) = (1/n)\mu_{n-1}(A \cap \Omega^{n-1})$ is valid. It should be recalled that the boundary ∂T of the convex cone T is a closed set. Therefore, the set $\partial T \cap \Omega^{n-1}$ is measurable and the equality cited above makes it possible to conclude that the measure of the intersection $\partial T \cap \Omega^{n-1}$ equals zero.

As a corollary of Lemma 4.6.4 we obtain the following result.

4.6.5. LEMMA 4.6.5. *The set $\gamma(T)$ of the tangential hyperplanes of the convex cone T in V^n is a set of zero measures in G_{n-1}^n.*

Proof. Let us consider the cone T^*. Let P be a tangential plane of the cone T, let ν be a unit normal vector of the plane P. In this case ν belongs to the intersection of the boundary of the dual cone T^* with the sphere Ω^{n-1}. In accordance with the lemma and due to the remarks preceding the lemma under discussion, we come to the conclusion that the unit normal vectors of the tangential hyperplanes of the cone T form a set of zero measure on the sphere Ω^{n-1}. If yields that the measure of the set $\nu[\gamma(T)] \subset G_{n-1}^n$ equals zero and, hence, the measure of the set $\gamma(T)$ in the manifold G_{n-1}^n also equals zero. The lemma is proved.

4.7. Length of a Spherical Curve as an Integral of the Function Equal to the Number of the Intersection Points

4.7.1. Let K be a curve on the sphere Ω^{n-1}. Let us define a certain function $\alpha_K(P)$ on the manifold G_{n-1}^n. Let $P \in G_{n-1}^n$. The plane P divides V^n into two closed semi-spaces, and in the case when the curve K lies in one of them entirely, we set $\alpha_K(P) = 0$. Let us assume that it is not the case. Then let us denote by $\alpha_K(P)$ the least upper boundary of the numbers m, such that on the curve K we can find a sequence of $m + 1$ points $X_0, X_1, \ldots, X_m$ for which $X_0 < X_1 < \cdots < X_m$ and at every $i = 1, 2, \ldots, m$ the points X_{i-1} and X_i lie on different sides of the plane P. If $\alpha_K(P) = m$, then, according to the definition of the curve, there is a sequence of $m + 1$ points, obeying the latter condition.

Let us say that the curve K on the sphere Ω^{n-1} is essentially intersected by the plane $P \in G_{n-1}^n$ at the point $Y \in K$, if any neighbourhood of the point Y on the curve K contains the points lying on different sides of P. If P contains no arcs of the curve K, the $\alpha_K(P)$ is equal to the number $\alpha_K^*(P)$ of the points where the curve K essentially intersects P. Indeed, let $X_0 < X_1 < \cdots < X_m$ be a sequence of the points of the curve K, such that the points X_{i-1} and X_i, $i = 1, 2, \ldots, m$, lie on different sides of the plane P, $m = \alpha_K(P)$, if $\alpha_K(P) < \infty$ and m is arbitrary in the opposite case. We can easily see that the arc $X_{i-1}X_i$ of the curve K is essentially intersected by the plane P at least at one point, and, hence $\alpha_K^*(P) \geqslant \alpha_K(P)$. Further, let $Y_1 < Y_2 < \cdots Y_m$ be the points of the essential crossing of the curve K with P, where $m = \alpha_K^*(P)$, if $\alpha_K^*(P)$ is finite, and m is arbitrary, if $\alpha_K^*(P) = \infty$. Let $U_1, U_2, \ldots, U_m$ be mutually exclusive neighbourhoods of the points $Y_1, Y_2, \ldots, Y_m$. Let X_0 and X_1, $X_0 < X_1$, be two points of the neighbourhood U_1, lying on different sides of the plane P. In the neighbourhood U_2 let us find a point X_2, such that the points X_1 and X_2 lie on different sides of P. Then, in the neighbourhood U_3 let us find a point X_3, such that the points X_2 and X_3 lie on different sides of P, and so on. As a result, we get a sequence of the points $X_0 < X_1 < \cdots < X_m$, any two neighbouring points of which lie on different sides of the plane P. This yields that $\alpha_K(P) \geqslant m$ and hence $\alpha_K(P) \geqslant \alpha_K^*(P)$. Therefore, $\alpha_K(P) = \alpha_K^*(P)$.

It should be noted that according to the corollary of Lemma 4.6.1, the set of those G_{n-1}^n which contains a certain arc of the curve K (the curve K is assumed to be non-degenerate) is a set of zero measure, which yields

$$\alpha_K(P) = \alpha_K^*(P)$$

for almost all $P \in G_{n-1}^n$.

Let us introduce one more quantity which characterizes the intersection of a curve and a hyperplane. Namely, let $n_K(P)$ be the number of points X of the curve K on the sphere Ω^{n-1}, lying on the plane $P \in G_{n-1}^n$.

LEMMA 4.7.1. *For almost all (in the sense of the measure $\mu_{n,n-1}$) planes $P \in G_{n-1}^n$ the equality $n_K(P) = \alpha_K^*(P)$ holds.*

Proof. It is obvious that at any P $n_K(P) \geqslant \alpha_K^*(P)$. Let us assume that for certain $P \in G_{n-1}^n$ $n_K(P) > \alpha_K^*(P)$. Since the subspaces $P \in G_{n-1}^n$, each of which contains the whole arc of the curve K, due to the corollary of Lemma 4.6.1, from a set of the zero measure, then it is sufficient to consider the case when P contains no arc of the curve K.

Let us set an arbitrarily normal parametrization $\{x(t), 0 \leqslant t \leqslant 1\}$ of the curve K

If $n_K(P) > \alpha_K^*(P)$, then there is a point $X \in K$, which is not a point of essential intersection of K and the plane P. This means that a certain neighbourhood of the point X on the curve K lies on one side of the plane P. Let $X = x(t_0)$. In this case there can be found rational t_1 and t_2, such that $t_1 \leqslant t_0 \leqslant t_2$ and the arc $x(t_1)x(t_2)$ lies on one side of P. Let us construct in V^n a closed convex hull of the arc $x(t_1)$ $x(t_2)$ and let $T(t_1, t_2)$ be a convex cone formed by the rays outgoing from the point 0 and passing through the points of the hull. The plane P is tangent to the cone $T(t_1, t_2)$. Therefore, if for a certain $P \in G_{n-1}^n$ $n_K(P) > \alpha_K^*(P)$, then the plane P is tangent to a certain cone $T(t_1, t_2)$, where t_1 and t_2 are rational numbers. In line with Lemma 4.6.5, the set of the planes $P \in G_{n-1}^n$, which are tangent to a convex cone, is a set of the zero measures. As far as the set of the cones $T(t_1, t_2)$, where t_1 and t_2 are rational, is not greater than countable, we come to the conclusion that the set of those $P \in G_{n-1}^n$ for which $n_K(P) > \alpha_K^*(P)$ is a set of zero measures. The lemma is proved.

THEOREM 4.7.1. *For any curve K on the sphere Ω^{n-1} the functions α_K, n_K are measurable on G_{n-1}^n and the equality:*

$$\int_{G_{n-1}^n} \alpha_K(P) \, d\mu_{n,n-1}(P) = \int_{G_{n-1}^n} n_K(P) \, d\mu_{n,n-1}(P) = \frac{1}{\pi} s(K) \qquad (28)$$

holds.

REMARK. Rectifiability of the curve K is not assumed in the theorem.

Proof. Let us set an arbitrary curve K on the sphere Ω^{n-1}. The proof is carried out in several steps. First, let us show measurability of the functions α_K and n_K. Since $\alpha_K = n_K$ almost everywhere, then it will be sufficient to show measurability of the function α_K alone.

Let us arbitrarily set $P_0 \in G_{n-1}^n$. Let us prove that for any sequence (P_m), $m = 1, 2, \ldots$, of the elements of G_{n-1}^n converging to P_0, the inequality

$$\alpha_K(P_0) \leqslant \varliminf_{m \to \infty \mathcal{L}} \alpha_K(P_m) \qquad (29)$$

holds. If $\alpha_K(P_0) = 0$, the inequality (29) is obvious. Let $\alpha_K(P_0) > 0$. Let us set an integer number $r \leqslant \alpha_K(P_0)$, $r > 0$. On the curve K let us find a sequence of the points $X_0 < X_1 < \cdots < X_r$, such that at every i the points X_{i-1} and X_i are located on different sides of the plane P_0. As $m \to \infty$ the planes P_m converge to P_0, there is an m_0, such that at every $m \geqslant m_0$ the points X_{i-1} and X_i will be located on different sides of the plane P_m for every $i = 1, 2, \ldots, r$. For $m \geqslant m_0$, obviously, $\alpha_K(P_m) \geqslant r$, which yields $\varliminf_{m \to \infty} \alpha_K(P_m) \geqslant r$. Since $r \leqslant \alpha_K(P_0)$ is arbitrarily chosen, we have (29). The function α_K is, thus, semi-continuous from below on the set G_{n-1}^n and is hence measurable.

For any curve K on the sphere a certain quantity

$$f(K) = \int_{G_{n-1}^n} \alpha_K(P) \, d\mu_{n,n-1}(P)$$

is defined. Let us prove that the function f, defined in such a way, obeys all the conditions of Theorem 2.2.5 on the set of all spherical curves.

(1) Let us check whether $f(K) = f(L)$ if the curves K and L are congruent. Let φ be the sphere Ω^{n-1} rotation around the point 0, transforming the curve K into the curve L. Let us arbitrarily choose $P \in G_{n-1}^n$ and show that

$$\alpha_K(P) \leqslant \alpha_L[\varphi(P)]. \qquad (30)$$

If $\alpha_K(P) = 0$, then the inequality (30) is obvious. Let $\alpha_K(P) > 0$. Let us choose an integer number $r > 0$, such that $r \leqslant \alpha_K(P)$, and let $X_0 \leqslant X_1 < \cdots <$

X_r be the points of the curve K such that at every $i = 1, 2, \ldots, r$ the points X_{i-1} and X_i are located on different sides of the plane P. Let Y_i, $i = 0, 1, 2, \ldots, r$, be the corresponding points of the curve L, i.e. $Y_i = \varphi(X_i)$. It is obvious that Y_{i-1} and Y_i are located on different sides of the plane $\varphi(P)$ at every i. Therefore, $\alpha_L[\varphi(P)] \geqslant r$ and, since $r \leqslant \alpha_K(P)$ is arbitrary chosen, the inequality (30) is proved. Integrating the inequality (30) term by term, we get

$$f(K) = \int_{G_{n-1}^n} \alpha_K(P) \, d\mu_{n,n-1}(P) \leqslant \int_{G_{n-1}^n} \alpha_L[\varphi(P)] \, d\mu_{n,n-1}(P) =$$

$$\int_{G_{n-1}^n} \alpha_L(P) \, d\mu_{n,n-1}(P) = f(L).$$

Here use has been made of the invariance property of the integral with respect to the measure G_{n-1}^n. Since K, in its turn, is obtained from L through a certain rotation of the sphere, then in line with the above proved, the inequality $f(L) \leqslant f(K)$ must also hold, which leads us to the conclusion that $f(L) = f(K)$.

(2) Let A and B be the end points of the curve K, $A < C < B$. The arc $[AC]$ will be denoted through K_1, the arc $[CB]$ through K_2. In this case, if the plane P does not pass through the point C, then obviously, $n_K(P) = n_{K_1}(P) + n_{K_2}(P)$. The set of the planes P, containing the point C, is a set of the zero measure. Therefore, for almost all $P \in G_{n-1}^n$ we have:

$$h_K(P) = n_{K_1}(P) + n_{K_2}(P).$$

Integrating this equality, we, obviously, get $f(K) = f([AC]) + f([CB])$.

(3) Let (K_m), $m = 1, 2, \ldots$, be an arbitrary sequence of the spherical curves converging to a certain curve K_0. Let us demonstrate that for any plane $P \in G_{n-1}^n$ the inequality

$$\alpha_{K_0}(P) \leqslant \varliminf_{m \to \infty} \alpha_{K_m}(P) \tag{31}$$

holds. If $\alpha_{K_0}(P) = 0$, then inequality (31) is obvious. Let $\alpha_{K_0}(P) > 0$. Let us arbitrarily set an integer $r \leqslant \alpha_K(P)$, $r > 0$, and on the curve K_0 construct a sequence of the points $X_0 < X_1 < \cdots < X_r$, such that at every i X_{i-1} and X_i are located on different sides of the plane P. Let $X_0^m < X_1^m < \cdots X_r^m$ be the points of the curve K_m, chosen in such a way that at $m \to \infty$ $X_i^m \to X_i$ for every $i = 1, 2, \ldots, r$. For sufficiently large m, $m \geqslant m_0$ the points X_i^m will be located with respect to the plane P in exactly the same position as the points X_i. It means that for $m \geqslant m_0$ $\alpha_{K_m}(P) \geqslant r$ and, hence $\varliminf_{m \to \infty} \alpha_{K_m}(P) \geqslant r$. Since $r \leqslant \alpha_L(P)$ is arbitrarily chosen, then the inequality (4) is obviously proved.

According to the known results from the Lebesgue theory of the integral, the above proved yields

$$f(K_0) = \int_{G_{n-1}^n} \alpha_{K_0}(P) \, d\mu_{n,n-1}(P) \leqslant \varliminf_{m \to \infty} \int_{G_{n-1}^n} \alpha_{K_m}(P) \, d\mu_{n,n-1}(P)$$

$$= \lim_{m \to \infty} f(K_m).$$

(4) Let us prove the fulfilment of the last condition of Theorem 2.2.5. Let K be an arc of a large circumference, joining two diametrically opposing points A and A' of the sphere Ω^{n-1}. In this case for any plane P, which does not contain the straight line AA', i.e. for almost all $P \in G_{n-1}^n$ $n_K(P) = 1$. Therefore,

$$f(K) = \int_{G_{n-1}^n} n_K(P) \, d\mu_{n,n-1}(P) = \mu_{n,n-1}(G_{n-1}^n) = 1$$

and, in particular, we get $0 < f(K) < \infty$.

All the condition of Theorem 2.2.5 are thus met. Hence $f(K) = Cs(K)$ for any spherical curve K. The value of the constant C can be found by way of choosing as K an arc of the circumference of a large circle joining two diametrically opposing points of the sphere. For such an arc, as has been shown above, $f(K) = 1$, $s(K) = \pi$ and, hence, $C = 1/\pi$.

The theorem is proved.

4.7.2. LEMMA 4.7.2. *Let K be a rectifiable curve on the sphere Ω^{n-1}, m be an integral number $1 \leqslant m \leqslant n-2$. The set E of all subspaces $X \in G_m^n$, such that $Q \cap |KL|$ is not empty, is a set of the zero measure.*

Proof. It is obvious that E is a closed set and hence that E is measurable. Let $P \in G_{n-1}^n$, $G_m(P)$ be the totality of all $Q \in G_m^n$, containing in P. The intersection $G_m(P) \cap E$ will be denoted through E_P. Let us prove that for almost all $P \in G_{n-1}^n$ $\mu_{P,m}(E_P) = 0$. Indeed, for almost all $P \in G_{n-1}^n$ $n_K(P) < \infty$, as the function $n_K(P)$ is integrable. Let $P \in G_{n-1}^n$ is such that $n_K(P) < \infty$. In this case the set $P \cap |K|$ is finite. Let $a_1, a_2, \ldots, a_r$ be all its points. The plane $Q \in G_m(P)$ belongs to E_P if and only if Q contains at least one of the points $a_1, a_2, \ldots, a_r$. The set of all $Q \in G_m(P)$, containing the point a_i is, in line with corollary 2 of Lemma 4.5.1, a set of zero measure, which yields that $\mu_{P,m}(E_P) = 0$. According to the corollary of Theorem 4.5.1, we get $\mu_{n,m}(E) = 0$, and therefore the lemma is proved.

4.8. Length of a Curve as an Integral of Lengths of its Projections

4.8.1. LEMMA 4.8.1. *Let a be an arbitrary vector in the space E^n, ad let $\pi_P(a)$ be its orthogonal projection on the k-dimensional subspace P of the space V^n. The function $P \mapsto \pi_P(a)$ is continous and the equality*

$$\int_{G_k^n} |\pi_P(a)| \mu_{n,k}(dP) = \frac{\sigma_n}{\sigma_k} |a| \tag{32}$$

holds, where σ_m, $m = 1, 2, \ldots$, is a constant, $\sigma_m = \Gamma(m/2)/\Gamma[(m+1)/2]$.

Proof. Continuity of the function $P \mapsto \pi_P(a)$ is obvious. Let $a = |a|e$ where $|e| = 1$. In this case $|\pi_P(a)| = |a| |\pi_P(e)|$ and the proof of equality (32) is reduced to the case when $a = e$ is a unit vector.

Let e be a unit vector. The integral

$$\int_{G_k^n} |\pi_P(e)| \, \mu_{n,k}(dP) \tag{33}$$

is independent of the choice of e. Indeed, let e' be another arbitrary unit vector in V^n, and let φ be an orthogonal transformation, such that $\varphi(e) = e'$. In this case $|\pi_P(e)| = |\pi_{\varphi(P)}(e')|$ and, hence, in view of the invariance property of the measure $\mu_{n,k}$ we have:

$$\int_{G_k^n} |\pi_P(e)| \mu_{n,k}(dP) = \int_{G_k^n} |\pi_{\varphi(P)}(e')| \mu_{n,k}(dP) =$$

$$\int_{G_k^n} |\pi_P(e')| \mu_{n,k}(dP).$$

Let us denote the integral (33) by $I_{n,k}$.

Let $k = 1$. In line with the remark made when concluding Section 4.4, we have:

$$I_{n,k} = \frac{1}{\omega_{n-1}} \int_{\Omega^{n-1}} |<e, \nu>| \mu(d\nu). \tag{34}$$

Let $S(e)$ be a cross-section of the sphere Ω^{n-1} formed by a hyperplane, passing through the point O perpendicular to the vector e. For the vector $\nu \in \Omega^{n-1}$, such that $\nu \neq \pm e$, let $\varphi = <\nu, e>$ and ζ be a unit vector lying in $S(e)$, which is colinear and has the same direction as the orthogonal projection of the vector ν onto the plane of the sphere $S(e)$. In this case $|<e, \nu>| = |\cos \varphi|$ and the right-hand part of (33) is equal to

$$\frac{1}{\omega_{n-1}} \int_{S(e)} \left[\int_0^\pi \sin^{n-2}\varphi \, |\cos\varphi| \, d\varphi \right] \mu_{n-2}(d\zeta)$$

$$= \frac{\omega_{n-2}}{\omega_{n-1}} \int_0^\pi \sin^{n-2}\varphi \, |\cos\varphi| \, d\varphi = \frac{\Gamma(\frac{n}{2})}{\sqrt{\pi} \, \Gamma(\frac{n+1}{2})},$$

as follows from the known expression for the quantity ω_m, $m = 1, 2, \ldots$.

To obtain the result for the general case, we shall calculate the integral (32), applying Theorem 4.5.1. Setting it $m = k$, $k = 1$ in (32) we get:

$$I_{n,1} = \int_{G_k^n} \left[\int_{G_1(P)} |\pi_\nu(e)| \, d\mu_{P,1}(\nu) \right] d\mu_{n,k}(P). \tag{35}$$

Let π_P denote the operation of orthogonal projection onto the plane P. For any vector $\nu \in P$, the projection e onto the straight line on which the vector ν lies, can be obtained in two steps. First we orthogonally project e onto the plane P, and then we orthogonally project the projection so obtained onto the given straight line. In other words, $\pi_\nu(e) = \pi_\nu[\pi_P(e)]$. Therefore, in line with what has been proved above, we have:

$$\int_{G_1(P)} |\pi_\nu(e)| \, d\mu_{P,1}(\nu) = \int_{G_1(P)} |\pi_\nu[\pi_P(e)]| \, d\mu_{P,1}(\nu)$$

$$= I_{k,1}|\pi_P(e)| = |\nu_P(e)| \frac{\Gamma(\frac{n}{2})}{\sqrt{\pi}\ \Gamma(\frac{n+1}{2})}.$$

Substituting this expression into (33) we get:

$$\frac{\Gamma(\frac{n}{2})}{\sqrt{\pi}\ \Gamma(\frac{n+1}{2})} = \frac{\Gamma(\frac{k}{2})}{\sqrt{\pi}\ \Gamma(\frac{k+1}{2})} \int_{G_k^n} |\pi_P(e)| d\mu_{n,k}(P),$$

which yields (32). The Lemma is proved.

Let K be an arbitrary curve in E^n. Let us fix an arbitrary point O in E^n. Let $P \in G_k^n$. Let us draw through O a k-dimensional plane P' which is parallel to the k-dimensional direction of P. The orthogonal projection of the curve K onto the plane P' will be denoted by K_P.

THEOREM 4.8.1. *For any curve K in E^n the function $P \mapsto s(K_P)$ is measurable in G_k^n and the equality*

$$\int_{G_k^n} s(K_P) d\mu_{n,k}(P) = \frac{\sigma_n}{\sigma_k} s(K)$$

holds.

Proof. In view of Lemma 4.8.1 the theorem is valid when the curve K is a segment of the straight line. This fact, obviously, yields the validity of the theorem in the case when the curve is a polygonal line.

Let us consider the general case. Let us construct a sequence of polygonal lines (L_m) inscribed into the curve K_P and converging to it at $m \to \infty$. At every $PL_{m,P}$ there is a polygonal line inscribed into the curve K_P and at $m \to \infty$ $L_{m,P} \to K_P$. It follows from this that $s(L_{m,P}) \to s(K_P)$ at $m \to \infty$ for all P. Since each of the functions $P \mapsto s(L_{m,P})$ is continuous, then, obviously, the function $P \mapsto s(K_P)$ is measurable. Since, for every m, $s(L_{m,P}) \leq s(K_P)$, and since, due to known theorems on the limiting transition under the sign of the Lebesgue integral, $s(L_{m,P}) \to s(K_P)$ as $m \to \infty$, we have:

$$\int_{G_k^n} s(L_{m,P}) d\mu_{n,k}(P) \to \int_{G_k^n} s(K_P) d\mu_{n,k}(P)$$

as $m \to \infty$. For every m

$$\int_{G_k^n} s(L_{m,P}) d\mu_{n,k}(P) = \frac{\sigma_n}{\sigma_k} s(L_m)$$

and, since $s(L_m) \to s(K)$ as $m \to \infty$, then we get:

$$\int_{G_k^n} s(K_P) d\mu_{n,k}(P) = \frac{\sigma_n}{\sigma_k} s(K).$$

The theorem is proved.

4.8.2. The theorem allows a certain analogy, concerning rectifiable curves on a sphere. The result obtained in this case will be used later. Let us introduce some preliminary notions.

Let $P \in G_k^n$, where $2 \leqslant k \leqslant n-1$, and let x be a point of the sphere Ω^{n-1}. Let us denote by $\eta_P(x)$ the nearest point to x of the $k-1$-dimensional sphere $\Omega^{n-1} \cap P$. The point $\eta_P(x)$ is not defined if $x \in \nu(P)$. If $x \notin \nu(P)$, then $\eta_P(x)$ can be obtained in the following way. Let $\pi = \pi_P$ be a mapping of the orthogonal projection onto the plane P. In this case

$$\eta_P(x) = \frac{\pi(x)}{|\pi(x)|} .$$

It shows, in particular, that the mapping η_P is continuous in the domain $\Omega^{n-1} \backslash \nu(P)$. For any spherical curve K, which does not intersect $\nu(P)$, the curve $\eta_P(K)$, which will be called the K projection onto the sphere $P \cap \Omega^{n-1}$, is defined. Let us assume that K is a rectifiable curve on the sphere and A is a set of those $P \in G_k^n$, where $2 \leqslant k \leqslant n-1$, for which the curve $\eta_P(K)$ is not defined. Then, let B be a set of all $Q \in G_{n-k}^n$, for which the intersection $Q \cap |K|$ is not empty. The plane $P \in A$ if and only if $Q = \nu(P) \in B$, so that $A = \nu(B)$ and hence $\mu_{n,k}(A) = \mu_{n,n-k}(B)$. Since $n-k \leqslant n-2$, then, as a consequence of Lemma 4.7.2, we get $\mu_{n,n-k}(B) = 0$ and, hence, $\mu_{n,k}(A) = 0$, too. Therefore, we have established that if the curve K is rectifiable, then the curve $\eta_P(K)$ is defined for almost all $P \in G_k^n$, where $2 \leqslant k \leqslant n-1$.

4.8.3. THEOREM 4.8.3. *For any rectifiable curve K on the sphere Ω^{n-1} for any integer k, where $2 \leqslant k \leqslant n-1$, the function $P \in G_k^n \mapsto s(\eta_P(K))$ is integrable and the equality*

$$\int_{G_k^n} s[\nu_P(K)] \, d\mu_{n,k}(P) = s(K)$$

is valid.

Proof. Let K be an arbitrary rectifiable spherical curve. The totality of those P for which $\nu(P) \cap |K| \neq \emptyset$ is a closed set of the zero measure. Consequently, the function $s[\pi_P(K)]$ is defined on the open set $U \subset G_k^n$, such that $\mu_{n,k}(U) = 1$. Let us choose arbitrarily $P_0 \in U$ and let (P_m), $m = 1, 2, \ldots$, be an arbitrary sequence of the elements of G_k^n, converging to P_0. In this case, beginning with a certain $m = m_0$, $P_m \in U$ and the curves $\eta_{P_m}(K)$ converge to the curve $\eta_{P_0}(K)$ at $m \rightarrow \infty$. This yields

$$s[\eta_{P_0}(K)] \leqslant \varliminf_{m \to \infty} s[\eta_{P_m}(K)].$$

The function $P \mapsto s[\eta_P(K)]$ is, therefore, semi-continuous from below on the set u and, hence, it is measurable.

For any rectifiable sperical curve K, therefore, a certain quantity

$$f(K) = \int_{G_k^n} s[\eta_P(K)] \, d\mu_{n,k}(P)$$

is defined. Let us prove that the function f obeys all the conditions of Remark 1 to Theorem 2.2.5. The class Σ is, in this case, the totality of all rectifiable curves on the sphere.

(1) Let us first check if $f(K) = f(L)$ in case K and L are congruent. Let φ be an orthogonal transformation of V^n, such that $L = \varphi(K)$. For an arbitrary point $X \in \Omega^{n-1}$, such that $X \notin \nu(P)$, where $P \in G_k^n$, we obviously have: $\varphi[\eta_P(X)]$ $= \eta_{\varphi(P)}(\varphi(X))$, which means that $\varphi[\eta_P(K)] = \eta_{\varphi(P)}(K)$ is defined. Therefore, $s[\eta_{\varphi(P)}(L)] = s[\varphi(\eta_P(K))]$ and

$$f(K) = \int_{G_k^n} s[\eta_{\varphi(P)}(L)] \, \mathrm{d}\mu_{n,k}(P)$$
$$= \int_{G_k^n} s[\eta_P(L)] \, \mathrm{d}\mu_{n,k}(P)$$

due to the invariance property of the integral with respect to the measure $\mu_{n,k}$.

(2) Let A be the origin of the curve K, let B be its end, $A < C < B$, and let C be a point of the curve K. In this case, obviously $s[\eta_P([AB])] = s[\eta_P([AC])]$ $+ s[\eta_P([CB])]$. Integrating this inequality, we get

$$f([AB])] = f([AC])] + f([CB])].$$

Therefore, the condition (2) of Theorem 2.2.5 is fulfilled.

(3) Let (K_m), $m = 1, 2, \ldots$, be an arbitrary sequence of rectifiable curves on the sphere, which converges to the curve K and let $P \in G_k^n$ be such that $\eta_P(K)$ is defined. In this case, obviously, the curve $\eta_P(K_m)$ will also be defined for all sufficiently large m and at $m \to \infty$ $\eta_P(K_m) \to \eta_P(K)$. It yields: $s[\eta_P(K)] \leqslant \underline{\lim}_{m\to\infty} s[\eta_P(K_m)]$. In accordance with the known properties of the Lebesgue integral, we, thus, get

$$f(K) = \int_{G_k^n} s[\eta_P(K)] \, \mathrm{d}\mu_{n,k}(P)$$
$$\leqslant \underline{\lim_{m\to\infty}} \int_{G_k^n} s[\eta_P(K_m)] \, \mathrm{d}\mu_{n,k}(P) = \underline{\lim_{m\to\infty}} f(K_m).$$

The condition (3) of Theorem 2.2.5 is therefore fulfilled.

(4) Now we have to check the fulfilment condition (4) of Theorem 2.2.5.

Let K be an arc of a large circumference, connecting two diametrically opposing points A and A' of the sphere Ω^{n-1}. In this case, as can be easily seen for almost all $P \in G_k^n$ the curve $\eta_P(K)$ is also a semi-circumference of a large circle, so that $s[\eta_P(K)] = \pi$ for all $P \in G_k^n$, for which the curve $\eta_P(K)$ is defined. This consequently yields

$$f(K) = \int_{G_k^n} s[\eta_P(K)] \, \mathrm{d}\mu_{n,k}(P) = \pi = s(K). \tag{36}$$

In particular, we get $0 < f(K) < \infty$ which is the required proof.

In line with Theorem 2.2.5, the above proof yields $f(K) = Cs(K)$, where $C = $ const for any rectifiable spherical curve. From the equality (36) we have $C = 1$. The theorem is proved.

4.9. Generalization of Theorems on the Mean Number of the Points of Intersection and Other Problems

4.9.1. Let $K = \{x(s), 0 \leqslant s \leqslant l\}$ (the parameter s is the arc length) be a rectifiable curve on the sphere Ω^{n-1}, and let $f(s)$ be a function defined on the segment $[0, l]$, $l = s(K)$. Let us denote by $\psi(f, K, P)$, where $P \in G_{n-1}^n$, the sum of the values of the function $f(s)$ for all s, at which the point $x(s) \in P$. (If the number of the points $x(s)$ is infinite, then $\psi(f, K, P)$ can be given any arbitrary value; in this case let us assume that $\psi(f, K, P) = \infty$).

The following theorem is a generalization of Theorem 4.7.1.

THEOREM 4.9.1. *If the function $f(s)$ is measurable (integrable) with respect to s, then $\psi(f, K, P)$ is a measurable (integrable) function of $P \in G_{n-1}^n$, and in the case when $f(s)$ is integrable too*

$$\int_0^l f(s)ds = \pi \int_{G_{n-1}^n} \psi(f, K, P) d\mu_{n,n-1}(P).$$

Proof. The theorem is obvious if $f(s)$ is a step-function, i.e. if the segment $[0, l]$ can be subdivided into a finite number of segments, on each of which $f(s)$ is constant.

In the general case the theorem is proved with a limiting transition, which is common for the theory of the Lebesgue integral. Let us consider here all the details of the proof in order to spare us the need to cite them any further.

It should be first of all noted that if $f_m(s) \rightarrow f(s)$ as $m \rightarrow \infty$ for all s, then $\psi(f, K, P) \rightarrow \psi(f, K, P)$ as $m \rightarrow \infty$ for any $P \in G_{n-1}^n$. It is also obvious that $|\psi(f, K, P)| \leq \psi(|f|, K, P)$ for any function f.

Let us assume that $f(s) = 0$ almost everywhere in $[0, l]$. Let us then prove that in this case also $\psi(f, K, P) = 0$ almost everywhere. Let us arbitrarily set $\varepsilon > 0$. To this ε we can find such an increasing sequence of non-negative step-functions $(f_m(S))$, $m = 1, 2, \ldots$, that $\int_0^l f_m(s)ds < \varepsilon/\pi$ for all m and for all $s \in [0, l]$ $|f(s)| \leq f_0(s) = \lim_{m \to \infty} f_m(s)$. For every m

$$\int_0^l f_m(s)ds = \pi \int_{G_{n-1}^n} \psi(f_m, K, P) d\mu_{n,n-1}(P).$$

The sequence of functions $\psi_m(P) = \psi(f_m, K, P)$ is increasing and for all P

$$|\psi(f, K, P)| \leq \psi(|f|, K, P) \leq \psi(f_0, K, P),$$

$$\lim_{m \to \infty} \psi_m(P) = \psi(f_0, K, P).$$

Therefore, an increasing sequence functions, integrable on G_{n-1}^n, has been constructed, such that for all P $|\psi(f, L, P)| \leq \lim_{m \to \infty} \psi_m(P)$ and at every m

$$\int_{G_{n-1}^n} \psi_m(P) d\mu_{n,n-1}(P) < \varepsilon.$$

In line with $\varepsilon > 0$ arbitrarity, it yields that $\psi(f, K, P) = 0$ almost everywhere in G_{n-1}^n. It means that the measure of the set of the planes $P \in G_{n-1}^n$, containing the points $x(s)$, where $s \in E$, equals zero.

Let $(f_m(s))$, $m = 1, 2, \ldots$, be an arbitrary sequence of the functions converging almost everywhere in $[0, l]$ to a certain function f. In this case the function $\psi(f_m, K, P)$ of the variable P converges almost everywhere to $\psi(f, K,$

P). Indeed, let $E \subset [0, l]$ be a set of those $s \in [0, l]$, for which the equality $f(s) = \lim_{m \to \infty} f_m(s)$ is not valid, let H be a totality of all planes P which contain the points $x(s)$, where $s \in E$. In line with the last remark, H is a set of zero measure. For every plane $P \notin H$ $\psi(f_m, K, P) \to \psi(f, K, P)$ at $m \to \infty$, i.e. we get that $\psi(f_m, K, P) \to \psi(f, K, P)$ for almost all $P \in G_{n-1}^n$.

Let $f(s)$, $0 \leqslant s \leqslant l$, be a measurable function. In this case there exists a sequence of step-functions $(f_m(s))$, $m = 1, 2, \ldots$, which converge to $f(s)$ almost everywhere. According to the above proof, at $m \to \infty$ $\psi(f_m, K, P) \to \psi(f, K, P)$ for almost all $P \in G_{n-1}^n$. The functions $\psi(f_m, K, P)$ of the variable P are integrable, which yields the measurability of $\psi(f, K, P)$.

Let us now consider the case when $f(s)$, $0 \leqslant s \leqslant l$, is a function integrable in the Lebesgue sense. In this case there is a sequence of step-functions $(f_m(s))$, $m = 1, 2, \ldots$, for which

$$\|f_m - f\|_{L_1} = \int_0^l |f_m(s) - f(s)| \, ds \to 0$$

as $m \to \infty$. Let (f_m), $m = 1, 2, \ldots$, be such a sequence. Let us consider it chosen in such a way that $\|f_m - f\|_{L_1} < 2^{-m}$. (The only requirement here is for the series $\sum_{m=1}^\infty \|f_m - f\|_{L_1}$ be convergent.) Then $f_m(s) \to f(s)$ nearly everywhere. Let us set $u_1(s) = f_1(s)$, $u_m(s) = f_m(s) - f_{m-1}(s)$ at $m > 1$, $U_m(P) = \psi(u_m, K, P)$. We have:

$$f(s) = \sum_{m=1}^\infty u_m(s) = \lim_{K \to \infty} \sum_{m=1}^K u_m(s)$$

which leads us to the conclusion

$$U(P) = \psi(f, K, P) = \lim_{K \to \infty} \sum_{m=1}^K U_m(P) = \sum_{m=1}^\infty U_m(P).$$

It should be now remarked that at $m > 1$

$$\|U_m\|_{L_1} = \int_{G_{n-1}^n} |U_m(P)| \, d\mu_{n,n-1}(P)$$

$$\leqslant \int_{G_{n-1}^n} \psi(|u_m|, K, P) \, d\mu_{n,n-1}(P) = \pi \int_0^l |u_m(s)| \, ds$$

$$= \pi \int_0^l |f_m(s) - f_{m-1}(s)| \, ds \leqslant \pi(\|f_m - f\|_{L_1} + \|f - f_{m-1}\|_{L_1}) < \frac{\pi}{2^{m-2}}.$$

The series $\sum_{m=1}^\infty \|U_m\|_{L_1}$ is thus convergent. Consequently, the function $U(P)$ is integrable and

$$\int_{G_{n-1}^n} U(P) \, d\mu_{n,n-1}(P) = \sum_{m=1}^\infty \int_{G_{n-1}^n} U_m(P) \, d\mu_{n,n-1}(P)$$

$$= \frac{1}{\pi} \sum_{m=1}^\infty \int_0^l u_m(s) \, ds = \frac{1}{\pi} \int_0^l f(s) \, ds.$$

The theorem is proved.

The theorem proved above states, in particular, that if the function $f(s)$ is measurable, then measurable is the function $\psi(f, K, P)$ of the variable $P \in G_{n-1}^n$. Our nearest aim is to prove the statement which is somewhat opposite to it. Namely, it will be stated below that if for any arc L of the curve K the function $\psi(f, L, P)$ is measurable, then $f(s)$ is a measurable function of s.

LEMMA 4.9.1. *Let $e_1, e_2, \ldots, e_n$ be an orthonormal system of vectors V^n, and let P_i be a hyperplane in V^n, orthogonal to e_i. Let us assume that on the sphere Ω^{n-1} there is given a certain arc L of a large circumference, not containing a single point e_i, φ is a length of this arc, φ_i is the length of the projection $\eta_{P_i}(L)$ of the arc L onto the sphere $\Omega^{n-1} \cap P_i$. In this case the inequality*

$$\varphi \leqslant \frac{1}{\sqrt{n-2}} \sum_{i=1}^{n} \varphi_i$$

holds.

Proof. Let $p, q \in \Omega^{n-1}$ be the ends of the arc L. Let us assume that $s(L) < \pi$, which is obviously what the general case is reduced to. In this case $s(L) = (p\frown q)$, and $\varphi_i = (p_i \frown q_i)$ where p_i, q_i are the projections of p and q, respectively, onto the plane P_i. Let us fix the vector p and rotate q in the plane of the vectors p and q. Let us set $q = p(\varphi)$, where $\varphi = (p \frown q)$.

In the space V^n let us introduce a Cartesian orthogonal system of coordinates, whose basis vectors are $e_1, e_2, \ldots, e_n$. For arbitrary vectors $a, b \in V^n$ the squared area of the parallelogram, constructed on them, equals

$$\begin{vmatrix} <a, a>, & <a, b> \\ <b, a>, & <b, b> \end{vmatrix} = \sum_{1 \leqslant K < l \leqslant n} \begin{vmatrix} a_k, & a_l \\ b_k, & b_l \end{vmatrix}^2, \tag{37}$$

where $a_1, \ldots, a_n$, $b_1, \ldots, b_n$ are the coordinates of the vectors a and b in the given system of coordinates. Let $a^{(i)}$, $b^{(i)}$ be the orthogonal projections of the vectors a and b onto the plane P_i. Then the kth coordinate $a_k^{(i)}$ of the vector $a^{(i)}$ at $k \neq i$ equals a_k, and at $k = i$ it equals zero. An analogous conclusion is true for the vector $b^{(i)}$. Let S_i be the area of the parallelogram, constructed on the vectors $a^{(i)}$ and $b^{(i)}$. In this case S_i^2 is expressed by the sum of the type (37) with the additional condition $k \neq i$, $l \neq i$. Let us consider the sum $\sum_{i=1}^{n} S_i^2$. Substituting it into the extended expression for each of the addends S_i^2, we find that the quantity $\begin{vmatrix} a_k, & a_l \\ b_k, & b_l \end{vmatrix}^2$ is a part of all the addends except for those two for which either $i = k$, or $i = l$. As a result, we get:

$$\sum_{n=1}^{n} S_i^2 = (n-2)S^2.$$

The squared area of an infinitely small parallelogram, constructed on the vectors $p(\varphi)$ and $p(\varphi+d\varphi)$ equals $d\varphi^2$. In line with what has been proved above, we have

$$d\varphi^2 = \frac{1}{n-2} \sum_{n=1}^{n} \alpha_i(v)d\varphi_i^2.$$

The coefficient $\alpha_i(\varphi)$ is equal here to the squared length of the projection of the vector $p(\varphi)$ onto the plane P_i and hence $0 \leqslant \alpha_i(\varphi) \leqslant 1$. Obviously, $d\varphi_i/d\varphi \geqslant 0$. We have:

$$\sum_{n=1}^{n} \frac{d\varphi_i}{d\varphi} \geqslant \sqrt{\sum_{n=1}^{n} \left(\frac{d\varphi_i}{d\varphi}\right)^2} \geqslant \sqrt{\sum_{n=1}^{n} \alpha_i(\varphi) \left(\frac{d\varphi_i}{d\varphi}\right)^2} = \sqrt{n-2}\, d\varphi.$$

By integration, we get from this:

$$\sum_{n=1}^{n} \varphi_i \geqslant \sqrt{n-2}\, \varphi$$

which is the required proof.

4.9.2. LEMMA 4.9.2. *If* $E \subset \Omega^{n-1}$ *is a set of* $(n-1)$-*dimensional zero measure, then there is an orthonormal frame* $\{e_1, e_2, \ldots, e_n\}$ *of the space* V^n, *such that* $e_i \notin E$ *at every* $i = 1, 2, \ldots, n$.

Proof. Let us first consider the case $n = 2$. In this case $S^{n-1} = S^1$ is a circumference. Let E' be a set, obtained from E by rotating about the angle $\pi/2$. In this case E' is the set of the zero measure. Let us set $E_0 = E \cup E'$. The measure of the set E_0 equals zero. Let us arbitrarily choose a vector $e_1 \notin E_0$ and let e_2 be a vector obtained from e_1 by rotating about the angle $\pi/2$ in the direction diametrically opposing that of the first rotation. The vector e_2 does not belong to E, because in the opposite case the vector e_0 belongs to the set $E' \subset E_0$, so that $\{e_1, e_2\}$ is the frame sought.

Let us assume that for a certain n the lemma is valid and let E be the set of zero measure on the sphere Ω^n in the space V^{n+1}. The corollary of Theorem 4.5.1 allows us to draw the conclusion that for almost all $e \in \Omega^n$ the n-dimensional subspace $\nu(e)$ which is orthogonal to e, intersects the set E through the set of $(n-1)$-dimensional zero measure. Let E' be the set of those vectors $e \in \Omega^n$ for which it is not the case. Let us choose an arbitrary vector $e \notin E \cup E'$ as e_1. Let P be an n-dimensional subspace of V^{n+1}, orthogonal to e_1, $\Omega_P = P \cap \Omega^n$, $E_P = P \cap E$. In view of the choice of e_1, E_P is a set of zero measure on the $(n-1)$-dimensional sphere Ω_P. Therefore, in line with the allowances made, in the plane P there is an orthogonal frame $\{e_2, \ldots, e_{n+1}\}$, such that $e_i \notin E_P$ at every $i = 1, 2, \ldots, n$. The system of vectors $e_1, e_2, \ldots, e_{n+1}$ is obviously the one sought. The lemma is proved.

THEOREM 4.9.2. *Let* $K \{x(s), 0 \leqslant s \leqslant l\}$ *be a rectifiable spherical curve (the parameter* s *is the arc length) and let* $f(s)$ *be a real function, defined on the segment* $[0, l]$. *Let* T *be a countable everywhere dense subset of the segment* $[0, l]$. *Let us assume that for any arc* $L = [x(p)x(q)]$, *where* $p \in T$, $q \in T$, *of the curve* K *the function* $\psi(f, L, P)$ *of the variable* $P \in G_{n-1}^n$ *is measurable on* G_{n-1}^n. *In this case the function* $f(s)$ *is also measurable.*

Proof. Let us first consider the case $n = 2$, when Ω^{n-1} is an ordinary circumference. Let us assume that the curve K is contained in one semi-circumference Γ, the latter considered semi-open. The general case, is obviously, reduced to the discussed one by subdividing the curve K into arcs. Let a be the end of the arc Γ which belongs to it. Let $p \in G_1^2$. The straight line p intersects Γ only at the point x. Assume $<a,\!\overset{\frown}{}x> = \xi$. The straight line p is uniquely defined by the number ξ, therefore, we write $p = p(\xi)$. For the arbitrary function $F(p)$ on G_1^2 we have:

$$\int_{G_1^2} F(p) \, d\mu_{2,1}(p) = \frac{1}{\pi} \int_0^\pi F[p(\xi)] \, d\xi.$$

In line with this remark, in this case the considerations are reduced to viewing the functions of the real variable.

Let us set $\xi(s) = (a,\!\overset{\frown}{}x(s))$ and prove two additional statements.

(a) Let $\varphi(x)$ be an arbitrary measurable function on the segment $[0, \pi]$. In this case the function $\varphi[\xi(s)]$ is measurable. Indeed, let us set an arbitrary number h and let

$$E_h = \{x \in [0, \pi) \, | \, \varphi(x) \leqslant h\}, \; E_h^* = \{s \in [0, l) \, | \, \varphi[\xi(s)] \leqslant h\}$$

The aim is to prove that E_h^* will be measurable at any h. Due to the measurability of f, the set E_h is measurable and hence for all integers $m > 0$, there are F_m and G_m such that $E_m \subset E_h \subset G_m$, the set F_m is closed, the set G_m is open with respect to the segment $[0, \pi]$, and $\mu_1(G_m \backslash F_m) < 1/m$. Let $F_m^* = \xi^{-1}(F_m)$, $G_m^* = \xi^{-1}(G_m)$. Because of the continuity of ξ, F_m^* is closed and G_m^* is open with respect to $[0, l)$. We have: $F_m^* \subset E_h^* \subset G_m^*$. Let $\chi_m(s)$ be an indicator of the set $G_m^* \backslash F_m^*$, i.e. $\chi_m(s) = 1$ at $s \in G_m^* \backslash F_m^*$ and $\chi_m(s) = 0$ if $s \notin G_m^* \backslash F_m^*$. For any $\xi \in [0, \pi)$ $\psi(\chi_m, K, p(\xi)) = 0$ if $\xi \notin G_m \backslash F_m$ and $\psi(\chi_m, K, p(\xi)) \leqslant n_K[p(\xi)]$ for all ξ. Setting $n = 2$ in Theorem 4.9.1, we get

$$\mu_1(G_m^* \backslash F_m^*) \leqslant \frac{1}{\pi} \int_0^\pi \psi(\chi_m, K, p(\xi)) \, d\xi \leqslant \frac{1}{\pi} \int_{G_m \backslash F_m} n_K[p(\xi)] \, d\xi \qquad (38)$$

In so far as the function $n_K[p(\xi)]$ is integrable, and $\mu_1(G_m \backslash F_m) \to 0$ at $m \to \infty$, then inequality (38) yields $\mu_1(G_m^* \backslash F_m^*) \to 0$ at $m \to \infty$, which proves the measurability of E_h^*.

(b) Let E_0 be the set of those $s \in [0, l]$, for which $n_K[p(\xi)] = \infty$. (For simplicity let us set $n_K(\xi) = n_K[p(\xi)]$). The set E_0, due to the supposition (a), is measurable; let us prove that $\mu_1(E_0) = 0$. If $\chi(s)$ is an indicator of the set E_0, then $\psi(\chi, K, p(\xi)) \leqslant n_K(\xi)$. The function $\psi(\chi, K, p(\xi))$ is equal to zero for all ξ, for which $n_K(\xi) < \infty$. The set $\{\xi \, | \, n_K(\xi) = \infty\}$ is a set of the zero measure, since the function $n_K(\xi)$ is integrable and, setting $n = 2$ and $f \equiv \chi$ in the equality of the theorem, we get

$$\mu(E_0) = \frac{1}{\pi} \int_0^\pi \psi(\chi, K, p(\xi)) \, d\xi = 0$$

Let us now pass to the proof proper of the required statement, restricting ourselves to the case $n = 2$.

At every $m = 1, 2, \ldots$ let us construct the division of the segment $[0, l]$ by the points $t_0^{(m)} = 0 < t_1^{(m)} < \cdots < t_{K_m}^{(m)} = l$, such that $t_i^{(m)} \in T$ at $1 \leqslant i \leqslant k - i$ and max $|t_i^{(m)} - t_{i-1}^{(m)}| < 1/m$. Let $L_{i,m} = [x(t_{i-1}^{(m)})x(t_i^{(m)})]$. Let us introduce the function $f_m(s)$ which is equal to $\psi[f, L_{i,m}, p(\xi(s))]$ at $t_{i-1}^{(m)} < s < t_i^{(m)}$, $s \notin T$ and equal to zero at the points of the set $T \cup \{0, l\}$.

The function $f_m(s)$ is measurable in view of the supposition on measurability of $\psi[f, L_{i,m}, p[\xi(s)]]$ and of the supposition (a). Let us show that at $s \notin T \cup \{0\} \cup \{l\} \cup E_0$ $\lim_{m \to \infty} f_m(s) = f(s)$. Indeed, at sufficiently large m the point s will belong to the only segment $[t_K^{(m)}, t_{K+1}^{(m)}]$.

Since $s \notin E_0$, then $n_K[\xi(s)] < \infty$, so that the number of the points $s' \neq s$, for which $\xi(s') = \xi(s)$, is finite, and at sufficiently large m the segment $[t_K^{(m)}, t_{K+1}^{(m)}]$ in which the point s lies, will contain not a single one of them s'. But in this case, obviously, $\psi(f, L_{K,n}, p[\xi(s(])) = f(s)$. In line with the supposition (b), $\mu(E_0) = 0$ and $f(s)$ is the limit of the almost everywhere converging sequence of measurable functions, which yields that f is measurable. For the case $n = 2$ the theorem is thus proved.

For the general case the validity of the theorem is stated by way of induction with respect to n^*. Let $n \geqslant 3$. Let us assume that the theorem is valid for the curves on the sphere in V^{n-1}. Let E_1 be the set of those one-dimensional directions p, for which $p \cap |K| \neq \emptyset$. It is obvious that E_1 is a set of zero measure. Then, let E_2 be the set of those $p \in G_1^n \setminus E_1$, for which the curve $\eta_{\nu(p)}(K)$ i.e. the projection of the curve K into the $(n-2)$-dimensional sphere $\nu(p)$ is unrectifiable. Let S be the set of those two-dimensional subspaces q of the space V^n, which contain the whole arc of the curve K. The set S is not greater than countable. Let us denote through E_3 the set of those $p \in G_1^n$, for which there exists $q \in S$, such that $p \subset q$. In line with Lemma 4.5.1, E_3 is the union of not more than countable set of the sets of zero measure, and, hence, E_3 is a set of the zero measure. Finally, let us define one more peculiar set of one-dimensional directions $p \in G_1^n$. Let us set an arbitrary arc $L = [x(\alpha)$ $x(\beta)]$ of the curve K, where $\alpha, \beta \in T$. The function $\psi_L(P) = \psi(f, L, P)$, due to the condition, is measurable on G_{n-1}^n. Therefore, the function $\psi_L(\nu(p))$, defined on G_1^n, is also measurable. In view of the corollary of Theorem 4.5.1, the function $\psi_L(\nu(p))$ will be measurable on $G_1(P)$ for almost all $p \in G_{n-1}^n$. Let us denote by E_L the set of those $p \in G_1^n$, for which the function $\psi_L[\nu(p)]$ is not measurable on $G_1[\nu(p)]$. E_L is a set of zero measure. Let us set $E_4 = \bigcup_L E_L$, where L runs through the set of all arcs $[x(\alpha)x(\beta)]$, where $\alpha, \beta \in T$. In this case E_4 is the set of zero measure, as the union of a countable number of sets of zero measure. Setting $E = \bigcup_{i=1}^4 E_i$, we have $\mu_{n-1,1}(E) = 0$. Let us denote through $\tilde{E}$ the set of all points $x \in \Omega^{n-1}$ lying on the straight lines $p \in E$. Then $\tilde{E}$ is a set of zero measure and, hence, according to Lemma 4.9.2, there is an orthonormal system of the vectors $e_1, e_2, \ldots, e_n$, such that $e_i \notin \tilde{E}$ at every $i = 1, 2, \ldots, n$. Let $p_i \in G_1^n$ be a straight line whereon the vector e_i, P_i

$= \nu(p_i) \in G_{n-1}^n$, lies. Let us denote by K_i the projection $\eta_{P_i}(K)$ of the curve K. The curve K_i is defined because $p_i \notin E_1$, and, since $p_i \notin E_2$, then K is rectifiable.

Let $\{x(s), 0 \leqslant s \leqslant l\}$ be the parametrization of the curve K, where s is the arc length, $x_i(s) = \eta_{P_i}[x(s)]$ is the projection of the point $x(s)$ onto the sphere $P_i \cap S^{n-1}$. Let $\sigma_i(s)$ be the length of the arc $[x_i(0)x_i(s)]$ of the curve K_i. The function $\sigma_i(s)$ is continuous and non-decreasing. It is not constant on any interval $(\alpha, \beta) \subset [0, l]$, $\alpha < \beta$. Indeed, since $p_i \notin E_3$, then no arc of the curve is projected into a point. Let $s_i(\sigma)$, $0 \leqslant \sigma \leqslant l_i$ be a function inverse to the function σ_i. Let us set $f_i(\sigma) = f[s_i(\sigma)]$. Therefore, on each of the curves K there is a certain function f_i determined. We have: $f(s) = f_i[\sigma_i(s)]$. Let $T_i = \sigma_i(T)$ be the totality of all $\sigma = \sigma_i(s)$, where $s \in T$. The set T_i is dense everywhere on the interval $[0, l_i]$.

Let L be the arc $[x_i(\alpha)x_i(\beta)]$ of the curve K_i. In this case on the set $G_{n-2}(P_i)$ of the $(n-2)$-dimensional subspace V^n contained in P_i, the function $\psi_{i,L}(Q) = \psi(f_i, L, Q)$ is defined. This function equals ∞ if the set of the points of the arc L lying in Q is infinite, and equals the sum of the function f_i values at the points of this arc which lie in Q. Let us prove that each of the functions $\psi_{i,L}$ is measurable. Let $p = \tilde{\nu}(Q)$ be a one-dimensional direction, orthogonal to Q and lying in the plane P_i. Since $p_i \notin E_4$, then the function $\Phi_L(p) = \psi(f, L, \nu(p))$ is measurable on $G_1(P_i)$. Let $p \in G_1(P_i)$, $P = \nu(p)$. The plane P passes throught the straight line p and intersects P_i through the $(n-2)$-dimensional plane Q. Let $x(s_j)$, $j = 1, 2, \ldots, k$ be the points of the intersection of the arc L with the plane P (we consider the case when the number of such points is finite). In this case the points $x_i(s_i) = \eta_p[x(s_i)]$ lie in the plane Q, and there are obviously no other points lying in Q on the curve K_i. Therefore, $\psi_{i,L}(Q) = \Phi_L(P) = \Phi_L(\tilde{\nu}(Q))$, since $p = \tilde{\nu}(q)$. Due to the measurability of the function Φ_L on $G_1(P_i)$ we conclude that the function $\psi_{i,L}$ is also measurable on the set $G_{n-2}(P_i)$. Therefore, for every of the planes P_i the function $\psi(f_i, L, Q)$ is measurable on $G_{n-2}(P_i)$ for any arc $L = [x_i(\alpha) x_i(\beta)]$, where $\alpha, \beta \in T$. Hence, due to the induction assumption, the measurability of each of the functions f_i, $i = 1, 2, \ldots, n$ follows.

Let $L = [x(\alpha)x(\beta)]$ be an arbitrary arc of the curve K, $L_i = [x_i(\alpha)x_i(\beta)]$ be the corresponding arc of the curve K_i. Then, in line with Lemma 4.9.1,

$$s(L) \leqslant \frac{1}{\sqrt{n-2}} \sum_{n=1}^{n} s(L_i). \tag{39}$$

Let $c \in \mathbb{R}$ arbitrarily. Let us set $A_c = \{x \in [0, l] \mid f(s) \leqslant c\}$ and prove that the set A_c is measurable. As far as, due to above proved, the function $f_i(\sigma) = f[s_i(\sigma)]$ is measurable, then the set $A_{ic} = \{\sigma \in [0_i l_i] \mid f_i(\sigma) \leqslant c\}$ is measurable. Obviously, $A_{ic} = \sigma_i(A_c)$.

Let us arbitrarily set $\varepsilon > 0$ and find the sets F_i and G_i, such that $F_i \subset A_{ic} \subset G_i$, F_i is closed, G_i is open with respect to $[0, l_i]$, and $\mu(G_i \backslash F_i) < \varepsilon$. Let $F_0 = \bigcup_{i=1}^{n} s_i(F_i)$, $G_0 = \bigcap_{i=1}^{n} s_i(G_i)$. It is easily seen that $F_0 \subset A_c \subset G_0$. At every i, $\sigma_i(F_0) \supset F_i$, $\sigma_i(G_0) \subset G_i$, so that $\sigma_i(G_0)\backslash\sigma_i(F_0) \subset G_i\backslash F_i$. The set $G_0\backslash F_0$ can be represented as a combination of mutually disjoint intervals. Applying the relation (39) to the arc of the curve K, corresponding to each of these intervals and its projection, we get:

$$\mu(G_0\backslash F_0) \leqslant \frac{1}{\sqrt{n-2}} \sum_{n=1}^{n} \mu[\sigma_i(G_0)\backslash\sigma_i(F_0)] < \frac{n\varepsilon}{\sqrt{n-2}}.$$

Due to the arbitrariness of $\varepsilon > 0$ the set A_c is measurable and, hence, f is measurable, since $c \in R$ was chosen arbitrarily. The theorem is proved.

CHAPTER V

Turn or Integral Curvature of a Curve

5.1. Definition of a Turn. Basic Properties of Curves of a Finite Turn

5.1.1. To any curve K in the space E^n there can correspond a certain number $\kappa(K)$, such that $0 \leqslant \kappa(K) \leqslant \infty$, which will be hereafter referred to as a turn or an integral curvature of the given curve K. In a regular case, which is an object of investigation in differential geometry, the turn of a curve is equal to the integral of the curve's curvature with respect to the arc length. Let us first define the concept of a turn for polygonal lines. In this case it refers, essentially, to elementary geometry. For arbitrary curves a turn is defined by way of approximating a curve with polygonal lines (see below). The fact that in a regular case a turn, in the sense of the definitions given here, is really equal to the integral of the curvature with respect to the arc length will be stated below. There is another possible way of constructing the theory of a curve turn, when the notion of a turn is defined by way of approximating an arbitrary curve by regular curves.

Let L be a polygonal line in the space E^n, $X_0 < X_1 < \cdots < X_{m-1} < X_m$ be its sequential vertices, X_0 be its origin, X_m be its end. Let a_i be the vector $\overline{X_{i-1}X_i}$. The angle $\varphi_i = (a_i, \widehat{} a_{i+1})$ is called a turn at the vertex X_i of the polygonal line L. The sum

$$\sum_{i=1}^{m-1} \varphi_i = \sum_{i=1}^{m-1} (a_i, \widehat{} a_{i+1})$$

is known as a turn of the polygonal line L and is denoted by the symbol $\kappa(L)$.

If L is a closed polygonal line in E^n, $a_1, a_2, \ldots, a_m$ are its links, then a turn of the polygonal line L is the sum

$$\kappa(L) = (a_m, \widehat{} a_1) + (a_1, \widehat{} a_2) + \cdots + (a_{m-1}, \widehat{} a_m) = \sum_{i=1}^{m} (a_{i-1}, \widehat{} a_i).$$

(In the sum in the right-hand part we assume $a_0 = a_m$).

LEMMA 5.1.1. *Let L and M be polygonal lines in E^n. In this case if L is insribed to M, then $\kappa(L) \leqslant \kappa(M)$. If L and M are closed polygonal lines in E^n, in which case L is inscribed in M, then $\kappa(L) \leqslant \kappa(M)$ as well.*

Proof. The considerations are the same for both closed and ordinary lines. Let us make use of the inequality of a triangle for angles. For any three non-zero vectors a, b, c in E^n, $(a, \widehat{} c) \leqslant (a, \widehat{} b) + (b, \widehat{} c)$. With no reduction of generality, we can obviously consider each vertex of the polygonal line L to be an apex of M. Let us first assume that M originates from L by way of substituting its line $A_{i-1}A_i = a_i$ by the two links.

$$A_{i-1}A' = a', \qquad A'A_i = a''.$$

It is obvious that $\kappa(M)$ is obtained if in the sum

$$\sum_{k=1}^{m} (a_k \frown a_{k+1})$$

the addends $(a_{i-1} \frown a_i) + (a_i \frown a_{i+1})$ will be substituted for by the sum

$$(a_{i-1} \frown a') + (a' \frown a'') + (a'' \frown a_{i+1}).$$

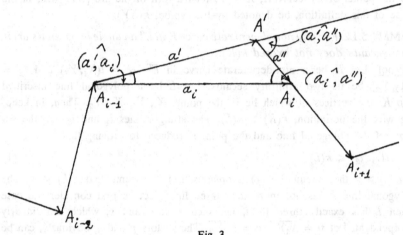

Fig. 3.

From the triangle $A_{i-1}A'A_i$ we have (see Fig. 3):

$$(a'' \frown a'') = (a'' \frown a_i) + (a_i \frown a'').$$

From this we get:

$$(a_{i-1} \frown a') + (a', a'') + (a'', a_{i+1})$$

$$= (a_{i-1} \frown a') + (a' \frown a_i) + (a_i \frown a'') + (a'' \frown a_{i+1})$$

$$\geqslant (a_{i-1} \frown a_i) + (a_i \frown a_{i+1})$$

and hence $\kappa(M) \geqslant \kappa(L)$. In a general case the polygonal line M can obviously be obtained from L in a finite number of steps each of which consists in substituting a separate segment of the polygonal line by two segments. In keeping with the above proof, at each step the turn does not decrease, whence we find that $\kappa(M) \geqslant \kappa(L)$, which is the required proof.

COROLLARY. *A turn of every polygonal (closed polygonal) line in E^n is a least upper boundary of the turns of the polygonal (respectively, closed) lines inscribed in it.*

Let K be an arbitrary non-degenerate curve in the space E^n. A turn of the curve K is the least upper boundary $\kappa(K)$ of the turns of the polygonal lines inscribed into the curve K. The value $\kappa(K) = \infty$ is admissible. In an analogous way, if K is a non-degenerate closed curve in E^n, then a turn of K is a least upper boundary $\kappa(K)$ of the turns of the closed polygonal lines inscribed into it.

If the curve K in E^n degenerates into a point, then let us consider that $\kappa(K) = 0$.

A turn of an arbitrary arc of the curve K in E^n (a closed curve K in E^n) will be a turn of this arc viewed as an independent curve in E^n. If X and Y are the points of the curve K, $X < Y$, then a turn of the arc $[XY]$ will, in the sense of this definition, be denoted by the symbol $\kappa(XY)$.

LEMMA 5.1.2. *For any non-degenerate curve K in E^n an angle between its arbitrary secants does not exceed $\kappa(K)$.*

Proof. Let K be a non-degenerate curve in E^n, $l_1 = l(X_1, Y_1)$ and $l_2 = l(X_2, Y_2)$ be its two arbitrary secants. Let L be a polygonal line inscribed into K, the vertices of which lie in the points X_1, Y_1, X_2, Y_2. Then, in keeping with the definition, $\kappa(K) \geqslant \kappa(L)$. The straight lines l_1 and l_2 are the secants of the polygonal line and the point is reduced to proving that

$$(l_1 \widehat{} l_2) \leqslant \kappa(L). \tag{1}$$

If $\kappa(L) \geqslant \pi$ then inequality (1) is obvious. Let us assume that $\kappa(L) < \pi$. The polygonal line L has not more than three links. Let us first consider the case when L has exactly three links, i.e. vectors a, b and c, which are linearly independent. Let $p = \overline{X_1 Y_1}$, $q = \overline{X_2 Y_2}$. The vectors p and q, obviously, can be expressed in terms of a, b and c as $p = \alpha_1 a + \beta_1 b + \gamma_1 c$, $q = \alpha_2 a + \beta_2 b + \gamma_2 c$, where the numbers α_i, β_i, γ_i, $i = 1, 2$ are non-negative. A set of all vectors x of the type $x = \alpha a + \beta b + \gamma c$, where $\alpha \geqslant 0$, $\beta \geqslant 0$, $\gamma \geqslant 0$ is a certain three-edged angle. Its intersection with the sphere $\Omega^{n-1} = \{u \in V^n \mid |u| = 1\}$ is a spherical triangle ABC. Let $A = a/|a|$, $B = b/|b|$, $C = c/|c|$, $P = p/|p|$, $Q = q/|q|$. The points P and Q lie in the spherical triangle ABC and, hence, the length of the shortest line PQ on the sphere is not greater than the triangle's diameter. The latter does not exceed the sum of the lengths of the sides AB and BC, i.e., $(p, \widehat{} q) \leqslant (a, \widehat{} b) + (b, \widehat{} c) = \kappa(L)$, which is the required proof.

In the case when the vectors a, b and c are linearly dependent, the set Γ of all vectors $x = \alpha a + \beta b + \gamma c$ where $\alpha \geqslant 0$, $\beta \geqslant 0$, $\gamma \geqslant 0$ is a plane angle since, by supposition, $\kappa(L) = (a, \widehat{} b) + (b, \widehat{} c) < \pi$. The value of this angle equals one of the numbers $(a, \widehat{} b)$, $(b, \widehat{} c)$ or $(a, \widehat{} c)$, and hence is not greater than $\kappa(L)$. The vectors p and q are directed inside the angle Γ, and hence (p, q) is not greater than the value of this angle, which yields $(p, \widehat{} q) \leqslant \kappa(L)$. In an analogous way we consider the case when the polygonal line L has only two links. The Lemma is proved.

5.1.2. THEOREM 5.1.1. *If the curves (K_m), $m = 1, 2, \ldots$, at $m \to \infty$ converge to* *the curve K, then $\kappa(K) \leqslant \underline{\lim}_{m \to \infty} \kappa(K_m)$. Analogously, if (K_m), $m = 1, 2, \ldots$ is a sequence of closed curves converging to a closed curve K, then $\kappa(K) \leqslant \underline{\lim}_{m \to \infty}$ $\kappa(K_m)$.*

Proof. Let L be an arbitrary polygonal line inscribed into K, and let $X_0 <$ $X_1 < \cdots X_r$ be its sequential vertices. On K_m let us find at every m a sequence of the points $X_0^m < X_1^m < \cdots < X_r^m$, such that at $m \to \infty$ $X_k^m \to X_k$ at all k and let L_m be a polygonal line inscribed into K_m with the vertices at these points. It is obvious that at $m \to \infty$ $\kappa(L_m) \to \kappa(L)$. But $\kappa(L_m) \leqslant \kappa(K_m)$, and, hence, $\underline{\lim}_{m \to \infty} \kappa(K_m) \geqslant \kappa(L)$. In view of the arbitrariness of the polygonal line L it yields that $\underline{\lim}_{m \to \infty} \kappa(K_m) \geqslant \sup \kappa(L) = \kappa(K)$. For the case of closed curves the considerations are the same.

COROLLARY 1. *Let K be a curve (closed curve) in E^n, (L_m), $m = 1, 2, \ldots$, be an arbitrary sequence of the polygonal (closed polygonal) lines inscribed into it converging to the curve K at $m \to \infty$. Then $\kappa(K) = \lim_{m \to \infty} \kappa(L_m)$.*

Indeed, if the sequence of the polygonal lines (L_m) obeys the condition of the corollary, then, in keeping with the theorem

$$\kappa(K) = \varliminf_{m \to \infty} \kappa(L_m). \tag{2}$$

On the other hand, at every m, $\kappa(K) \geqslant \kappa(L_m)$, whence

$$\kappa(K) \geqslant \varlimsup_{m \to \infty} \kappa(L_m). \tag{3}$$

Comparing (2) and (3), we get:

$$\kappa(K) = \lim_{m \to \infty} \kappa(L_m)$$

which is the required proof.

COROLLARY 2. *Let K be a curve in E^n, let A be its beginning, and let B be its end. Then for every point X of the curve K the inequality*

$$\kappa(AX) + \kappa(XB) \leqslant \kappa(AB) = \kappa(K)$$

is valid.

Proof. Let (L_m), $m = 1, 2, \ldots$, be a sequence of polygonal lines inscribed into the curve K such that, at $m \to \infty$, $\lambda(L_m) \to 0$, in which case the point X is a vertex of each of the polygonal lines L_m. The point X divides L_m into two arcs L_m' and L_m''. At $m \to \infty$ $\kappa(L_m) \to \kappa(K)$, $\kappa(L_m') \to \kappa(AX)$, $\kappa(L_m'') \to \kappa(XB)$, in keeping with Corollary 1. At every m we obviously have: $\kappa(L_m) = \kappa(L_m') + \kappa(L_m'')$ $+ \varphi_m \geqslant \kappa(L_m') + \kappa(L_m'')$. (Here $\varphi_m \geqslant 0$ is a turn of the polygonal line L_m at the apex X). Passing in this inequality to the limit at $m \to \infty$, we get the result sought.

COROLLARY 3. *Let K be a non-degenerate curve in E^n, such that $\kappa(K) < \infty$, X is an arbitrary point of the curve K. In this case a turn of the arc $[XY]$ ($[YX]$) of the curve K tends to zero when Y tends to the point X along the curve K on the right (on the left, respectively).*

Proof. Let A be the beginning and B the end of the curve K. It is, obviously, enough to consider the case when X is one of the points A or B. For the sake of definitness let $X = B$. Let us arbitrarily set a sequence (Y_m) of the points of the curve K, such that $Y_m \rightarrow B$ on the left of the curve K. The corollary will be proved if we show that $\kappa(Y_m B) \rightarrow 0$ at $m \rightarrow \infty$. Let us denote by K_m an arc $[AY_m]$ of the curve K. In this case at $m \rightarrow \infty$ the curves K_m are converging to the curve K and, hence, in keeping with the theorem, $\kappa(K) \leqslant \underline{\lim}_{m \to \infty} \kappa(K_m)$. On the other hand, Corollary 2 allows us to conclude that at every m

$$\kappa(K) \geqslant \kappa(K_m) \tag{4}$$

which obviously yields

$$\kappa(K) \geqslant \overline{\lim_{m \to \infty}} \kappa(K_m) \tag{5}$$

From (4) and (5) we get:

$$\kappa(K) = \lim_{m \to \infty} \kappa(K_m)$$

Then, according to Corollary 2, at every m we have:

$$\kappa(K) - \kappa(K_m) \geqslant \kappa(Y_m B) \geqslant 0$$

whence we get $\kappa(Y_m B) \rightarrow 0$ at $m \rightarrow \infty$, which is the required proof.

THEOREM 5.1.2. *Any curve of a finite turn in E^n is one-sidedly smooth.*

Proof. Let K be a non-degenerate curve in E^n. Let us assume that $\kappa(K) < \infty$. Let us choose an arbitrary point X of the curve K. Let us suppose that X is not the end of the curve K, and arbitrarily set $\varepsilon > 0$. In keeping with Corollary 3 of Theorem 5.1.1, there is a point Y of the curve K, such that $X < Y$ and $\kappa(XY) < \varepsilon$. By Lemma 5.1.2, the angle between any two secants of the arc (XY) is less than ε. Since $\varepsilon > 0$ is chosen arbitrarily, then the curve K has a right-hand tangent in a strong sense at the point X. In an analogous way we can establish that if X is not the beginning of the curve K, then at the point X there exists a left-hand tangent in a strong sense. The Theorem is proved.

COROLLARY 1. *All the curves of a finite turn are rectifiable.*

COROLLARY 2. *Any curve of a finite turn is a conglomeration of a finite number of simple arcs.*

THEOREM 5.1.3. *Let $K = AB$ be an arbitrary curve, and let X be an internal point of the curve K. Then, if a turn of the curve K is finite, then the turns of the arcs AX and XB are also finite, and vice versa, if $\kappa(AX) < \infty$, $\kappa(XB) < \infty$, then also $\kappa(AB) < \infty$. In this case, if $\kappa(K) < \infty$, then*

$$\kappa(AB) = \kappa(AX) + \kappa(XB) + (t_l(X), \widehat{} t_r(X)).$$

Let Q be a closed curve in E^n, and let K be a loop obtained by a cross-section of Q. In this case for a turn of the curve Q to be finite, it is necessary and sufficient that the quantity $\kappa(K)$ be finite. In this case if $\kappa(Q) < \infty$, then the equality

$$\kappa(Q) = \kappa(K) + (t_l(B) \widehat{} t_r(A))$$

holds, where A is the origin and B is the end of the loop K.

Proof. Let $(L_{1,m})$ and $(L_{2,m})$, $m = 1, 2, \ldots$, be polygonal lines inscribed into the arcs AX and XB, respectively, and having general terminal points with these arcs, in which case at $m \to \infty$ $\lambda(L_{1,m}) \to 0$ and $\lambda(L_{2,m}) \to 0$. The polygonal lines $L_{1,m}$ and $L_{2,m}$ jointly comprise the only polygonal line L_m inscribed into the arc AB. Let $t_{l,m}$ and $t_{r,m}$ be the left and the right tangential unit vectors of the polygonal line L_m at the point X. Then at all m we have: $\kappa(L_m)$ $= \kappa(L_{1,m}) + \kappa(L_{2,m}) + (t_{l,m} \widehat{} t_{r,m})$. At $m \to \infty$ $\kappa(L_m) \to \kappa(AB)$, $\kappa(L_{1,m}) \to$ $\kappa(AX)$, $\kappa(L_{2,m}) \to \kappa(XB)$, and the angle $(t_{l,m} \widehat{} t_{r,m})$ does not exceed π at all m, and at $m \to \infty$ it converges to the angle between the left and right tangential unit vectors of the curve at the point X, provided these unit vectors exist. From this we can obviously derive all the statements referring to common curves.

Let Q be a closed curve, and let K be a loop obtained by a cross-section of Q. Let us construct a sequence (L_m), $m = 1, 2, \ldots$, of the polygonal lines inscribed in the curve K, such that $\lambda(L_m) \to 0$ at $m \to \infty$ and at every m the terminal points of L_m coincide with those of the curve K. Let S_m be a closed polygonal line resulting from the polygonal line L_m. Then $\kappa(S_m) = \kappa(L_m) +$ $(t_m(A) \widehat{} t_m(B))$, where $t_m(A)$ and $t_m(B)$ are the tangential unit vectors at the points A and B of the polygonal line L_m. At $m \to \infty$ $\kappa(L_m) \to \kappa(K)$, and at every m, $\kappa(L_m) \leqslant \kappa(S_m) \leqslant \kappa(L_m) + \pi$. This demonstrates that if one of the limits $\lim_{m \to \infty} \kappa(L_m)$ and $\lim_{m \to \infty} \kappa(S_m)$ is finite, then the other is also finite, which proves that $\kappa(Q)$ is finite if and only if $\kappa(K)$ is finite. Let us assume that $\kappa(Q) < \infty$. In this case $\kappa(K) < \infty$ and at $m \to \infty$ the vectors $t_m(A)$ and $t_m(B)$ converge to the tangential unit vectors at the points A and B of the curve K. Whence, passing to the limit, we get

$$\kappa(Q) = \kappa(K) + (t_r(A) \widehat{} t_l(B)).$$

The theorem is proved.

Let us introduce the following notion for Theorem 5.1.3. Let K be an arbitrary one-sided smooth curve, and let X be an internal point of the curve K. The angle $(t_l(X) \widehat{} t_r(X))$ is called a turn of the curve K at the point X and is denoted by $\kappa_K(X)$. The subscript K in this notation will be omitted in the cases when its absence will result in no ambiguity.

5.1.3. THEOREM 5.1.4. If a turn of a non-degenerate curve equals zero, then this curve is a simple arc lying in one straight line.

Proof. Let the curve K be non-degenerate and let $\kappa(K) = 0$, A be the origin, and B the end of the curve K; let (L_m), $m = 1, 2, \ldots$, be a sequence of the polygonal lines inscribed into K, such that $\lambda(L_m) \to 0$ at $m \to \infty$, in which case A is the beginning, and B is the end of the polygonal line L_m at every m. Since, at any m, $\kappa(L_m) \leqslant \kappa(K)$, then the turns of all polygonal lines L_m are equal to zero. Let $a_1, a_2, \ldots, a_p$ be sequential links of the polygonal line L_m. In this case $(a_{i-1}\widehat{}a_i) = 0$ at every $i = 2, \ldots, p$, whence it follows that the polygonal line L_m lies in one straight line and is a simple arc with the ends A and B, which means that the polygonal lines L_m coincide with the segment AB. Since $L_m \rightarrow K$ at $m \rightarrow \infty$, it leads one to the conclusion that the curve K also coincides with this segment. The theorem is proved.

LEMMA 5.1.3. *If a non-degenerate curve K lies in one straight line and its turn is finite, then we can find on the curve K a sequence of points $X_0 < X_1 < \cdots < X_\mu$, where $X_0 = A$ is the beginning, $X_\mu = B$ is the end of the curve K, such that each of the arcs $[X_{i-1}, X_i]$ is simple and the angle between the vectors $\overline{X_{i-1}X_i}$ and $\overline{X_iX_{i+1}}$ is equal to π. In this case the number μ equals $\kappa(K)/\pi + 1$. If Q is a closed curve lying in one straight line and $\kappa(Q) < \infty$, then the curve Q can be divided into $\mu \geqslant 2$ arcs $[X_\mu X_1], [X_1 X_2], \ldots, [X_{\mu-1}X_\mu]$, each of which is a rectilinear segment, in which case the vectors $\overline{X_{i-1}X_i}$ and $X_i X_{i+1}$ are oppositely directed at every $i = 1, 2, \ldots, \mu$. (Here we set $X_0 = X_\mu$, $X_{\mu+1} = X_1$). In this case μ is an even number and equals $\kappa(Q)/\pi$.*

Proof. Let K be a curve lying in one straight line such that $\kappa(K) < \infty$. Let (L_m), $m = 1, 2, \ldots$, be a sequence of the polygonal lines inscribed into K, such that $\lambda(L_m) \rightarrow 0$ at $m \rightarrow \infty$ and the terminal points of the polygonal line L_m coincide with those of the curve K at all m. In this case $\kappa(L_m) \rightarrow \kappa(K)$ at $m \rightarrow \infty$. As far as each of the polygonal lines L_m lies in one straight line, then $\kappa(L_m) = \pi\mu_m$, where μ_m is an integer. This shows that $\kappa(K) = \pi\mu$, where μ is also an integer. Since $\mu_m \rightarrow \mu$ at $m \rightarrow \infty$ and the numbers μ_m and μ are integers, then $\mu_m = \mu$, beginning with a certain m. Let us assume that $\mu_m = \mu$ for all m. Let $X(t)$, $a \leqslant t \leqslant b$, be a normal parametrization of the curve K, $t_{0m} = a < t_{1m} < \cdots < t_{K_m m} = b$ be the values of the parameter corresponding to the consequent vertices of the polygonal line L_m, $X_m(t)$, $a \leqslant t \leqslant b$, be a parametrization of L_m, such that $X_m(t_{im}) = X(t_{im})$ for every $i = 0, 1, \ldots, k_m$. The vector-functions $X_m(t)$ uniformly converge to $X(t)$ on the segment $[a, b]$. Let $\tau_{1m} < \tau_{2m} < \cdots < \tau_{\mu m}$ be the values of the parameter t corresponding to those vertices of the polygonal line L_m, at which turn of L_m is other than zero. Let us assume that at every $i = 1, 2, \ldots, \mu$ there exists a limit $\lim_{m \to \infty} \tau_{im} = \tau_i$. (This can be always achieved by way of passing to a subsequence.) Obviously, $a \leqslant \tau_1 \leqslant \tau_2 \leqslant \cdots \leqslant \tau_\mu \leqslant b$. Let us set $\tau_0 = a$, $\tau_{\mu+1} = b$. If, for a certain i, $\tau_i < \tau_{i+1}$, then at $m \rightarrow \infty$ the arc $L_m^i = \{X_m(t), \tau_{im} \leqslant t \leqslant \tau_{i+1,m}\}$, of the polygonal line L_m converges to the arc $K^{(i)} = \{X(t), \tau_1 \leqslant t \leqslant t_{i+1}\}$ of the curve K. But the arc $L_m^{(i)}$ is a rectilinear segment, and hence $K^{(i)}$ is also a segment. We therefore come to the conclusion that the curve K consists of not more than $\mu + 1$ rectilinear segments, i.e., it is polygonal. Since $\kappa(K) = \pi\mu$, then K consists of exactly $\mu + 1$ segments and thus, the statement of the Lemma concerning common curves is proved.

Let Q be a non-degenerate closed curve lying in one straight line and such that $\kappa(Q) < \infty$. The set $|Q|$ contains more than one point and is a connected closed subset of a certain straight line. Consequently, $|Q|$ is a rectilinear segment. Let A and B be two terminal points of this segment and K be a loop obtained by a cross-section of Q and such that its terminal points coincide with A, and let X be a point of the curve K, the carrier of which is point B. In keeping with Theorem 5.1.3, a turn of the curve K is finite. Therefore, in view of the above proof, the curve K is polygonal. This shows that Q is a closed polygonal line. We obviously get $\kappa(Q) = \pi + \kappa(K)$. The point X is a point of return of K and hence $\kappa(K) \geqslant \pi$ and $\kappa(Q) \geqslant 2\pi$. The curve K falls into $\mu = \kappa(K)/\pi + 1 = \kappa(Q)/\pi$ of rectilinear segments. Let X_0 be the beginning, X_μ the end of the curve K, $X_1 < X_2 < \cdots < X_{\mu-1}$ be the points the return of K. The vectors $\overline{X_i X_{i+1}}$ and $\overline{X_{i-1} X_i}$ are oppositely directed. Herefrom we can easily deduce that the vectors $\overline{X_0 X_1}$ and $\overline{X_{i-1} X_i}$ are identically directed when i is uneven, and oppositely directed when i is even. From this we can easily deduce that the vectors $\overline{X_0 X_1}$ and $\overline{X_{\mu-1} X_\mu}$ are oppositely directed, then μ is even.

THEOREM 5.1.5. *For any closed curve Q in E^n the inequality $\kappa(Q) \geqslant 2\pi$ is valid. In this case the sign of equality is valid only when Q is either a plane convex curve or a closed polygonal line consisting of two links.*

REMARK 1. In a regular case the corresponding theorem is known as the Fenchel theorem.

REMARK 2. A closed curve Q is called a plane convex curve if Q lies in one two-dimensional plane, is a simple closed curve and its carrier $|Q|$ is the boundary of a plane convex domain.

Proof. Let Q be an arbitrary closed curve in E^n. Then, in keeping with the definition, for any closed polygonal line L which is inscribed into Q, we have: $\kappa(Q) \geqslant \kappa(L)$. Let L be a two-link polygonal line inscribed into Q. In this case $\kappa(L) = 2\pi$ and we thus get: $\kappa(Q) \geqslant 2\pi$ which proves the inequality of the Theorem.

Let us assume that for a given closed curve Q $\kappa(Q) = 2\pi$. Then for any closed polygonal line L inscribed into Q we have: $\kappa(Q) \geqslant \kappa(L) \geqslant 2\pi$ and, since $\kappa(Q) = 2\pi$, then $\kappa(L) = 2\pi$. Let us prove that any closed polygonal line with its turn equal to 2π is either a plane convex polygon or it consists of two simple arcs lying in one straight line. The proof will be carried out by the method of induction with respect to the number m of the links of a polygonal line.

The statement being proved is valid when $m = 2$ or $m = 3$. Let us assume that it has been proved for the case of m-links of the polygonal lines, and that the polygonal line L has $(m+1)$-links, in which case let $\kappa(L) = m + 1$. Let us choose three consecutive vertices B, C and D of the polygonal line L. Let A be the vertex immediately preceding B, and let E be that immediately following D. Let us set $a = \overline{AB}$, $b = \overline{BC}$, $c = \overline{CD}$, $d = \overline{DE}$, $b' = \overline{BD}$. If the turn of the polygonal line L at the vertex C equals zero, then the links b and c can be united into one. As a result, we get an m-link polygonal line. By induction, we come to the conclusion that in this case L is either a plane convex polygon or it consists of two simple arcs lying in one straight line.

Let us assume that $(b,\hat{}c) \neq 0$. Substituting the links BC and CD with the vector BD, we get a closed m-link polygonal line L' inscribed into L. When going over from $\kappa(L)$ to $\kappa(L')$, the sum of the angles $(a,\hat{}b) + (b,\hat{}c) + (c,\hat{}d)$ is substituted by the sum $(a,\hat{}b') + (b',\hat{}d)$. Since $\kappa(L) = \kappa(L')$, then, consequently,

$$(a,\hat{}b) + (b,\hat{}c) + (c,\hat{}d) = (a,\hat{}b') + (b',\hat{}d). \qquad (6)$$

And then

$$(b,\hat{}c) = (b,\hat{}b') + (b',\hat{}c), \qquad (7)$$

$$(a,\hat{}b) + (b,\hat{}b') \geqslant (a,\hat{}b'), \qquad (8)$$

$$(b',\hat{}c) + (c,\hat{}d) \geqslant (b',\hat{}d) \qquad (9)$$

Comparing relations (6) - (9), we obtain

$$(a,\hat{}b) + (b,\hat{}b') = (a,\hat{}b'), \qquad (b',\hat{}c) + (c,\hat{}d) = (b',\hat{}d) \qquad (10)$$

Let us first assume that L' consists of two simple arcs lying in one straight line l. Then if the point C lies in l, the polygonal line L also lies in it. Since $\kappa(L) = 2\pi$, then L has exactly two points of return and, thus, it consists of two simple arcs lying in one straight line. Let C not lie in the straight line l. Let us denote by F and G the points of return of the polygonal line L'. In this case neither of the points B and D belongs to the segment FG. Indeed, if, for instance, the point B were to lie inside the segment FG, then the angle $(a,\hat{}b')$ would equal 0, and hence the angle $(a,\hat{}b)$ would also equal zero, i.e., the point C would lie in the straight line l. Therefore, the points B and D coincide with the ends of the segment F and G, and the polygonal line L coincides in this case with the perimeter of the triangle.

Let us now assume that L' is a plane convex polygon. Let us extend its sides AB and DE beyond the points B and D, respectively. In this case in keeping with equalities (10) we get that the point C lies in the domain limited by the extensions of the sides AB and DE and the segment BD. Hence, L is a plane convex polygon.

It follows from the above proof that, if $\kappa(Q) = 2\pi$, then any polygonal line inscribed into L is either a twice covered segment or a plane convex polygon. Let us construct a sequence of the polygonal lines (L_m), $m = 1, 2, \ldots$, inscribed into Q, such that at every m the polygonal line L_m is inscribed into L_{m+1} and $\lambda(L_m) \to 0$ at $m \to \infty$. Going over to the limit at $m \to \infty$ we conclude that Q is a twice covered simple arc lying in one straight line, if all polygonal lines L_m are curves of such kind. In the case when Q does not lie in one straight line, the polygonal line L_m, beginning with a certain $m = m_0$, is a plane convex polygon, which affords that Q is a plane convex curve. The theorem is proved.

5.2. Definition of a Turn of a Curve by Contingencies

5.2.1 Let K be an arbitrary curve in E^n. Let us place in order the set of all its contingencies assuming that $q_l(X) < q_r(X)$ throughout and that if $q(X)$ and $q(Y)$ are some contingencies at different points $X, Y \in K$, then $q(X) < q(Y)$ at $X < Y$.

Let $\xi = \{q_1, q_2, \ldots, q_m\}$, where $q_1 < q_2 < \cdots < q_m$ is an arbitrary finite system of the contingencies of the curve K. Let us arbitrarily choose a straight line l_i in the contingency q_i. Let us set

$$\kappa^*(\xi) = \sup_{l_i \in q_i} \sum_{i=1}^{m-1} (l_i \widehat{} l_{i+1})$$

where the least upper boundary is taken by all possible sequences of the straight lines $l_1, l_2, \ldots, l_m$, such that $l_i \in q_i$ at every $i = 1, 2, \ldots, m$.

The least upper boundary of the quantity $\kappa^*(\xi)$ chosen by all the sequences of the contingencies of the curve K, will be denoted by $\kappa^*(K)$.

The basic results of this paragraph can be expressed in the following theorems.

THEOREM 5.2.1. *For any curve K in E^n the value of $\kappa^*(K)$ is equal to a turn of this curve.*

THEOREM 5.2.2. *A turn of any one-sidedly smooth curve is equal to the length of its indicatrix of the tangents.*

The theorems formulated above are the corollaries of the lemmas which will be proved below.

LEMMA 5.2.1. *If $\kappa^*(K) < \infty$, then the curve K is one-sidedly smooth.*

Proof. Let K be an arbitrary curve in E^n, and let X be a point of the curve K. Let us denote by $Q_l(X)$ $(Q_r(X))$ the totality of all straight lines p passing via the point X, for each of which we can construct a sequence (X_m) of the points of the curve K, converging to X from the left (right), and a sequence of the straight lines p_m, such that, at every m, $p_m \in q_l(X_m) \cup q_r(X_m)$ and $p_m \to p$ at $m \to \infty$. If $Q_l(X)$ $(Q_r(X))$ consists of the only straight line p, then the latter is the limit on the left (respectively, on the right) at the point X of the contingencies of the curve K in the sense of the definition of Section 3.3 and hence, in this case, according to Theorem 3.3.1 the curve K has a left (right) tangent at the point X in a strong sense.

Let us assume that for the curve K in E^n the value of $\kappa^*(K)$ is finite. Let us prove that at its every point X of the set of the straight lines, $Q_l(X)$ and $Q_r(X)$ consist of a single straight line each. Let us assume, on the contrary, that at a certain point $X \in K$ one of these sets, for instance, $Q_l(X)$, contains two different straight lines a and b. In this case on the curve K there is a sequence of points $Y_0 < Y_1 < \cdots < Y_m < \cdots X$, such that the contingency q_m at the point Y_m contains a straight line l_m which forms an angle less than $\varepsilon = \frac{1}{3}(a \widehat{} b) > 0$ with the straight line a for all even m, and with the straight line b for all uneven m, respectively.

Let us consider a system of contingencies $\xi_m = \{q_0, q_1, \ldots, q_m\}$. It is easily seen that $\kappa^*(\xi_m) \geqslant \sum_{i=1}^{m} (l_{i-1} \overset{\frown}{,} l_i) \geqslant m\varepsilon$. Due to the arbitrariness of m, we have $\kappa^*(K) = \infty$, which contradicts the condition. The Lemma is thus proved.

LEMMA 5.2.2. *For any one-sidedly smooth curve the value of $\kappa^*(K)$ equals the length of its indicatrix of the tangents.*

Proof. Let $\xi = \{t_1, t_2, \ldots, t_m\}$ be an arbitrary finite sequence of the curve's tangents numbered in order of their succession. Let us plot the unit vectors of the tangents t_i from the centre of a unit sphere Ω^{n-1} and connect their ends in succession with the shortest lines on the sphere. As a result, we get a spherical polygonal line L inscribed into the indicatrix of the tangents Q of the curve K. In this case, obviously, $s(Q) \geqslant s(L) = \kappa^*(\xi)$ and, since ξ is arbitrary, then $s(Q) \geqslant \sup_\xi \kappa^*(\xi) = \kappa^*(K)$.

Let us prove the inverse inequality: $s(Q) \leqslant \kappa^*(K)$. Let $Y_0 < Y_1 < \cdots < Y_m$ be an arbitrary finite sequence of the points of the curve Q. The curve Q can contain the arcs of the great circle corresponding to the angular points of the curve K. Let us add the ends of all those arcs on which there are points Y_i of the given sequence, and exclude from it the points Y_i lying inside these arcs. As a result, we get a new finite sequence $Y_0', Y_1', \ldots, Y_p'$ of the points of the curve Q. Connecting the points of the sequences $\{Y_i\}$ and $\{Y_i'\}$ by the arcs of great circles in the order of their location on Q, we get two spherical polygonal lines L and L' which are inscribed into Q. In this case the polygonal line L is inscribed into L', so that $s(L) \leqslant s(L')$. All the points of the sequence Y_i' correspond to the tangents of the curve forming a certain sequence $\xi = (t_i)$. We have:

$$s(L) \leqslant s(L') = \kappa^*(\xi) \leqslant \kappa^*(K).$$

Since L is an arbitrary polygonal line inscribed into Q, then $s(Q) = \sup s(L) \leqslant \kappa^*(K)$. The Lemma is proved.

Let K be a one-sidedly smooth curve in E^n. Let us assume that the tangents of the curve K are directed into on semi-space if there exists a vector ν, such that for any tangential unit vector t of the curve K the inequality $<t, \nu> \, > 0$ is valid. In this case the vector ν is called a vector of the internal normal of the semi-space into which the tangents of the curve K are directed.

LEMMA 5.2.3. *Let K be a one-sidedly smooth curve in E^n, such that all its tangential vectors are directed into one semi-space (denoting by ν the vector of the internal normal of this semi-space). Let $h = \overline{AB}$, where A is the beginning and B is the end of the curve K. In this case the vector h is other than zero and allows the presentation*

$$h = k_1 t_1 + k_2 t_2 + \cdots + k_r t_r$$

where $r \leqslant n$, $k_1, k_2, \ldots, k_r$ are positive numbers, and each vector t_i is ei-

ther a left or a right tangential unit vector of the curve K.

Proof. Let us assume that the curve K obeys the Lemma conditions. Let us denote by T a set of all vectors $t \in V^n$, each of which is either a left or a right tangential unit vector of the curve K. The set T is closed.

Let $\{x(s), 0 \leqslant s \leqslant l\}$ be a parametrization of the curve K, where the parameter s is an arc length. In this case

$$h = \int_0^l x'(s)\mathrm{d}s.$$

For almost all s $<x'(s), \nu> \ > 0$. Integrating this inequality term by term, we get $<h, \nu> \ > 0$, which, in particular, allows one to conclude that $h \neq 0$.

Let us denote by H the least convex cone in V^n, containing the set T. According to the known results of the theory of convex bodies the cone H can be defined in two ways. Firstly, H is an intersection of all possible closed semi-spaces containing T and such that their boundary hyper-planes pass via the point O. Secondly, H coincides with the set of all vectors x, which can be presented as

$$x = k_1 t_1 + k_2 t_2 + \cdots + k_n t_n$$

where the numbers $k_1, k_2, \ldots, k_n$ are non-negative, and $t_1, t_2, \ldots, t_n$ are the elements of the set T.

Let us prove that $h \in H$. For this purpose let us make use of the first definition of H. Let P be an arbitrary closed semi-space containing T and such that O belongs to the boundary P. Obviously, P is a set of all $x \in V^n$, such that $<\mu, x> \ \geqslant 0$, where μ is a vector of the internal normal P. For every s, for which the derivative $x'(s)$ is defined (i.e., for almost all s) the vector $x'(s) \in T$. We therefore find that $<\mu, x'(s)> \ \geqslant 0$ for almost all $s \in [0, l]$. Integrating this inequality, we get $<\mu, h> \ \geqslant 0$, i.e., $h \in P$. Therefore, any closed semi-space containing T contains the vector h. Consequently, h belongs to the intersection of all these semi-spaces, i.e., $h \in H$.

The required result follows from the second definition of H. Since $h \in H$, then, according to this definition, there are vectors $t_1, t_2, \ldots, t_n$ and numbers $k_1 \geqslant 0, k_2 \geqslant 0, \ldots, k_n \geqslant 0$ such that

$$h = k_1 t_1 + k_2 t_2 + \cdots + k_n t_n.$$

Preserving the addends other than zero in the sum, we get the result sought.

The Lemma is proved.

LEMMA 5.2.4. *For any non-degenerate curve K in E^n $\kappa^*(K) \leqslant \kappa(K)$.*

Proof. Let K be an arbitrary non-degenerate curve in E^n. If $\kappa(K) = \infty$, then the required inequality is obvious. Let us assume that $\kappa(K)$ is finite. In this case, according to Theorem 5.1.2, the curve K is one-sidedly smooth. Let $\xi = \{t_1, t_2, \ldots, t_m\}$ be an arbitrary sequence of the tangents of the curve K, numbered in the order of their location and let $X_1 \leqslant X_2 \leqslant \cdots \leqslant X_m$ be the points at which these tangents are taken. Let us construct a sequence of mutually non-intersecting semi-neighbourhoods $U_1, U_2, \ldots, U_m$ of the points X_1, $X_2, \ldots, X_m$, such that U_i is the right-hand semi-neighbourhood of the point X_i, if $t_i = t_r(X_i)$ and the left-hand one, if $t_i = t_l(X_i)$. Let us arbitrarily choose $\varepsilon > 0$ and find a secant $l_i = l(X_i, Y_i)$, where $Y_i \in U_i$, such that $(t_i, l_i) < \varepsilon/2m$. Connecting the points X_i, Y_i in the order of their location on the curve K, we get the polygonal line L inscribed into K. It is obvious that

$$\kappa(K) \geqslant \kappa(L) \geqslant \sum_{i=1}^{m-1} (l_i, \widehat{} l_{i+1}) \geqslant \sum_{i=1}^{m-1} (t_i, \widehat{} t_{i+1}) - \frac{2m-2}{2m} \varepsilon.$$

From this, due to the arbitrariness of $\varepsilon > 0$, we have

$$\sum_{i=1}^{m-1} (t_i, \widehat{} t_{i+1}) \leqslant \kappa(K)$$

i.e., $\kappa^*(\xi) \leqslant \kappa(K)$ and, since ξ is arbitrary, $\kappa^*(K) \leqslant \kappa(K)$, which is the required proof.

LEMMA 5.2.5. *For any one-sidedly smooth curve K $\kappa^*(K) \geqslant \kappa(K)$.*

Proof. Let us divide the curve K into arcs in such a way that the indicatrix of the tangents of each of them is contained inside the semi-sphere. This possibility results from Theorem 5.1.1 in keeping with which, at every point of the curve, there exist left-hand and right-hand semi-neighbourhoods, such that all the tangents of these semi-neighbourhoods form an angle of less than $\pi/2$ with the tangent $t_l(X)$ for the left-hand semi-neighbourhood and with $t_r(X)$ for the right-hand one. These semi-neighbourhoods form a certain neighbourhood of a point of the curve. By the Borel lemma, covering the curve K with a finite number of such neighbourhoods and dividing each of them by the point into two arcs, we obviously get the required subdivision.

Let $L_1, L_2, \ldots, L_m$ be an arbitrary sequence of polygonal lines inscribed into the curve K, such that $\lambda(L_m) \to 0$ at $m \to \infty$, and all the points taking part in the subdivision discussed above are the vertices of the polygonal lines L_m.

Let spherical curves Q and M be given. In this case let us say that the curve M is a simple extension of Q, provided M is obtained from Q in the following way. First we choose a certain arc Q' of the curve Q and on it a certain finite system of arcs $[X_i Y_i]$, $i = 1, 2, \ldots, m$, no single pair of which has general internal points. (In particular, m can be equal to zero). The arc $[X_i Y_i]$ is substituted by the shortest arc of a great circumference connecting the points X_i and Y_i. (This arc, in particular, can degenerate into a point.) It is the curve M that is formed from Q as a result of these substitutions. Let us say that M is an extension of the curve Q if we can find such a finite sequence of the curves $Q_0 = Q, Q_1, Q_2, \ldots, Q_s$ that at every $i = 1, 2, \ldots, s$ the curve Q_i is a simple extension of the curve Q_{i-1}.

It is seen from the definition that any spherical polygonal line inscribed into the spherical curve Q is its simple extension. Any spherical curve is a simple extension of itself. It also follows from the definition that if the curve M extends the curve Q, then $s(M) \leqslant s(Q)$.

Let Q be a spherical indicatrix of the tangents of the curve K, M_m is an indicatrix of the tangents' polygonal lines L_m. We have: $s(Q) = \kappa^*(K)$, $s(M_m) = \kappa(L_m)$. At $m \to \infty$ $\kappa(L_m) \to \kappa(K)$. The lemma will be proved if we establish that, at every m, $s(M_m) \leqslant s(Q)$. For this purpose it is sufficient to show that at every m the curve M_m is an extension of the curve Q.

Let L be one of the polygonal lines L_m, and let M be its indicatrix of the tangents. Let $X_0 < X_1 < \cdots < X_r$ be the consecutive vertices of the polygonal line L. The indicatrix of the tangents of the arc $[X_{i-1}X_i]$ of the curve K, which is considered to be independent due to the construction, lies in one open semi-sphere and is a certain arc $[Y_iZ_i]$ of the curve Q. If the symbols x, y, z etc. denote the non-zero vectors in V^n, then let X, Y, Z etc. be the points of the sphere Ω^{n-1}, such that $\overline{OX} = x/|x|$, $\overline{OY} = y/|y|$ and so on. Let us say that X, Y, Z are spherical images of the vectors x, y and z, respectively. Let $h_i = \overline{X_{i-1}X_i}$. According to Lemma 5.2.3, $h_i \neq 0$ and there are vectors t_i, $t_2, \ldots, t_{m_i}$, $1 \leqslant m \leqslant n$, each of which is either a left-hand or a right-hand tangential unit vector of the arc $[X_{i-1}X_i]$, and there are positive numbers k_1, $k_2, \ldots, k_{m_i}$, such that $h_i = k_1 t_1 + k_2 t_2 + \cdots + k_{m_i} t_{m_i}$. Let us assume that the unit vectors t_j, $j = 1, 2, \ldots, m$, are numbered in the order of the sequence of the corresponding tangents of the curve K. Let $T_1 < T_2 < \cdots < T_{m_i}$ be the points of the arc $[Y_iZ_i]$, which are spheric images of the vectors t_i, $i = 1, 2, \ldots, m_i$, H_i be a spheric image of the vector h_i. If the vectors a and b are such that $(a,\hat{\,}b) < \pi$, $c = ka + lb$, where $k > 0$, $l > 0$, then a point of the sphere C lies on the shortest arc of the great circumference connecting the points A and B. Let us present the expressions of h_i in the following form:

$$h_i = (((k_1 t_1 + k_2 t_2) + k_3 t_3) + \cdots +) + k_{m_i} t_{m_i}.$$

Using what has been written above, we come to the conclusion that the point H_i can be constructed in the following way. The points T_1 and T_2 are connected on the sphere with the shortest line, on which a point T' is chosen. The latter is connected via the shortest line with the point T_3 and a certain point T'' is chosen on the arc $T'T_3$, and so forth. At the last step we obtain a certain shortest line $T^{(m-2)}T_m(m = m_i)$ on which the point H_i lies. This means that H_i lies on the curve which is obtained from the arc $Q_i = [Y_iZ_i]$ in several steps, at each of which the arc constructed at the preceding step of the curve is substituted for with the shortest arc of the large circumference which connects into a certain curve which is an extension of the curve Q, i.e. M itself is an extension of the curve Q, which is the required proof.

The lemma is proved.

Proof of Theorem 5.2.1. If $\kappa^*(K) = \infty$, then, according to Lemma 2.5.4, $\kappa(K) = \infty$ too, so that $\kappa^*(K) = \kappa(K)$. If $\kappa^*(K) < \infty$, then, in keeping with Lemma 5.2.1, the curve K is one-sidedly smooth. Due to Lemma 5.2.4 we have: $\kappa^*(K) \leqslant \kappa(K)$. In this case, in line with Lemma 5.2.5 we also have $\kappa(K) \leqslant \kappa^*(K)$ and, consequently, $\kappa^*(K) = \kappa(K)$ in all cases.

Proof of Theorem 5.2.2. Let K be a one-sidedly smooth curve. In this case, according to Lemma 5.2.2, the value of $\kappa^*(K)$ is equal to the length of the indicatrix of the tangents of K. However, in keeping with Theorem 5.2.1, we have $\kappa^*(K) = \kappa(K)$ in all cases. We, therefore, come to the conclusion that for any one-sidedly smooth curve in E^n $\kappa(K)$ is equal to the length of the indicatrix of the tangents of the curve K, which is the required proof.

5.3. Turn of a Regular Curve

5.3.1. Let K be an arbitrary curve in the space. The curve K is referred to as a curve of the class C^r, where $r \geqslant 1$ is an integer, if it allows the parametrization $\{x(t), a \leqslant t \leqslant b\}$ such that the function $x(t)$ has a continuous derivative $x^{(r)}(t)$ in which case the first derivative $x'(t) \neq 0$ at all $t \in [a, b]$. If K belongs to the class C^r, where $r \geqslant 2$ and the parametrization $x(t)$, $a \leqslant t \leqslant b$, obeys the conditions listed here, then we shall say that the curve K is regular and $x(t)$ is its regular parametrization.

Let K be a curve of the class C^r, $r \geqslant 2$, $\{x(t), a \leqslant t \leqslant b\}$ be its regular parametrization. Let us set

$$\tau(t) = \frac{x'(t)}{|x'(t)|}, \quad a \leqslant t \leqslant b.$$

The vector-function τ defined in such a way is the parametrization of the indicatrix of the tangents of the curve K. In line with known results of differential geometry, for every $t_0 \in [a, b]$ there exists a limit

$$\lim_{t \to t_0} \frac{(\widehat{\tau(t), \tau(t_0)})}{s(t, t_0)} = k(t_0)$$

where $s(t, t_0)$ is the length of the arc with its ends at the points $x(t)$ and $x(t_0)$ of the curve K. The quantity $k(t_0)$ is the curvature at the point $x(t_0)$ of the curve K.

Let K be a regular curve, which will be parametrized by choosing the arc length as the parameter. Let $x(s)$, $0 \leqslant s \leqslant l$, be the parametrization obtained, where $k(s)$ denotes the curvature at the point $x(s)$. It can easily be checked that if K belongs to the class C^r, where $r \geqslant 1$, then the function $x(s)$ belongs to the same class C^r. In this case $|x'(s)| = 1$ for all $x \in [0, l]$. The function $x'(s)$, $0 \leqslant s \leqslant l$, is the parametrization of the indicatrix of the tangents Q of the curve K. The function $x'(s)$ belongs to the class C^1. Using the results of Chapter II we find that the length of the indicatrix of the tangents of the curve K is

$$\int_0^l |x''(s)|\,ds.$$

But, on the other hand, in accordance with known results of differential geometry, $|x''(s)| = k(s)$, where $k(s)$ is the curvature at the point $x(s)$ of the curve K. Therefore, the length of the indicatrix of the tangents of the curve K is

$$\int_0^l k(s)\,ds.$$

Taking into account Theorem 5.2.1, we find that for any regular curve K

$$\kappa(K) = \int_0^l k(s)\,ds.$$

For regular curves, therefore, the turn of a curve coincides with its integral curvature, i.e. with the integral of the curvature at a point of the curve with respect to the arc length.

When determining a turn of the curve we have made use of the approximation of the curve by a polygonal line. Had we initially started with the approximation of an arbitrary curve by regular curves, this would have had certain advantages. The point is that the constructions made above are essentially based on the specificity of Euclidean space which is, for instance, revealed in the fact that if the polygonal line L_1 is inscribed into the polygonal line L_2, then the turn of L_1 does not exceed that of L_2, $\kappa(L_1) \leqslant \kappa(L_2)$. If we wish to extend the notion of a turn, for instance, to the case of curves in a Riemanian space, we would be unable to use anything of the kind. Let us now demonstrate the way of introducing the notion of the turn of a curve by way of approximating the curve with regular curves. Let us first prove the following supposition.

LEMMA 5.3.1. *Let L be an arbitrary polygonal line in E^n, $A_0, A_1, \ldots, A_m$ be its sequential vertices. In this case for any $\varepsilon > 0$ there exists a polygonal line L', which has no points of return, with its vertices at the points A_0', A_1', A_m' and such that $k = 0, 1, 2, \ldots, m \ |A_k - A_k'| < \varepsilon$.*

Proof. Let us set $A_0' = A_0$, $A_1' = A_1$ and assume that the points $A_0', A_1', \ldots$, A_k' have already been constructed. The point A_{k+1}' is chosen in such a way that we have $|A_{k+1}' - A_{k+1}| < \varepsilon$ and the point A_{k+1}' does not lie on the straight line $A_{k-1}' A_k'$. Continuing this construction we get, as a result, the points A_0', A_1', $\ldots, A_m'$, and by connecting them in a sequential manner with a line segment we get a certain polygonal line L'. At every $k \geqslant 1$ the point A_{k+1}' does not lie on the straight line $A_{k-1}' A_k'$ and hence the point A_k' cannot be a point of return of L'. The Lemma is proved.

THEOREM 5.3.1. *For every curve K in E^n there exists a convergent sequence of regular curves, such that at $m \to \infty \ \kappa(K_m) \to \kappa(K)$.*

Proof. Let us first suppose that the curve K is a polygonal line having no points of return, and let $A_0, A_1, \ldots, A_m$ be its consequent vertices. A curve of the class C^r, $r \geqslant 2$ which approximates the polygonal line K will be obtained by way of rounding up the angles at the polygonal line vertices in the way described below. Let $h > 0$ be a number which is less than half of the length of each of the links of the polygonal line K. Let us choose an arbitrary internal vertex A_i. On the links radiating from it let us plot the segment A_iB_i and A_iC_i, $B_i < A_i < C_i$, of length h. The arc $[B_iC_i]$ will be denoted by U_h^i. In the plane $A_iB_iC_i$ let us construct an r-times differentiable ($r \geqslant 2$) convex arc lying in the triangle $B_iA_iC_i$, with terminal points B_i and C_i, which has a tangency with the links of the polygonal line K of order higher than r at these points. Let us denote this arc by V_h^i. It is obvious that a turn of the arc V_h^i is equal to that of the polygonal line K at the vertex A_i. Substituting each of the arcs U_h^i by the corresponding arc V_h^i, we get a certain curve. This curve, as is readily seen, belongs to the class C^r, and its turn is equal to that of the polygonal line K. At $h \to \infty$ the curve $K_h \to K$.

The Theorem is thus proved for polygonal lines.

Let K be an arbitrary curve, $L_1, L_2, \ldots, Lm, \ldots$ be a sequence of polygonal lines inscribed into it such that, at $m \to \infty$, $L_m \to K$. In this case, at $m \to \infty$, $\kappa(L_m) \to \kappa(K)$.

The polygonal line L_m can, generally speaking, have points of return. By an arbitrarily small shift of the vertices, we can get from it a polygonal line L_m' having no points of return (Lemma 5.3.1). In this case the polygonal line L_m' can be chosen in such a way that

$$\rho(L_m, L_m') < \frac{1}{m}, \qquad |\kappa(L_m) - \kappa(L_m')| < \frac{1}{m}.$$

Now let K_m be a curve of class C^r, such that

$$\rho(K_m, L_m) < \frac{1}{m}, \qquad |\kappa(L_m') - \kappa(K_m)| < \frac{1}{m}.$$

Obviously, at $m \to \infty$, the curves K_m converge to the curve K, and their turns converge to the turn of the curve K. The Theorem is thus proved.

5.4. Analytical Criterion of Finiteness of a Curve Turn

5.4.1. In this section a particularly important criterion of the finiteness of a curve turn is established.

THEOREM 5.4.1. *Let K $\{X(s),\ 0 \leqslant s \leqslant l\}$ be a rectifiable curve (the parameter s is an arc length). In this case, if the curve K has a finite turn, then at all s there exist left and right derivatives $X_l'(s)$ and $X_r'(s)$, and these derivatives are functions of a finite variation.*

Proof. The existence of the derivatives $X_l'(s)$ and $X_r'(s)$ results from Theorem 5.1.2. If $0 \leqslant s_1 \leqslant s_2 \leqslant \cdots \leqslant s_m \leqslant l$, then

$$\sum_{i=1}^{m-1} |X_l'(s_{i+1}) - X_l'(s_i)| \leqslant \sum_{i=1}^{m-1} (X_l'(s_{i+1}), \widehat{} X_l'(s_i)) \leqslant \kappa(K).$$

From this we have: $V_0^l\, X_l'(s) \leqslant \kappa(K)$ and the theorem is proved.

THEOREM 5.4.2. *Let K be a curve in E^n. Let us assume that the curve K allows the parametrization $\{X(t), a \leqslant t \leqslant b\}$ which obeys the following conditions:*

(1) *the vector-function $X(t)$ is one-sidedly smooth, i.e., at any $t < b$ there exists a derivative $X'_r(t)$, at $t > a$ there exists a derivative $X'_l(t)$, in which case for any $t > a$ $X'_l(t) = \lim_{\tau \to t-0} X'_l(\tau) = \lim_{\tau \to t-0} X'_r(\tau)$ and, for $t < b$, $X'_r(t) = \lim_{\tau \to t+0} X'_l(\tau) = \lim_{\tau \to t+0} X'_r(\tau)$;*

(2) *for every $t > a$ $|X'_l(t)| > 0$ and for any $t < b$ $|X'_r(t)| > 0$;*

(3) *$X'_l(t)$ $(X'_r(t))$ is a function of bounded variation.*

In this case K is a curve of a finite turn in E^n.

Proof. Let us assume that the curve K in E^n allows the parametrization obeying conditions (1) and (2) and such that $X'_l(t)$ is a function of bounded variation. The case when the function of bounded variation is the right derivative is considered in an anologous way. Let us define $X'_l(t)$ additionally setting $X'_l(a) = X'_r(a) = \lim_{t \to a} X'_l(t)$. In this case $X'_l(t)$ obviously not only remains a function of bounded variation, but also the value of its variations remains unchanged.

Let us prove that there exists a number $\gamma > 0$ such that $\gamma < |X'_l(t)|$ for all $t \in [a, b]$. Indeed, let us assume that $\inf_{t \in [a,b]} |X'_l(t)| = 0$. In this case there can be found a sequence of values (t_m), $m = 1, 2, \ldots$, such that $|X'_l(t_m)| \to 0$ at $m \to \infty$. Withough sacrifice of generality, we can consider that the sequence (t_m) converges, and only from one side, to a certain point $t_0 \in [a, b]$. The $X'_l(t_m) \to X'(t_0)$, where $X'(t_0) = X'_l(t_0)$, if $t_m \to t_0$ from the left, $X'(t_0) = X'(t_m)$, if $t_m \to t_0$ from the right. In this case $|X'_l(t_m)| \to |X'(t_0)|$, i.e., we come to a contradiction since, according to the condition, $|X'_l(t)| \neq 0$, $|X'_r(t)| \neq 0$ for all $t \in [a, b]$. Hence, $\inf |X'_l(t)| \neq 0$, which is the required proof.

Let us now make use of the results of Theorem 5.2.1. Let $p_0 < p_1 < \cdots < p_m$ be an arbitrary sequence of the tangents of the curve. Substituting, in the case of necessity, the right-hand tangential unit vectors with the left-hand tangent at neighbouring points in such a way that the sum $\sum_{i=1}^{m-1}(p_i, \frown p_{i+1})$ changes by not greater than an arbitrarily small $\varepsilon > 0$, we will have:

$$\sum_{i=1}^{m} (p_{i-1}, \frown p_i) = \sum_{i=1}^{m} (X'(s_{i-1}), \frown X'(s_i))$$

For arbitrary non-zero vectors a, b in E^n, the angle between which equals φ, we have: $|a - b|^2 = |a^2| + |b^2| - 2|a||b|\cos\varphi = |a^2| + |b^2| - 2|a||b| + 2|a||b| (1 - \cos\varphi) \geqslant 4|a||b| \sin^2 \varphi/2$, and hence $|a - b| \geqslant 2\sqrt{|a||b|} \sin\varphi/2$. At $0 \leqslant \varphi \leqslant \pi$ we have: $2 \sin \varphi/2 \geqslant 2\varphi/\pi$. Whence we get

$$(X'_l(t_{i-1}), \frown X'_l(t_i)) \leqslant \frac{\pi}{2\gamma} |X'_l(t_{i-1}) - X'_l(t_i)|$$

and hence,

$$\sum_{i=1}^{m} (p_{i-1} \widehat{} p_i) \leqslant \frac{\pi}{2\gamma} \sum_{i=1}^{m} |X_i'(t_{i-1}) - X_i'(t_i)| \leqslant \frac{\pi}{2\gamma} \bigvee_{a}^{b} X_i'(t).$$

In line with Theorem 5.2.1, the above formula yields

$$\kappa(K) \leqslant \frac{\pi}{2\gamma} \bigvee_{a}^{b} X_i'(t) < \infty$$

and the Theorem is proved.

As a supplement to Theorems 5.4.1 and 5.4.2, let us prove the invariancy of the class of curves of finite turn with respect to a certain sufficiently broad set of transformations of E^n.

Let there be given an open set $U \subset E^n$ and a mapping $f : U \to E^m$. Let us also assume that in the spaces E^n and E^m Cartesian orthogonal systems of coordinates are introduced. Let us suppose that f is a mapping of the class $C^{1,1}$, if f belongs to the class C^1 and for any compact set $A \subset U$ one can find a number $L = L(A) < \infty$, such that for any $X^{(1)}, X^{(2)} \in A$ the inequality

$$\left| \frac{\partial f}{\partial X_i} (X^{(1)}) - \frac{\partial f}{\partial X_i} (X^{(2)}) \right| \leqslant L(A) \, |X^{(1)} - X^{(2)}|$$

holds for all $i = 1, 2, \ldots, n$.

THEOREM 5.4.3. *Let U be an open set in E^n, $f : U \to E^m$ be a mapping of the class $C^{1,1}$, K be a non-degenerate curve of finite turn contained in the domain U. In this case, if for any point X of the curve K the vectors $df_X[t_l(X)]$ and $df_X[t_r(X)]$ are other than zero, then $f(K)$ is a curve of a finite turn in E^m.*

Proof. Let us assume that the curve K and the mapping f obey all the conditions of the theorem. Let $X(s)$, $0 \leqslant s \leqslant l$, be a parametrization of the curve K, where the parameter s is the arc length. Let us set $Y(s) = f[X(s)]$. The function Y is a parametrization of the curve $f(K)$. Let us prove that $Y(s)$ obeys all the conditions of Theorem 5.4.2. Since the mapping f belongs to the class C^1, then the function Y has the derivatives $Y_l'(s)$ and $Y_r'(s)$ at every point $s \in [a, l]$. (We omit here all the remarks concerning the ends of the interval $[0, l]$). In this case

$$Y_l'(s) = df_{X(s)}[X_l'(s)] = \sum_{i=1}^{m} \frac{\partial f}{\partial X_i} [X(s)] \, X_{i,l}'(s),$$

$$Y_r'(s) = df_{X(s)}[X_r'(s)] = \sum_{i=1}^{n} \frac{\partial f}{\partial X_i} [X(s)] \, X_{i,r}'(s)$$

for all s. Since f belongs to the class C^1, then the derivatives $\partial f/\partial X_i$ are continuous, which shows that at $s > 0$ $Y_l'(s) = \lim_{\sigma \to s-0} Y_l'(\sigma) = \lim_{\sigma \to s-0} Y_r'(\sigma)$, and at $s < l$ $Y_r'(s) = \lim_{\sigma \to s+0} Y_l'(\sigma) = \lim_{\sigma \to s+0} Y_r'(\sigma)$, so that the vector-function $Y(s)$ is one-sidedly smooth. Furthermore, due to the condition of the theorem, $Y_l'(s) \neq 0$, $Y_r'(s) \neq 0$ for all $s \in [0, l]$. Let us prove that $Y_l'(s)$ is a function of bounded variation. The set $A = |K| \subset U$ is compact and, hence, there can be found a constant $L_1 < \infty$, such that for any

$$X^{(1)}, X^{(2)} \in A \quad \left| \frac{\partial f}{\partial X_i}(X^{(1)}) - \frac{\partial f}{\partial X_i}(X^{(2)}) \right| \leqslant L_1 |X^{(1)} - X^{(2)}|$$

at every $i = 1, 2, \ldots, n$.

Owing to the compactness of A, there is also a number $L_2 < \infty$, such that at

$$X \in A \quad \left| \frac{\partial f}{\partial X_i}(X) \right| < L_2 \quad \text{for all } i = 1, 2, \ldots, n.$$

Let us arbitrarily set the numbers $s_0, s_1, \ldots, s_m$, such that $0 \leqslant s_0 < s_1 < \cdots < s_m \leqslant l$. At every $j = 1, 2, \ldots, m$ we have:

$$|Y_i'(s_j) - Y_i'(s_{j-1})| = \left| \sum_{i=1}^{n} \frac{\partial f}{\partial X_i}[X(s_j)] X_{i,l}'(s_j) - \sum_{i=1}^{n} \frac{\partial f}{\partial X_i}[X(s_{j-1})] X_{i,l}'(s_{j-1}) \right|$$

$$\leqslant \left| \sum_{i=1}^{n} \left(\frac{\partial f}{\partial X_i}[X(s_j)] - \frac{\partial f}{\partial X_i}[X(s_{j-1})] \right) X_{i,l}'(s_j) \right|$$

$$+ \left| \sum_{i=1}^{n} \frac{\partial f}{\partial X_i}[X(s_{j-1})][X_{i,l}'(s_j) - X_{i,l}'(s_{j-1})] \right|$$

$$\leqslant L_1 \sqrt{n} \ |X(s_j) - X(s_{j-1})| + L_2 \sqrt{n} \ |X'(s_j) - X'(s_{j-1})|.$$

Summing with respect to j we get, allowing for the fact that $|X(s_j) - X(s_{j-1})| \leqslant s_j - s_{j-1}$:

$$\sum_{j=1}^{m} |Y_i'(s_j) - Y_i'(s_{j-1})| \leqslant L_1 l \sqrt{n} + L_2 \sqrt{n} \bigvee_0^l X_i'(s)$$

In line with Theorem 5.4.1, the right-hand part is finite and, since the sequence of values $s_o < s_1 < \cdots < s_m$ has been arbitrarily chosen, the above proof shows that the variation of the vector-function $Y_i'(s)$ is finite. In conformity with Theorem 5.4.2, the above proof also shows that $L = f(K)$ is a curve of a finite turn, which is the required proof.

Theorem 5.4.3 now makes it possible to define the notion of a curve of a finite turn in an arbitrary n-dimensional manifold of the class $C^{1,1}$ in a complete analogy with the way the notion of a rectifiable curve in an n-dimensional manifold of the class Lip has been defined earlier.

Let M be an n-dimensional manifold. In this case we shall say that there is a structure of a differential manifold of the class $C^{1,1}$ introduced in M, provided in M there is given an atlas $\mathfrak{U} = \{U_\alpha, \varphi_\alpha\}$, such that for any $\alpha, \beta \ \varphi_\beta \circ \varphi_\alpha^{-1}$ is a mapping of the class $C^{1,1}$. This definition is correct since a superposition of two mappings of the class $C^{1,1}$ is always a mapping of the class $C^{1,1}$.

Let K be an arbitrary curve in the manifold M of the class $C^{1,1}$. We shall say that K is a curve of finite turn in M, if for K we can find such a subdivision of the curve K into arcs $K_i = [X_{i-1}, X_i]$, where $X_0 = A < X_1 < \cdots < X_m = B$, A is the beginning and B is the end of the curve K, and the admissible charts $\varphi_i : U_i \to \mathbb{R}^n$, such that at every $i = 1, 2, \ldots, m$ the arc K_i is contained in U_i and $l_i = \varphi(K_i)$ is a curve of finite turn in $\mathbb{R}^n$. In an analogous way we can define the notion of a closed curve of finite turn in M.

Let K be a curve of finite turn in M, $L = [PQ]$ be its arbitrary closed arc, $\varphi:U \to \mathbb{R}^n$ is an admissible chart of the manifold M, such that $L \subset U$. In this case $\varphi(L)$ is a curve of finite turn in $\mathbb{R}^n$. Indeed, let $X_0 = A < X_1 < \cdots < X_m = B$ (A is the beginning, B is the end of K) be the points of the curve K, $\varphi_i:U_i \to \mathbb{R}^n$ be the admissible charts, such that $L_i = [X_{i-1}X_i] \subset U$ at all $i = 1, 2, \ldots, m$ and $\varphi_i(L_i) = S_i$ be a curve of finite turn. Let $G_i = \varphi_i(U_i)$, $\theta_i = \varphi \circ \varphi_i^{-1}$. The mapping θ_i belongs to the class $C^{1,1}$ and, hence, it transforms any curve of finite turn to that of a finite turn. The points X_i divide L into partial arcs. Let $L_i' = L \cap [X_{i-1}X_i]$. The mapping φ_i transforms L_i into a curve of finite turn. We have: $\varphi(L_i) = \theta_i[\varphi_i(L_i')]$ and, hence, $\varphi(L_i')$ is a curve of finite turn. We therefore find that $\varphi(L)$ is subdivided into not greater than m arcs, the turns of which are finite, and, hence $\kappa[\varphi(L)] < \infty$.

THEOREM 5.4.4. *Let K be a one-sidedly smooth curve in an n-dimensional manifold M of the class $C^{1,1}$. Then, for the curve K to be a curve of finite turn, it is necessary and sufficient that its any complete indicatrix be a rectifiable curve in the manifold $\hat{T}(M)$.*

Proof. Let us first consider the case when the curve K lies in the domain of the definition of a certain admissible chart $\varphi:U \to \mathbb{R}^n$ in the manifold M. Let $G = \varphi(U)$. The map φ defines a certain mapping $\hat{\varphi}:\hat{U}G \times \Omega^{n-1}$. In this case the mapping $\hat{\varphi}$ proves to be a Lipshitz one.

Let K be a one-sidedly smooth curve, $R = \varphi(K)$. The mapping $\hat{\varphi}$ maps $\hat{\tau}(K)$ onto a complete indicatrix of tangents $\sigma(R)$ of the curve R. Let us assume that K is a curve of finite turn. In this case R is a curve of finite turn in $\mathbb{R}^n$. Let $X(t)$, $e(t)$, $a \leqslant t \leqslant b$, be the consistent parametrizations of the curve R and its indicatrix of the tangents. Then $X(t)$ and $e(t)$ are the functions of bounded variation: $X(t)$ because the curve K is rectifiable, and $e(t)$ because the indicatrix of the tangents of the curve K is rectifiable. The function $t \mapsto (X(t), e(t)) \in \mathbb{R}^{2n}$ is a parametrization of the curve $\sigma(R)$. It is a function of bounded variation and, hence, the curve $\sigma(R)$ is rectifiable. The curve $\sigma(R)$ is transformed into $\hat{\tau}(K)$ through a Lipshitz transform of $\hat{\varphi}^{-1}$. Therefore, $\hat{\tau}(K)$ is a rectifiable curve in $\hat{T}(M)$.

Inversely, if $\hat{\tau}(K)$ is a rectifiable curve in $\hat{T}(M)$, then $\sigma(R) = \hat{\varphi}(\hat{\tau}(K))$ is a rectifiable curve in $G \times \Omega^{n-1}$. This curve is transformed into an indicatrix of the tangents R by the mapping $\pi_H:(X, l) \in G \times \Omega^{n-1} \mapsto e \in \Omega^{n-1}$. The mapping π_H is Lipshitzian and, hence, $\sigma(R) = \pi_H(\sigma(R))$ is a rectifiable curve. We come to the conclusion that $R = \varphi(K)$ is a curve of a finite turn. Therefore, K is also a curve of finite turn in M.

The general case is obviously reduced to the one considered by subdividing the curve into a finite number of arcs each of which is contained in the domain of the definition of a certain admissible chart of the manifold M.

The theorem is proved.

Comparing Theorem 5.4.4 with the classification of curves in a differentiable manifold (see 3.6.4), we come to the conclusion that the class of curves of a finite turn coincides with the class $R_2(M)$.

5.5. Basic Integro–Geometrical Theorem on the Curve Turn

5.1.1. Let us fix a point O in a space E^n in an arbitrary way. Let us also arbitrarily set a k-dimensional direction $P \in G_k^n$ and draw through O a k-dimensional plane which is parallel to P. For the sake of simplicity, let us denote this plane with the symbol P which is used to denote the given k-dimensional direction. Let K be an arbitrary curve in E^n. Let us denote by K_P its orthogonal projection onto the plane P in E^n. If the k-dimensional planes P_1 and P_2 are parallel, then the orthogonal projections of the curve K onto P_1 and P_2 are obtained from one another by way of a parallel transfer. Therefore, in the case when the point O is changed, it is only a parallel displacement that the curve K_P is subjected to.

Let a and b be arbitrary vectors in E^n. The symbols a_P and b_P will denote their orthogonal projections onto the plane $P \in G_k^n$.

LEMMA 5.5.1. *Let a and b be non-zero vectors in E^n. In this case for almost all $P \in G_k^n(O)$ the angle $(a_P, \hat{} b_P)$ is defined (i.e., the vectors a_P and b_P are both non-zero), the function $P \mapsto (a_P, \hat{} b_P)$ is measurable and the following equality is valid:*

$$\int_{G_k^n} (a_P, \hat{} b_P) \, \mathrm{d}\mu_{n,k}(P) = (a, \hat{} b).$$

Proof. Let E_0 be a totality of all planes $P \in G_k^n$, each of which is orthogonal to at least one of the vectors, a or b. According to Corollary 1 of Lemma 4.5.1, E_0 is a set of a zero measure. If $P \neq E_0$, then for any plane P' which is sufficiently close to P the vectors a_P and b_P are non-zero and hence the set $G_k^n \backslash E_0$ is open. The function $P \mapsto (a_P, \hat{} b_P)$ is obviously continuous on the set $G_k^n \backslash E_0$. Therefore this function is measurable.

Let us set:

$$\Phi(a, b) = \int_{G_k^n} (a_P, \hat{} b_P) \, \mathrm{d}\mu_{n,k}(P).$$

The value $\Phi(a, b)$ is independent of the lengths of the vectors a and b, and if a pair of vectors (a_1, b_1) can by way of rotating the space, be made coincident with the pair (a_2, b_2), then $\Phi(a_1, b_1) = \Phi(a_2, b_2)$. Indeed, let $\varphi \in O^n$ be such that $\varphi(a_1) = a_2$, $\varphi(b_1) = b_2$. In this case for any plane $P \in G_k^n$ we obviously have:

$$(a_{1,P}, \hat{} b_{1,P}) = (a_{2,\varphi(P)}, \hat{} b_{2,\varphi(P)}).$$

Integrating this equality term by term, we get

$$\Phi(a_1, b_1) = \int_{G_k^n} (a_{1,P}, \hat{} b_{1,P}) \, \mathrm{d}\mu(P) = \int_{G_k^n} (a_{2,\varphi(P)}, \hat{} b_{2,\varphi(P)}) \, \mathrm{d}\mu(P)$$

$$= \int_{G_k^n} (a_{2,P} \widehat{} b_{2,P}) \, d\mu(P) = \Phi(a_2, b_2).$$

Here we make use of the invariance of the masure $\mu = \mu_{n,k}$ in G_k^n. It follows from the above proof that Φ depends only one the angle between the vectors a and b, i.e.,

$$\Phi(a, b) = f[(a, \widehat{} b)].$$

Let us show that the function f is linear. It is defined in the segment $[0, \pi]$ and is non-negative. Let the numbers x and y be such that $0 \leqslant x \leqslant \pi$, $0 \leqslant y \leqslant \pi$ and $x + y \leqslant \pi$. Let us choose three vectors a, b, c lying in the same plane such that $(a, \widehat{} b) = x$, $(b, \widehat{} c) = y$ and $(a, \widehat{} c) = x + y$. Then for nearly all $P \in G_k^n$ we also have $(a_P, \widehat{} b_P) + (b_P, \widehat{} c_P) = (a_P, \widehat{} c_P)$. Integrating this equality term by term, we get: $\Phi(a; b) + \Phi(b, c) = \Phi(a, c)$ and hence $f(x) + f(y) = f(x+y)$. The function f therefore satisfies the functional Cauchy equation and, hence, $f(x) = Cx$, where C is a constant. The constant C can be found by setting $(a, \widehat{} b) = \pi$. In this case for nearly all $P(a_P, \widehat{} b_P) = \pi$, which yields $\Phi(a, b) = \pi$ and, hence, $C = 1$. The Lemma is proved.

COROLLARY. *For any polygonal line L in E^n the function $P \mapsto \kappa(L_P)$ is measurable in G_k^n, in which case*

$$\int_{G_k^n} \kappa(L_P) \, d\mu_{n,k}(P) = \kappa(L).$$

REMARK. In the case when $k = 1$, the elements of the manifold G_k^n are straight lines of the vector space V^n, passing through the point O. For any straight line $p \in G_1^n$ the angle $(a_P, \widehat{} b_P)$ assumes only the values 0 and π. In this case Lemma 5.4.1 can be proved by a direct description of a set of those p for which $(a_P, \widehat{} b_P) = \pi$. For this purpose it would be advisable to make use of the remark made in Section 4.4.3 by which an integration with respect to G_1^n reduced to one with respect to the sphere.

THEOREM 5.5.1. *For any curve K in E^n the function $P \rightarrow \kappa(K_P)$ is measurable in G_k^n and*

$$\kappa(K) = \int_{G_k^n} \kappa(K_P) \, d\mu_{n,k}(P). \tag{11}$$

Proof. For simplicity, let us limit ourselves to the case of ordinary curves, the considerations for closed curves being analogous. Let (L_m), $m = 1$, $2, \ldots$, be a sequence of the polygonal lines inscribed into the curve K, such that $\lambda(L_m) \rightarrow 0$ at $m \rightarrow \infty$. At every $P \in G_k^n$ $L_{m,P}$ is a polygonal line inscribed into the curve K and at $m \rightarrow \infty$ $L_{m,P} \rightarrow K_P$. Therefore, in conformity with Corollary to Theorem 5.5.1, we have $\kappa(L_{m,P}) \rightarrow \kappa(K_P)$ at $m \rightarrow \infty$ for all $P \in G_k^n$. It makes it possible to conclude that the function $P \mapsto \kappa(K_P)$ is measurable as a limit of a sequence of measurable functions.

Applying Corollary to Lemma 5.5.1 to the polygonal line L_m, we get that at every m

$$\int_{G_k^n} \kappa(L_{m,P}) \; \mathrm{d}\mu_{n,k}(P) = \kappa(L_m) \tag{12}$$

In line with the known theorems on a limiting transition under the sign of an integral, the following inequality is valid:

$$\int_{G_k^n} \kappa(K_P) \; \mathrm{d}\mu_{n,k}(P) \leqslant \lim_{m \to \infty} \int_{G_k^n} \kappa(L_{m,P}) \; \mathrm{d}\mu(P).$$

In line with (12) the integral in the right-hand part equals $\kappa(L_m) \longrightarrow \kappa(K)$ at $m \longrightarrow \infty$ and at the transition to a limit at $m \longrightarrow \infty$ we get:

$$\int_{G_k^n} \kappa(K_P) \; \mathrm{d}\mu_{n,k}(P) \leqslant \kappa(K) \tag{13}$$

On the other hand, since $\kappa(L_{m,P}) \leqslant \kappa(K_P)$ at all m, then we have:

$$\int_{G_k^n} \kappa(K_P) \; \mathrm{d}\mu(P) \geqslant \int_{G_k^n} \kappa(L_{m,P}) \; \mathrm{d}\mu(P) = \kappa(L_m) \tag{14}$$

at all m. At $m \longrightarrow \infty$ $\kappa(L_m) \longrightarrow \kappa(K)$ and, passing to a limit in inequality (14), we obtain:

$$\int_{G_k^n} \kappa(K_P) \; \mathrm{d}\mu(P) \geqslant \kappa(K). \tag{15}$$

From inequalities (13) and (15) we obviously get (11). The Theorem is proved.

Theorem 5.5.1 was first obtained by I. Fari [6, 7] and somewhat later by J. Milnor [19, 20]. These authors formulated the theorem for regular curves since they did not have at their disposal the notion of the turn of a curve in a general form. I. Fari considers only the case when $k = n - 1$, and J. Milnor does the cases when $k = 1$ and $k = n - 1$, and at $k = 1$ J. Milnor considers instead of $\kappa(K_P)$ another particular quantity, which coincides in fact with $\kappa(K_P)$.

5.5.2. Theorem 5.5.1 proves an effective tool for studying properties of the curves related to the notion of a turn. Examples of the application of this theorem will be given in subsequent paragraphs of the present chapter. Let us see in what way one can deduce some of the results of the preceding section from Theorem 5.5.1.

<u>Proof of one-sided smoothness of the curves of a finite turn.</u> Let the curve K in E^n be such that $\kappa(K) < \infty$. In this case according to Theorem 5.5.1 for almost all $p \in G_1^n$

$$\int_{G_k^n} \kappa(K_p) \; \mathrm{d}\mu(p) < \infty.$$

and, hence, $\kappa(K_p) < \infty$ for almost all p. Let us arbitrarily set $\varepsilon > 0$. Let us assume that the unit vectors e_i, $i = 1, 2, \ldots, N$ form an $\varepsilon/2$-net on the sphere Ω^{n-1} in V^n. For any i there can be found a straight line $p_i \in G_1^n$ which makes with e_i an angle less than $\varepsilon/2$. Let ν_i be a unit vector lying on the straight line p_i and such that $(e_i, \widehat{} \nu_i) < \varepsilon/2$. The vectors ν_i form an ε-net on the sphere Ω^{n-1}. For any $i = 1, 2, \ldots, N$ the curve K_{p_i} has a finite turn and hence, in conformity with Lemma 5.1.3, it consists of a finite number of straight lines. This fact, in particular, affords the definition of a one-sidedly smooth curve. Therefore, for any $\varepsilon > 0$ there exists a finite ε-net on the sphere Ω^{n-1}, such that the projections of the curve K onto the straight lines, which are parallel to the vectors of this net, are one-sidedly smooth curves. In line with Theorem 3.2.1, this shows that the curve K is one-sidedly smooth, which is the required proof.

Proof of Lemma 5.2.5. This lemma plays a central part in proving Theorems 5.2.1 and 5.2.2. The latter themselves refer to a number of basic results of the theory under discussion since, in particular, it is with their help that the relation with the notions known from differential geometry is established.

Let K be a curve in E^n. Let us prove that $\kappa(K) \leqslant \kappa^*(K)$. In the case when $\kappa^*(K) = \infty$ there is nothing to prove. Let us assume that $\kappa^*(K)$ is finite. In this case, in view of Lemma 5.2.1, the curve K is one-sidedly smooth and, in view of Lemma 5.2.2, $\kappa^*(K)$ is equal to the length of the indicatrix of the tangents Q of the curve K.

For any $(n-1)$-dimensional direction P in E^n, as shown in Section 4.7.1 there is a certain number $\alpha_Q(P)$ defined and, in view of Theorem 4.7.1

$$s(Q) = \int_{G_{n-1}^n} \pi\alpha_Q(P) \, d\mu_{n,n-1}(P).$$

Let $\nu: G_1^n \to G_{n-1}^n$ be a canonical correspondence of the Grassman manifolds G_1^n and G_{n-1}^n. For any $p \in G_1$ $\nu(p)$ is an $(n-1)$-dimensional direction orthogonal to p. In line with Theorem 5.5.1,

$$\kappa(K) = \int_{G_1^n} \kappa(K_p) \, d\mu_{n,1}(p).$$

Lemma 5.2.5 will be established provided we can prove the following supposition.

LEMMA 5.5.2. *Let K be a curve of a finite turn in E^n, and Q be its indicatrix of the tangents. Then for almost all $p \in G_1^n$ the following inequality holds:*

$$\kappa(K_p) \leqslant \pi\alpha_Q(\nu(p)) \tag{16}$$

Proof. Let $X(s)$, $0 \leqslant s \leqslant 1$, be the parametrization of the curve K, where the parameter s is the arc length. Let us arbitrarily take $p \in G_1^n$ and let K_p be an orthogonal projection of the curve K onto a straight line parallel to the direction p. If the curve K_p is degenerate then $\kappa(K_p) = 0$, and in this case the inequality (16) is obvious. Let us consider K_p to be a non-degenerate curve. Then L is an arbitrary polygonal line inscribed into K_p. We can obviously consider that L has no vertex wherein the turn equals zero. Let $X_0 < X_1 < \cdots < X_m < X_{m+1}$ be successive vertices of L (X_0 is the beginning, X_{m+1} the end). In this case $\kappa(L) = m\pi$. Let $Y_0 < Y_1 < \cdots < Y_m < Y_{m+1}$ be the vertices of the curve K which are projected into the points $X_0, X_1, \ldots, X_{m+1}$, respectively. Let us arbitrarily set the vector e colinear to the direction p. Let $Y_i = X(s_i)$, $i = 0, 1, 2, \ldots, m+1$. Let us consider the arc $[Y_{i-1}Y_i]$ of the curve K. Let us assume that the vector $\overline{X_{i-1}X_i}$ is directed in the same direction as the vector e. Let us prove that in this case there is the value $\sigma_i \in (S_{i-1}, S_i)$, such that $\langle X_i'(\sigma_i), e \rangle > 0$. Indeed, let us, on the contrary, assume that

$<X'(\sigma), e> \leqslant 0$ for all $s \in (S_{i-1}, S_i)$. Integrating this inequality term by term, we get

$$<\overline{Y_{i-1}Y_1}, e> = <X(s_i) - X(s_{i-1}), e> \leqslant 0.$$

This obviously contradicts the fact that the vector $\overline{X_{i-1}X_i}$, which is an orthogonal projection of the vector $\overline{Y_{i-1}Y_i}$ onto p, is directed in the same direction as e. Analogously we see that if the vector $\overline{X_{i-1}X_i}$ opposes e in direction, then there is a $\sigma_i \in (s_{i-1}, s_i)$, such that $<X'_i(\sigma_i), e> < 0$. The vector $X'_i(\sigma_i)$ defines a certain point Y_i of the curve Q. In this case, obviously, $Y_1 < Y_2 < \cdots < Y_{m+1}$. At every $i > 0$, $i < m + 1$, the vectors $\overline{X_{i-1}X_i}$ and $\overline{X_iX_{i+1}}$ are oppositely directed and hence the scalar products have different signs. It means that the points Y_i and Y_{i+1} lie to different sides of the plane, $P = \{y | <y, e> = 0\}$. This plane is nothing other than $\nu(p)$. In view of the definition of the number $\alpha_Q(P)$, we get that $\alpha_Q(P) \geqslant m = 1/\pi \kappa(L)$. Thus, $\kappa(L) \leqslant \pi\alpha_Q(P) = \pi\alpha_Q[\nu(p)]$. Since the polygonal line L inscribed into the curve K_p has been arbitrarily chosen, then $\kappa(K_p) \leqslant \pi\alpha_Q[\nu(p)]$, and the Lemma is proved.

COROLLARY 1. *For any one-sidedly smooth curve K in E^n the inequality*

$$\kappa(K) \leqslant s(Q)$$

is valid.

To prove this statement, it is sufficient to integrate inequality (16) term by term and to use the results of Theorems 5.5.1 and 4.7.1.

Corollary 1 yields a new proof of Lemma 5.2.5.

COROLLARY 2. *For any curve K in E^n such that $\kappa(K) < \infty$ the following equality is valid for almost all $p \in G_1^n$*

$$\kappa(K_p) = \pi\alpha_Q[\nu(p)].$$

where Q is the indicatrix of the tangents of curve K.

Proof. According to the Lemma, for almost all p, we have $\kappa(K_p) \leqslant \pi\alpha_Q[\nu)p)]$. In line with Theorem 5.2.2, $\kappa(K) = s(Q)$. Therefore,

$$\int_{G_1^n} \kappa(K_p)\mathrm{d}\mu(p) = \kappa(K) = s(Q) = \int_{G_1^n} \pi\alpha_Q[\nu(p)]\mathrm{d}\mu(p)$$

and, hence, $\kappa(K_p) = \pi\alpha_Q[\nu(p)]$ for almost all ν, which is the required proof.

Let us give two more examples of a purely illustrative character where Theorem 5.5.1 is used to prove some of the results obtained above.

Proof of Corollary 3 to Theorem 5.1.1. Let K be a non-degenerate curve in E^n, such that $\kappa(K) < \infty$. Let us arbitrarily choose a point X of the curve K and assume that X is not the end of K. The task is to prove that $\kappa(XY) \rightarrow 0$ when $Y \rightarrow X$ from the right. Let us arbitrarily set a sequence (Y_m), $m = 1, 2, \ldots,$ of the points of the curve K converging to X from the right. Let $p \in G_1^n$, let K_p be an orthogonal projection of the curve K onto p, and let X_p and $Y_{m,p}$ be the projections of the points X and Y_m onto p. In this case $Y_{m,p} \rightarrow X_p$ from the right along the curve K_p at $m \rightarrow \infty$. In conformity with Theorem 5.5.1, we have

$$\kappa(XY_m) = \int_{G_1^n} \kappa(X_p Y_{m,p}) \, d\mu_{n,1}(p).(p).$$

Let $p \in G_1^n$ be such that $\kappa(K_p) < \infty$. In this case, according to Lemma 5.1.3, the curve K_p is a polygonal line. Hence, at sufficiently large m the arc $[X_p Y_{m,p}]$ is a segment of a straight line and $\kappa(X_p Y_{m,p}) = 0$. We therefore find that $\kappa(X_p Y_{m,p}) \to 0$ at $m \to \infty$. At every p we have: $\kappa(X_p Y_{m,p}) \leqslant \kappa(K_p)$. On the basis of the Lebesgue theorem on a limiting transition, the above proof yields

$$\int_{G_1^n} \kappa(X_p Y_{m,p}) \, d\mu_{n,1}(p) \to 0.$$

at $m \to \infty$, i.e., $\kappa(XY_m) \to 0$ at $m \to \infty$, which is the required proof.

Integro-geometrical proof of Lemma 5.1.2. Let $l_1 = l(X_1, Y_1)$ and $l_2 = L(X_2, Y_2)$ be two secants of the curve K. Let us arbitrarily set a one-dimensional direction $p \in G_1^n$ and consider the curve K_p. Let us assume that p is orthogonal to none of the straight lines l_1 and l_2. In view of Corollary 1 of Lemma 4.5.1 such are almost all the elements p of the set G_1^n. Let $l_{1,p}$ and $l_{2,p}$ are the projections of the directed straight lines l_1 and l_2 onto the straight line p. It is obvious that $l_{1,p}$ and $l_{2,p}$ are the secants of the curve K_p. If $\kappa(K_p) = 0$ then, according to Theorem 5.1.4 the curve K_p is a straight line segment. In this case obviously $(l_{1,p}, \widehat{} l_{2,p}) = 0$ and, therefore, $(l_{1,p}, \widehat{} l_{2,p}) \leqslant \kappa(L_p)$. If $\kappa(K_p) > 0$, then $\kappa(K_p) \geqslant \pi \geqslant (l_{1,p}, \widehat{} l_{2,p})$. We therefore find that for almost all $p \in G_1^n$

$$(l_{1,p}, \widehat{} l_{2,p}) \leqslant \kappa(K_p).$$

Integrating the above inequality with respect to the variable p and applying Lemma 5.5.1 and Theorem 5.5.1, we obtain the required inequality.

5.6. Some Estimates and Theorems on a Limiting Transition

5.6.1. THEOREM 5.6.1. *For any curve K in E^n the following inequality is valid:*

$$s(K) \leqslant C_n d(K) [\kappa(K) + \pi] \tag{17}$$

where $d(K)$ is the diameter of the curve K, C_n is a constant,

$$C_n = \frac{\Gamma\left(\frac{n+1}{2}\right)}{\sqrt{\pi} \; \Gamma\left(\frac{n}{2}\right)}.$$

For any closed curve Q in E^n

$$s(Q) \leqslant C_n d(Q) \kappa(K). \tag{18}$$

Proof. Let K be a curve, Q be a closed curve in E^n. Let us arbitrarily set a straight line $p \in G_1^n$ passing via a fixed point $O \in E^n$. Let Q_p and K_p be orthogonal projections of the curves Q and K onto p. Then the diameters of the curves Q_p and K_p are not greater than $d(Q)$ and $d(K)$, respectively. Let us prove that

$$s(Q_p) \leqslant \tfrac{1}{\pi} d(Q_p) \kappa(Q_p), \tag{19}$$

$$s(K_p) \leqslant \tfrac{1}{\pi} d(K_p) [\kappa(K_p) + \pi]. \tag{20}$$

When $\kappa(Q_p) = \infty$ inequality (19) is obvious. In the same way, when $\kappa(K_p) = \infty$ inequality (20) is fulfilled.

Let $\kappa(Q_p) < \infty$. Then

$$\kappa(Q_p) = 2mp\pi$$

where $m \geqslant 1$ is an integer. In this case, in line with Lemma 5.1.3, the curve Q_p consists of $2m$ rectilinear segments, the lengths of which do not exceed $d(Q_p)$ and hence $s(Q_p) \leqslant 2md(Q_p) = [\kappa(Q_p)/\pi]d(Q_p)$, and inequality (19) is proved.

If $\kappa(K_p) < \infty$, then $\kappa(K_p) = m\pi$, where $m \geqslant 0$ is an integer and hence according to Lemma 54.1.3, the curve K_p consists of $(m+1)$ rectilinear segments, the length of each of which does not exceed $d(K_p)$. This affords

$$s(K_p) \leqslant d(K_p)(m+1) = \tfrac{1}{\pi} d(K_p)[\kappa(K_p) + \pi]$$

and inequality (20) is thus also proved.

Substituting in (19) the quantity $d(Q_p)$ with $d(Q) \geqslant d(Q_p)$ and integrating the inequality obtained with respect to p, we get

$$\int_{G_1^n} s(Q_p) d\mu(p) \leqslant \frac{d(Q)}{\pi} \int_{G_1^n} \kappa(Q_p) d\mu(p).$$

In line with Theorem 4.8.1, the left-hand part of the latter inequality equals $(\sigma_n/\sigma_1)s(Q)$. Due to Theorem 5.5.1, the integral in the right-hand part equals $\kappa(Q)$ and inequality (18) is thus proved. Inequality (17) is deduced in an analogous way, i.e., integrating inequality (20) term by term. The theorem is proved.

When $n = 2$, Theorem 5.5.1 tells us that for any plane curve the following inequality is valid:

$$s(K) \leqslant \frac{d}{2} [\kappa(K) + \pi].$$

For the curves in E^3 Theorem 5.5.1 yields the following estimate:

$$s(K) \leqslant \frac{2}{\pi} d[\kappa(K) + \pi].$$

For the closed curves on the plane we get the estimate:

$$s(Q) \leqslant \frac{d}{2} \kappa(Q)$$

for the closed curves in E^3:

$$s(Q) \leqslant \frac{2d}{\pi} \kappa(Q).$$

The latter estimate was obtained by I. Fari [6, 7] by a method analogous to that discussed above, with only one exception: in [6, 7] d denotes the diameter of the smallest sphere containing a curve. There will be no drastic changes, however, in the considerations of I. Fari, if we assume that d is a curve diameter. Further on, employing some other means, we will obtain an exact estimate of the length of a non-closed curve through its turn and the diameter of the sphere which contains the curve.

COROLLARY. *Any sequence of closed curves* (K_m), $m = 1, 2, \ldots$, *in* E^m, *the turns of which do not exceed a certain number* $M < \infty$, *all lying in a bounded domain* G *of the space* E^n, *can yield a converging subsequence.*

Proof. Let d be the diameter of the domain G. Then in view of Theorem 5.6.1 the lengths of the curves K_m do not exceed a number $A < \infty$, which is equal to $C_n d[M + \pi]$ for ordinary curves and to $C_n dM$ for closed curves. The statement being proved, therefore, results from Theorem 2.1.5.

5.6.2. THEOREM 5.6.2. *For any two rectifiable curves* K *and* L:

$$|s(K) - s(L)| \leqslant 2C_n \rho(K, L) \, [\kappa(K) + \kappa(L) + \pi],$$

where C_r *is the constant from Theorem 5.5.1.*

If K *and* L *are closed rectifiable curves, the valid is the estimate*

$$|s(K) - s(L)| \leqslant 2C_n \rho(K, L) \, [\kappa(K) + \kappa(L)\pi].$$

Proof. Let us first consider the case when the curves K and L lie in one straight line. Let us prove that in this case the following estimates are valid:

$$|s(K) - s(L)| \leqslant \frac{2}{\pi} \, \rho(K, L) \, [\kappa(K) + \kappa(L) + \pi] \tag{21}$$

for non-closed curves and

$$|s(K) - s(L)| \leqslant \frac{2}{\pi} \, \rho(K, L) \, [\kappa(K) + \kappa(L)] \tag{22}$$

for the case when K and L are closed curves.

If at least one of the values $\kappa(K)$ and $\kappa(L)$ equals ∞, then inequalities (21) and (22) are obvious. Let us assume that $\kappa(K) < \infty$ and $\kappa(L) < \infty$. In this case, in conformity with Lemma 5.1.3, the curves K and L are polygonal lines. Let us arbitrarily set $\varepsilon > 0$, and with its help find the parametrizations $X(t)$, $0 \leqslant t \leqslant 1$, and $Y(t)$, $0 \leqslant t \leqslant 1$, of the curves K and L, such that $|X(t) - Y(t)| < \rho(K, L) + \varepsilon$ for all $t \in [0, 1]$. Let us first consider the case of non-closed curves. Let $t_1, t_2, \ldots, t_m$ and $u_1, u_2, \ldots, u_l$ be the values of the parameter which correspond to those points of the curves K and L, respectively, the turn of which equals π. We have $m = (1/\pi)\kappa(K)$, $l = (1/\pi)\kappa(L)$. The points t_i, u_j divide the interval $[0, 1]$ into no more than $l + m + 1$ segments. Let $[\alpha, \beta]$ be one of them. Each of the arcs $X(\alpha)X(\beta)$ and $Y(\alpha)Y(\beta)$ of the curves K and L is a segment of a straight line. As far as $|X(\alpha) - Y(\alpha)| < \rho(K, L) + \varepsilon$, $|X(\beta) - Y(\beta)| < \rho(K, L) + \varepsilon$, then the lengths of these segments differ by less than $2[\rho(K, L) + \varepsilon]$. Therefore, the curves K and L are divided into no more than $l + m + 1$ segments in such a way that the lengths of the corresponding segments differ by not greater than $2[\rho(K, L) + \varepsilon]$. This means that the lengths of the curves themselves differ by less than $2[\rho(K, L) + \varepsilon](l + m + 1)$ and, since $\varepsilon > 0$ has been arbitrarily chosen, then

$$|s(K) - s(L)| \leqslant 2\rho(K, L)(l+m+1) = \tfrac{2}{\pi} \rho(K, L) \,[\kappa(K) + \kappa(L) + \pi].$$

In the case when K and L are closed curves, the considerations are analogous and, as a result, we find that the curves K and L are divided into $r \leqslant m + l$ straight line segments in such a way that the lengths of the corresponding segments differ by not more than $2[\rho(K, L) + \varepsilon]$. Here $m = \kappa(K)/\pi$, $l = \kappa(L)/\pi$. In this case, therefore, we have

$$|s(K) - s(L)| \leqslant \tfrac{2}{\pi} \rho(K, L) \,[\kappa(K) + \kappa(L)].$$

Let us now consider the case when the curves K and L are not assumed to be lying in one straight line. Let us orthogonally project K and L onto an arbitrarily straight line $p \in G_1^n$. If $X(t)$ and $Y(t)$ are arbitrary parametrizations of the curves K and L, such that the interval $[0, 1]$ is a domain of the changes of t, and $X_p(t)$ and $Y_p(t)$ are the projections of the points $X(t)$ and $Y(t)$ onto p, then

$$\rho(K_p, L_p) \leqslant \max_{0 \leqslant t \leqslant 1} |X_p(t) - Y_p(t)| \leqslant \max_{0 \leqslant t \leqslant 1} |X(t) - Y(t)|$$

and in line with the arbitrariness of the parametrizations of $X(t)$ and $Y(t)$, we conclude that

$$\rho(K_p, L_p) \leqslant \inf_{X, Y} \max_{0 \leqslant t \leqslant 1} |X(t) - Y(t)| = \rho(K, L).$$

In accord with what has been proved above, we get:

$$|s(K_p) - s(L_p)| \leqslant \tfrac{2}{\pi} \rho(K_p, L_p) \,[\kappa(K_p) + \kappa(L_p) + \pi]$$
$$\leqslant \tfrac{2}{\pi} \rho(K, L) \,[\kappa(K_p) + \kappa(L_p) + \pi].$$

Let us integrate this inequality term by term with respect to P. Applying Theorems 4.8.1 and 5.6.2 we get:

$$\frac{\sigma_n}{\sigma_1} |s(K_p) - s(L_p)|$$
$$= \left| \int_{G_1^n} [s(K_p) - s(L_p)] \, d\mu_{n,1}(p) \right| \leqslant \int_{G_1^n} |s(K_p) - s(L_p)| \, d\mu_{n,1}(p) |$$
$$\leqslant \tfrac{2}{\pi} \rho(K, L) \,[\kappa(K) + \kappa(L) + \pi].$$

The theorem is proved.

COROLLARY. *If at $m \to \infty$ the (closed) curves K_m, $m = 1, 2, \ldots$, converge to the (closed) curve K, and there exists a constant $M < \infty$, such that, at all m, $\kappa(K_m) \leqslant M$, then the curve K is rectifiable and the lengths of the curves K_m converge to the length of the curve K at $m \to \infty$.*

<u>Proof.</u> In accordance with Theorem 5.5.1, $\kappa(K) \leqslant \varliminf_{m\to\infty} \kappa(K_m) \leqslant M$. In line with Corollary 1 of Theorem 5.1.2, this yields that the curve K is rectifiable. In conformity with Theorem 5.6.2, at every m we have, for non-closed curves:

$$|s(K_m) - s(K)| \leqslant 2C_n \rho(K, K_m) \,[\kappa(K) + \kappa(K_m) + \pi]$$

$$\leqslant 2C_n[2M + \pi] \, \rho(K, K_m) = M_1\rho(K, K_m) \qquad (23)$$

where $M_1 = \text{const} < \infty$. In the case of closed curves the theorem yields the estimate

$$|s(K_m) - s(K)| \leqslant M_2\rho(K, K_m) \qquad (24)$$

where $M_2 = 4C_nM$. Since at $m \to \infty$ the curves K_m converge to the curve K, then $\rho(K, K_m) \to 0$ at $m \to \infty$. From this, in view of inequalities (23) and (24), it follows that $|s(K_m) - s(K)| \to 0$ at $m \to \infty$, and the corollary is proved.

5.7. Turn of a Curve as a Limit of the Sum of Angles Between the Secants

5.7.1. Let K be an arbitrary non-degenerate curve in E^n. Let us refer to a double chain ξ of the curve K as any sequence of the pairs of the curve points $\{X_1, Y_1; X_2, Y_2; \ldots; X_m, Y_m\}$, such that $X_1 \leqslant X_2 \leqslant \cdots \leqslant X_m$, $Y_1 \leqslant Y_2 \leqslant \cdots Y_m$ and at every i, $X_i < Y_i$, the points X_i, Y_i not coincide spatially. The secant $l(X_i, Y_i)$ will be, for simplicity, denoted by l_i. Let us call it an i-th secant of the double chain ξ.

To every double chain ξ let us put into correspondence the two numbers: $\nu(\xi)$ and $\lambda(\xi)$. The former is the greatest integer ν, such that on the curve K there can be found a point X belonging simultaneously to ν open arcs (X_iY_i), $i = 1$, $2, \ldots, m$. The quantity $\nu(\xi)$ is referred to as the multiplicity of the double chain ξ.

Let us denote by $\lambda(\xi)$ the largest of the diameters of the arcs

$$[AX_1], \quad [X_iY_{i+1}], \quad i = 1, 2, \ldots, m-1, \quad [Y_mB],$$

where A and B are the terminal points of the curve K. The quantity $\lambda(\xi)$ characterises the density of the location of the points of the double chain on the curve and, at the same time, the smallness of the arcs (X_iY_i) tends to zero. Let us refer to $\lambda(\xi)$ as to the modulus of the double chain ξ.

The sum $\prod(\xi) = \sum_{i=1}^{m-1} (l_i \frown l_{i+1})$ is called the turn of a double chain.

The basic result of the present section is the following theorem on the approximation of a curve turn by turns of its double chains.

THEOREM 5.7.1. *Let K be an arbitrary non-degenerate curve of finite turn in E^n, and let (ξ_m), $m = 1, 2, \ldots$, be a sequence of double chains of the curve K. In this case if at $m \to \infty$ $\lambda(\xi_m) \to 0$ and there exists a constant $N < \infty$, such that*

$$\nu(\xi_m) < N$$

at all m, then at $m \to \infty$ the turns of the double chains ξ_m converge to the turn of the curve K.

The proof of the Theorem is based on the integro-geometrical Theorem 5.5.1 and no other proof is available. Let us first prove certain lemmas on the curves lying in one straight line.

LEMMA 5.7.1. *Let L be an arbitrary polygonal line lying in one straight line. In this case a turn of any double chain ξ of the polygonal line L, such that* $\lambda(\xi) < \delta/2$, *where δ is the least of the lengths of the links of the polygonal line L, is equal to the turn of the polygonal line L.*

Proof. Let $A_0, A_1, \ldots, A_m$ be the vertices of the polygonal line L, and $\xi = \{X_1, Y_1; \ X_2, Y_2; \ \ldots; \ X_p, Y_p\}$ be an arbitrary double chain of the polygonal line, in which case $\lambda(\xi) < \delta/2$. On each of the links of the polygonal line L there lies at least one pair of the points of the double chain. Indeed, it is obvious that $X_1 \in (A_0 A_1)$, $Y_p \in (A_{m-1} A_m)$. The arcs $[X_i Y_{i+1}]$ cover the whole arc $[X_1 Y_p]$ and, therefore, on each link $A_j A_{j+1}$ there lies at least one of the points $X_i, \ Y_i, \ i = 1, 2, \ldots, p$ since in the opposite case the link $A_j A_{j+1}$ would be entirely contained within one arc $[X_i Y_{i+1}]$, which is impossible, since the diameter of each such arc is less than $\delta/2$. If on the arc $[A_j A_{j+1}]$ there are points X_i, then among them there is the utmost left one - let it be the point X_k. In this case, obviously, $X_{k-1} \notin (A_j A_{j+1})$, and $Y_k \in [A_j A_{j+1}]$. If on the arc $[A_j A_{j+1}]$ there are the points Y_i, then let Y_k be the utmost right one among them. In this case, obviously, $Y_{k+1} \notin (A_j A_{j+1})$ and $X_k \in [A_j A_{j+1}]$, so that in this case we can also state that on the link $A_j A_{j+1}$ there lies a pair of the points $(X_i Y_i)$ of the double chain ξ.

Let (X'_j, Y'_j) be the utmost right pair of the points of the chain ξ, such that $X'_j < A_j$ and the secant $l'_j = \overline{X'_j Y'_j}$ is directed in the same way as the link $A_{j-1} A_j$. Then for the secant pair l''_j which immediately follows (X'_j, Y'_j) we have: $(l'_j \frown l''_j) = \pi$, $j = 1, 2, \ldots, m-1$, while the remaining angles between the secants of the successive pairs equal zero. Therefore, a turn of the double chain ξ

$$\Pi(\xi) = \sum_{j=1}^{m-1} (l'_j \frown l''_j) = (m - 1)\pi = \kappa(K)$$

which is the required proof.

LEMMA 5.7.2. *If K is a polygonal line lying in one straight line, and ξ is an arbitrary double chain of the polygonal line, then*

$$\Pi(\xi) \leqslant [\nu(\xi) + 1] \kappa(K).$$

Proof. Let us consider the secants l_i and l_{i+1} of the double chain ξ. If $(l_i \frown l_{i+1}) = 0$, then $(l_i \frown l_{i+1}) \leqslant \kappa(X_i Y_{i+1})$. If $(l_i \frown l_{i+1}) = \pi$, then a turn of the arc $[X_i Y_{i+1}]$, as is easily seen, is other than zero and, hence

$$\kappa(X_i Y_{i+1}) \geqslant \pi = (l_i \frown l_{i+1}).$$

Therefore, $(l_i \frown l_{i+1}) \leqslant \kappa(X_i Y_{i+1})$ in all cases. We, thus, come to the conclusion that

$$\Pi(\xi) = \sum_{i=1}^{p-1} (l_i \frown l_{i+1}) \leqslant \sum_{i=1}^{p-1} \kappa(X_i Y_{i+1}). \tag{25}$$

Each point of the curve K can belong to as many as $\nu(\xi) + 1$ arcs $(X_i Y_{i+1})$. Indeed, let us choose an arbitrary point X of the curve K. Let i_1 and i_2 be the least and the largest values of the index i, such that $X \in (X_i Y_{i+1})$. It is obvious that at $i_1 \leqslant i \leqslant i_2$

$$X_{i_1} \leqslant X_i \leqslant X_{i_2} < X < Y_{i_1+1} \leqslant Y_{i+1} \leqslant Y_{i_2+1}.$$

If $i_1 = i_2$, then the sought statement is obvious. Let us assume that $i_1 < i_2$. Let $i_1 < i \leqslant i_2$. Then we have: $X_i < X$, $Y_i \geqslant Y_{i+1} > X$ and, hence, the point X belongs to all the arcs $(X_i Y_i)$ for $i_1 < i \leqslant i_2$. The total number of the latter arcs is not greater than $\nu(\xi)$, according to the definition of the multiplicity; the number of the arcs $(X_i Y_{i+1})$, for which $X_i < X < Y_{i+1}$, does not, therefore, exceed $\nu(\xi) + 1$, which is the required proof.

In the sum in the right-hand part of inequality (25) the turn at each of the vertices of the polygonal line K is allowed for not more than $\nu(\xi) + 1$ times, which results in

$$\sum_{i=1}^{p-1} \kappa[(X_i Y_{i+1})] \leqslant [\nu(\xi) + 1] \kappa(K),$$

which is the required proof.

Proof of Theorem 5.7.1. Let $\xi_1, \xi_2, \ldots, \xi_m, \ldots$ be an arbitrary sequence of the double chains of the curve K which obeys all the conditions of the Theorem, and let $\kappa(K) < \infty$.

Let E_1 be a set of all the directions p for which $\kappa(K_p) = \infty$, E_2 be a set of all p which are perpendicular to at least one of the secants of one of the double chains ξ_m.

It is obvious that E_1 is a set of zero measure, as a result of the summability of the function $\kappa(K_p)$. The set E_2 is also that of a zero measure in G_1^n, as a result of Lemma 4.5.1.

Let $p \notin E_1 \cup E_2$. The projections of the points of the double chain ξ_m form a double chain $\xi_{m,p}$ of the curve K_p. As a result of the Corollary to Lemma 5.4.1, we have

$$\Pi(\xi_m) = \int_{\Omega^{n-1}} \Pi(\xi_{m,p}) \, d\mu(p).$$

At $m \to \infty$ $\lambda_{K_n}(\xi_{m,p}) \to 0$. The curve K_p is a curve of finite turn (by condition, $p \notin E_1$). Therefore, at $m \to \infty$ $\Pi(\xi_{m,p}) \to \kappa(K_p)$, in view of Lemma 5.6.1. According to Lemma 5.6.2, at all m:

$$\Pi(\xi_{m,p}) \leqslant (N + 1) \kappa(K_p).$$

The functions $\Pi(\xi_{m,p})$ of the variable $p \in G_1^n$ are, thus, majorated by the summable function $(N + 1) \kappa(K_p)$ and, hence, in line with the known Lebesgue theorem,

$$\lim_{m \to \infty} \int_{\Omega^{n-1}} \Pi(\xi_{m,p}) \, d\mu(p) = \int_{\Omega^{n-1}} \kappa(K_p) \, d\mu(p),$$

i.e., at $m \to \infty$

$$\Pi(\xi_m) \to \kappa(K)$$

which is the required proof.

REMARK 1. It is still unknown whether the limitation $\nu(\xi_m) \leqslant N < \infty$ can be eliminated from Theorem 5.7.1. As is seen from the proof of the theorem, this limitation has been used only for ensuring the legality of a limiting transition under the sign of the Lebesgue integral.

REMARK 2. For the curves having an infinite turn, Theorem 5.7.1 is, generally speaking, invalid.

REMARK 3. By way of integrating we can easily derive from Lemma 5.7.2 an estimate of the turn of double chain through its multiplicity and the turn of the curve:

$$\Pi(\xi) \leqslant [\nu(\xi) + 1] \, \kappa(K)$$

which is valid for any spatial curve. The problem of the exactness of this estimate has not been considered in this book. It should be noted that in any case the quantity $\nu(\xi) + 1$ in the estimate presented cannot be replaced with a number independent of the double chain ξ, as can be shown by the example of a curve lying in one straight line.

5.8. Exact Estimates of the Length of a Curve

5.8.1. In a preceding paragraph a certain estimate of the length of a curve through its turn and diameter has been obtained (Theorem 5.5.1). Here we are going to obtain two more analogous estimates: an estimate of the length of a curve through the distance between the terminal points and the turn (Theorem 5.8.1) under the assumption that the turn of a curve is less than π, and an estimate of the length through the turn and the diameter of the ball, containing the curve (Theorem 5.8.2). The estimates of Theorems 5.8.1 and 5.8.2 are exact ones.

THEOREM 5.8.1. *Let K be a curve in E^n, such that $\kappa(K) < \pi$ and let r be the distance between its terminal points. In this case the following estimate is valid:*

$$s(K) \leqslant \frac{r}{\cos \dfrac{\kappa(K)}{2}}. \tag{26}$$

The sign of equality in relation (26) is valid if and only if K is a polygonal line consisting of two links equal in length.

Proof. Let the curve K in E^n be such that $\kappa(K) < \pi$ and $X(s)$, $0 \leqslant s \leqslant l$, be its parametrization, where the parameter s is the arc length, $l = s(K)$. Let us construct an indicatrix of the tangents Q of the curve K. Let e be a point of the curve Q which divides its length in half. In this case for any point $p \in Q$ we have: $(\widehat{p, e}) \leqslant \frac{1}{2}s(Q) = \frac{1}{2}\kappa(K)$. As a result, for any $s \in [0, l]$

$$\langle X_l'(s), e \rangle \geqslant \cos(\tfrac{1}{2}\kappa(K)), \quad \langle X_r'(s), e \rangle \geqslant \cos(\tfrac{1}{2}\kappa(K)).$$

Let $Y = X(l) - X(0)$. Then we have:

$$r = |Y| \geqslant <Y, e> = \int_0^l <X_i'(s), e> \, ds \geqslant l \cos \frac{\kappa(K)}{2} \qquad (27)$$

and inequality (26) is obviously proved.

Let us now see in what cases relation (26) turns into an equality. We can easily prove that this is valid when K is a polygonal line made of two equal links. The task is to prove that the equality is impossible in any other case. Hence, let K be a curve such that $r = l \cos [\kappa(K)/2]$. Let us then assume that $\kappa(K) > 0$, since in the case when $\kappa(K) = 0$ there is nothing to prove. We have:

$$|Y| = <Y, e>, \int_0^l <X_i'(s), e> \, ds = l \cos \frac{\kappa(K)}{2}$$

and, since $<X_i'(s), e> \geqslant \cos [\kappa(K)/2]$ for all s, then $<X_i'(s), e> = \cos [\kappa(K)/2]$ for almost all s. In view of the considerations of continuity this affords $<X_i'(s), e> = \cos [\kappa(K)/2]$ for all $s = [0, l]$. In the preceding considerations the left derivative can be replaced by the right one and, hence, we also have: $<X_r'(s), e> = \cos [\kappa(K)/2]$. It follows from this that both the left and the right tangents at every point of the curve K form with the vector e an angle equal to $\kappa(K)/2$. We see that $e \neq X_i'(s)$ and $e \neq X_r'(s)$ for all $s \in [0, l]$ and hence e belongs to the arc $\tau[X(s_0)]$ of the indicatrix of the tangents of K, which corresponds to a certain angular point $X(s_0)$. We have: $(e \widehat{} X_i'(s_0)) = (e \widehat{} X_r'(s_0)) = \kappa(K)/2$. Since the point e lies inside the arc of a great circumference which connects the points $X_i'(s_0)$ and $X_r'(s_0)$ on the sphere S^{n-1}, then it follows from what has been proved above that the length of this curve equals $\kappa(K)$. We come to the conclusion that a turn of the curve K at the point $X(s_0)$ is equal to $\kappa(K)$. Therefore, a turn of its arcs $[X(0)X(s_0)]$ and $[X(s_0)X(l)]$ equals zero, i.e., each of these arcs is a length of a straight line. We see that if $l = r \cos [\kappa(K)/2]$, then the curve K is a two-link polygonal line ABC (Fig. 4). In this case, as is seen from (27), the equality $|Y| = <Y, e>$ must be valid, which is possible only when $e = Y/|Y|$. The vector e is parallel to the bisector of the external angle at the vertex B of the triangle ABC, $Y = \overline{AC}$. Therefore, the side AC of the triangle ABC is parallel to the bisector of the external angle at the point B. The triangle ABC is hence isosceles. The theorem is completely proved.

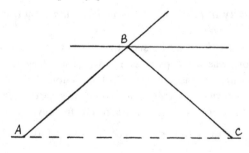

Fig. 4.

REMARK. For a regular case the estimate of the theorem was obtained by E. Schmidt [34] and the considerations used for obtaining inequality (27) confirm the conclusion drawn by E. Schmidt.

LEMMA 5.8.1. *Let ABCD be a plane non-convex tetragon with a non-convex angle C (Fig. 5). Let us assume that the tegragon ABCD is continuously deformed in such a way that the angle at its apex A increases, while the lengths of its sides remain constant. In this case the angles B and D also increase.*

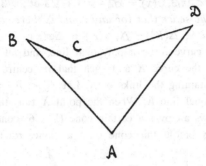

Fig. 5.

Proof. The lengths of the sides given, the angle $\hat{A} = \varphi$ completely defines the tetragon $ABCD$. Let us assume that $\hat{B} = \beta(\varphi)$, $\hat{D} = \delta(\varphi)$. Let us arbitrarily set a φ value, such that there exists a non-convex tetragon with the given lengths of the sides and the angle $\hat{A} = \varphi$, and let us give to φ an arbitrary but sufficiently small increment $h > 0$. If h is small, then the tetragon $ABCD$ with the angle $\hat{A} = \varphi + h$ does exist. Let us denote it by $A_1B_1C_1D_1$. The tetragon $A_1B_1C_1D_1$ is divided by the diagonal line A_1C_1 into two triangles: $A_1B_1C_1$ and $A_1B_1D_1$. The difference of their angles at the vertex A_1 is, obviously, less than φ, provided h is sufficiently small. Due to this fact we can carry out the following constructions.

Let us bend the tetragon $A_1B_1C_1D_1$ in the space about the diagonal line A_1C_1 in such a way that the angle $B_1A_1D_1$ be equal to φ, and transfer the figure obtained as a whole, so that the points A_1, B_1 and D_1 be coincident with the points A, B and D, respectively. Therefore, we obtain a certain triangle angle in the space, the two sides of which are the triangles $A_1B_1C_1$ and $A_1B_1D_1$, and the third side is the tetragon $ABCD$. Let us draw the plane BDC_1, and let p be the plane dividing in half the angle between this plane and the plane $ABCD$ and intersecting the segment AC_1 at a certain point E. Let the triangles DEC_1 and BEC_1 be mirror-reflected in the plane P. In this case the point C_1 will coincide with the point C, the segments BC_1 and C_1D with the segments BC and CD. Let us now consider a non-convex pyramid, the basis of which is the tetragon $ABCD$, and the vertex -E. Obviously, $\angle ABC + \angle EBC = \hat{B}_1 = \beta(\varphi+h)$, which affords $\hat{B}_1 \geqslant \angle ABC = \hat{B}$, and, analogously, $\hat{D}_1 > D$. Therefore, $\beta(\varphi+h) > \beta(\varphi)$, $\delta(\varphi+h) > \delta(\varphi)$; and the Lemma is thus proved.

Let K be an arbitrary curve in the space. Among all the balls containing the curve K, there exists a ball of least radius. Let us denote the diameter of this ball by $\Delta(K)$.

THEOREM 5.8.2. *For any curve K the following inequality is valid*

$$s(K) \leqslant \Delta(K) \; \varphi[\kappa(K)] \qquad (28)$$

where $\varphi(x)$ is the function which equals $1/\cos(x/2)$ at $0 \leqslant x \leqslant \pi/2$, $\varphi(x) = 2\sin(x/2)$ at $\pi/2 \leqslant x \leqslant 2\pi/3$, $\varphi(x) = x/2 - \pi/3 + \sqrt{3}$ at $2\pi/3 \leqslant x < \infty$. Inequality (28) is exact in the sense that for any x and Δ there can be found a curve K for which $\kappa(K) = x$, $\Delta(K) = \Delta$, $s(K) = \Delta\varphi(\kappa)$.

Proof. (a) Let the curve K be a polygonal line, and let S be a ball of the radius R, containing the curve K and such that its centre O lies in none of the straight lines containing the links of K. Let $A_0 < A_1 < \cdots < A_m$ be the vertices of the polygonal line K. When the point X runs through the curve K, the segment OX draws all over a certain cone (Fig. 6) composed of the triangles A_iOA_{i+1}. Let us unfold this cone onto a plane, rotating the component

Fig. 6.

triangles around the segment OA_i in such a way that after the unfolding the triangles adjacent to the side OA_i had no common internal points. The polygonal line K, as a result, will go over to a certain plane polygonal line $L = A_0'A_1'\dots A_m'$, lying inside a plane circle S' of the radius R and centred at O.

Let us say that the plane curve K has a star-like location with respect to the point O if it does not pass through this point and at a monotonous travelling of the point X along the curve K the ray OX monotonously rotates around the point O. The polygonal line L obviously has a star-like location with respect to the point O.

It is clear that $s(K) = s(L)$. Let us prove that $\kappa(K) \geqslant \kappa(L)$. For this purpose it is sufficient to demonstrate that, at the transition from K to L, there is no increase in the turn at every vertex of the polygonal line K. Let us consider the vertex A_i, to which the links $A_{i-1}A_i$ and A_iA_{i+1} of the polygonal line K join (Fig. 7). Let us rotate the triangle OA_iA_{i+1} around the straight line OA_i in such a way that it was lying in the plane of the triangle $OA_{i-1}A_i$ on the other side of the straight line OA_i. Let B be the point into

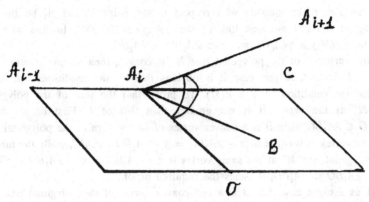

Fig. 7.

which the vertex A_{i+1} transfers in this case, and let C be the point on an extension of the straight line $A_{i-1}A_i$ beyond the point A_i. The angle $\angle CA_iA_{i+1}$ is equal to the turn of the polygonal line K at the point A_i and the angle $\angle CA_iB$ to the turn of the polygonal line L at the vertex A_i'. The task is to prove that $\angle CA_iB \geqslant \angle CA_iA_{i+1}$. Let us consider a trihedral angle formed by the rays A_iC, A_iO and A_iA_{i+1}. We have: $\angle CA_iA_{i+1} + \angle A_{i+1}A_iO \geqslant \angle CA_iO = \angle CA_iB + \angle BA_iO$. As, however, $\angle A_{i+1}A_iO = \angle BA_iO$, then $\angle CA_iA_{i+1} \geqslant \angle CA_iB$, which is the required proof.

(b) The vertex A_i' of the polygonal line L will be called convex if A_i' is a convex angle in the tetragon $OA_{i-1}'A_i'A_{i+1}'$ and non-convex in the opposite case. Let us porve that for L there exists a polygonal line M lying in the circle S' and having a star-like location with respect to its centre, such that $s(M) = s(L)$, and $\kappa(M) \leqslant \kappa(L)$, in which case the polygonal line M has no non-convex vertices.

If BC is the last link of the polygonal line L, then let $\gamma(L) = \angle BCO$, $r(L) = |OC|$. Let us prove by way of induction with respect to the number of the links n of the polygonal line L the existence of a polygonal line M for which, along with the above listed conditions, the relation $\gamma(L) \leqslant \gamma(M)$ and $r(L) = r(M)$ hold. For the case when $n = 1$ the statement (b) is obvious: it is sufficient to take $M = L$. Let this statement be proved for any $(n-1)$-link polygonal line, and let the polygonal line L have n links. Let us denote by A_1, B and C the three last vertices of the polygonal line L, and its origin by A_0. The arc (A_0B) is an $(n-1)$-link polygonal line. Let M' be a polygonal line corresponding to (A_0B) due to the assumption of induction. Let us, by rotating the polygonal lline M' around the point O, make its end coincident with the point B; it is possible in view of the condition $r(M') = r(A_0B)$. The polygonal line M' together with the segment BC forms a new polygonal line M''. The last link of the polygonal line M' and of the segment BC will be considered located at different sides of the straight line OB (it can be obtained by mirror-reflecting the polygonal line M' in the straight line OB). The polygonal line M'' there-

fore has a star-like location with respect to the point O. Let A_1' be the beginning of the last but one link of the polygonal line M''. In line with the
condition $\gamma(M') \geqslant \gamma(A_0B)$, we have $\angle A_1'BO \geqslant \angle A_1BO$.

If the vertex B of the polygonal line M'' is convex, then we can assume that
$M'' = M$. Indeed, in this case it is obvious that all the conditions are fulfilled but the condition $\kappa(L) \geqslant \kappa(M)$. To prove that the turn of the polygonal
line M'' at the vertex B is not greater than that of L. First of all, since
$\angle A_1BO \leqslant \angle A_1'BO$, then B is a convex vertex of L. The turn of the polygonal line
L at the vertex B is equal to $\pi - \angle A_1BC = \pi - \angle A_1BO - \angle OBC$, while the turn of
the polygonal line M'' at the same vertex is $\pi - \angle A_1'BC = \pi - \angle A_1'BO - \angle OBC$.
Since $\angle A_1BO \leqslant \angle A_1'BO$, it yields the required proof.

Let us assume now that B is a non-convex vertex of the polygonal line M''.
Let us continuously deform the tetragon $OA_1'BC$, preserving the lengths of the
sides, in such a way that the side OA_1' remains motionless, while the angle at
its vertex O increases. Let us continue this deformation until the turn at one
of the vertices A_1', B becomes equal to zero. As a result, we get a certain new
polygonal line M'''. It is obvious that $r(M''') = r(M'')$, $s(M''') = s(M'')$. In line
with Lemma 5.8.1, the angles at the vertices A_1' and C of the tetragon $OA_1'BC$ do
not decrease; therefore, $\kappa(M''') \leqslant \kappa(M')$, $\gamma(M''') \geqslant \gamma(M'') = \gamma(L)$. It is also obvious that the polygonal line M''' lies inside the circle S'. The polygonal line
M''' has at least one link less than the polygonal line L. By the supposition of
induction, for M''' there can be found a polygonal line M which obeys all the
conditions required. This polygonal line is the one sought.

(c) Let us now transform the polygonal line M when preserving its turn and
increasing its length. Let us prove that the polygonal line M can be transformed in this manner into a curve N of one of the following types: (a) N is a
two-link polygonal line, (b) N is a curve consisting of three arcs AB, BC and
CD, where AB and CD are the chords of the circle S', and BC is the arc of the
circumference of this circle (to be more exact, it is a curve formed by a
point moving monotonously along the circumference of the circle S').

Let $B_0, B_1, B_2, \ldots, B_m$ be the subsequent vertices of the polygonal line M,
$b_1, b_2, \ldots, b_m$ be its links, $b_i = \overline{B_{i-1}B_i}$. Let us extend the segments b_1 and
b_3 beyond the vertices B_1 and B_2. In this case two situations are possible:
these extensions either intersect outside the circle S', or inside it.

Let us first consider the former case. Let the extension of the vector b_1
intersect the circumference of the circle S' at a certain point B_1', and let
the extension of the vector b_3 intersect at a certain point B_2' (Fig. 8). Let
us denote by $B_1'B_2'$ an arc of the circumference of the circle S', which has the
ends B_1', B_2', in which case $B_1'B_2'$ and the links b_1, b_2 of the polygonal line M
lie at different sides of the straight line B_1B_2. Let us replace the segment
B_1B_2 of the polygonal line M with an arc which consists of the segment B_1B_1',
the arc $B_1'B_2'$ and the segment $B_2'B_2$. As a result, we get a certain curve M_1. A
turn of the arc $B_1B_1'B_2'B_2$ of this curve, as is seen, is equal to the sum of the

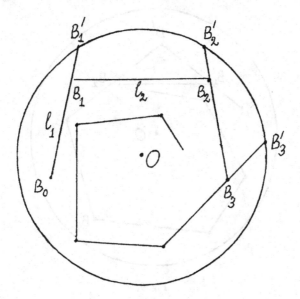

Fig. 8.

turns at the vertices B_1 and B_2 of the polygonal line M. Therefore, $\kappa(M_1) = \kappa(M)$. At the same time it is clear that $s(M_1) > s(M)$. If M has one more link b_4, let us extend b_4 up to its crossing with the circumference of the circle S' at a certain point B_3'. The segment $B_2 B_3$ included in the curve M_1 will be replaced with an arc consisting of the arc $B_2' B_3'$ of the circumference of the circle S', which is directed in the same way as $B_1' B_2'$, and of the segment $B_2' B_3$. As a result, we get a new curve M_2 for which, obviously, $\kappa(M_2) = \kappa(M_1)$, $s(M_2) > s(M_1)$. Continuing this process, we get, as a result, a curve N which consists of two straight segments adjoining its terminal ends and of an arc of the circumference of the circle S', in which case $\kappa(N) = \kappa(M)$, and $s(N) > s(M)$. The terminal ends of the curve N can here be considered to be the chords of the circle S', since it can always be attained by extending the segments.

Let us now consider the second case. Let the extensions of the first and third links of the polygonal line M intersect in the sphere S' at a certain point B_1' (Fig. 9). Substituting the segment $B_1 B_2$ of the polygonal line M with an arc which consists of the segments $B_1 B_1'$ and $B_1' B_2$, we get a new polygonal line M_1, for which, obviously, $\kappa(M_1) = \kappa(M_2)$, $s(M_1) > s(M_2)$. The polygonal line M_1 has one link less than the polygonal line M. Let us consider the first and third links of the polygonal line M_1. In the case when their extensions beyond the ends of second link do not intersect in the circle S', then, by analogy with the case considered above, we get a curve which consists of two segments and an arc of the circumference. If these extensions do intersect inside the circle S', then, by analogy with what has been done for M, from M_1 we get a new polygonal line M_2, for which $\kappa(M_2) = \kappa(M_1)$, and $s(M_2) > s(M_1)$. Con-

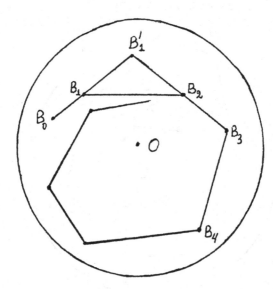

Fig. 9.

sidering the polygonal line M_2 in the same way and continuing this process, we finally get a curve of the type (a) or (b), which will be denoted by N.

(d) To prove inequality (28) it is sufficient to show that it is valid for the curve N. Indeed, as follows from the construction of the curve N, $\kappa(N) \leqslant \kappa(K)$, $s(N) \geqslant s(K)$. Consider that it has already been proved that $s(N) \leqslant 2R\varphi[\kappa(N)]$; since $\varphi(x)$ is a non-decreasing function, this affords

$$s(K) \leqslant s(N) \leqslant 2R\varphi[\kappa(N)] \leqslant 2R\varphi[\kappa(K)].$$

(e) Let us now consider the curves N of the first and second kind with the given $\kappa(N)$ values and seek among them the curve with the maximum length.

Let us first consider the following problem: to find a polygonal line with the maximum length among all the two-link polygonal lines with the given turn κ, contained in the circle S'. Let ABC be an arbitrary two-link polygonal line contained in S', with the turn κ, the terminal points of which lie on the boundary of the circle.

Let us construct a circumference constructed around the triangle ABC. The arc ABC of this circumference obviously lies entirely inside S'. Let B' be the middle of this arc. The triangle $AB'C$ is isosceles. A turn of the polygonal line $AB'C$ is also equal to κ, and the length of $AB'C$, as will be shown below, is not less than that of the polygonal line ABC and equals it only when $B = B'$.

We can easily prove this in the following way (Fig. 10). Let us assume that B lies between B' and C. Let the segment AB intersect $B'C$ at the point D. The triangles $AB'D$ and DBC are similar. In this case $BC < B'C = AB'$, since the arc $B'C$ is less than the arc BC; therefore, $DC < DA$, $DB < DB'$. Let us plot the segments $DK = DC$ and $DL = DB$ from the point D on the segments DB' and DA.

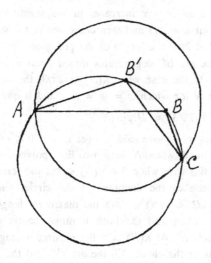

Fig. 10.

Let us draw through L a straight line $LM \parallel BA$, where M lies on AB'. In this case

$$AB + BC = AK + KD + DB + BC = AK + DC + DL + KL$$
$$= ML + LC + AM < AB' + B'C$$

which is the required proof.

The required extremum line therefore consists of two links that are equal in length, and its length will be maximum only if the distance between the end is the greatest possible at a given κ.

Let us assume that $\kappa = \kappa(ABC) \leqslant \pi/2$. Then, if ABC is an extremal two-link polygonal line, then A and C are the ends of the diameter of the circle S'. The point B, in line with the condition $\kappa \leqslant \pi/2$, will in this case belong to S', and, as is seen, we get:

$$s(N) \leqslant \frac{2R}{\cos \frac{\kappa(N)}{2}} = 2R\varphi(\kappa(N)).$$

If $\kappa(N) > \pi/2$, then the chord AC is seen to be maximum if and only if all the three vertices of the polygonal line N lie on the boundary of the circle S, and hence in this case $s(N) = 4R \sin \kappa/2 = 2R\varphi(\kappa)$.

(f) Let us now consider the curve N consisting of the arcs AB, BC and CD, where AB and CD are the chords of the circle, and BC is an arc of its circumference (possibly multiply bypassing the circumference). Let φ and ψ be the turns at the points B and C of the curve N. A turn of the arc BC is $\kappa - \varphi - \psi$. The length of the curve N, as can easily be calculated, is equal to $2R \sin \varphi + 2R \sin \psi + R\{(2 \sin \varphi - \varphi) + (2 \sin \psi - \psi) + \kappa\} = s(\varphi, \psi, \kappa)$. In the segment

$[0, \pi]$ the function $2 \sin x - x$ increases in the segment $[0, \pi/3]$, reaches its maximum at the point $x = \pi/3$ and then decreases in the segment $[\pi/3, \pi]$. This affords that at $\kappa(N) \leqslant 2\pi/3$ the length of the polygonal line N will be maximum in the case when the arc BC degenerates into a point and N turns into a two-link polygonal line. In the case when $\kappa(N) > 2\pi/3$, the maximum length is attained by the curve N for which $\varphi = \psi = \pi/3$, and it equals

$$R(\kappa(N) - \tfrac{2}{3}\pi + 2\sqrt{3}) = 2R\varphi[\kappa(N)]/$$

By way of concluding the above said, we get that at $0 \leqslant \kappa(N) \leqslant 2\pi/3$ the maximum length is attained when N is a two-link polygonal line with its links equal in length. In this case when $0 \leqslant \kappa(N) \leqslant \pi/2$ the terminal ends of the polygonal line are those of the diameter of the circle, and its vertex lies inside the circle. At $\pi/2 < \kappa(N) \leqslant 2\pi/3$ the maximum length is attained by the polygonal line the vertex and the both terminal points of which lie on the boundary of the circle S'. At $\kappa(N) \geqslant \pi$ the maximum length is attained by the curve which consists of the chord AB, the arc BC and the chord CD of the circle S', in which case the turns at the vertices B and C are $\pi/3$. When $2\pi/3 < \kappa(N) < \pi$, the maximum is attained by one of the two curves, of which the first is a two-link polygonal line, and the second is a curve of the same type as the maximum curve at $\kappa(N) \geqslant \pi$. Comparing the values of their lengths, we see, however, that a two-link polygonal line cannot in this case attain a maximum length.

We therefore get $s(K) \leqslant 2R\varphi[\kappa(K)]$. Now let K be an arbitrary curve. Let us construct a sequence L_m, $m = 1, 2, \ldots$, of the polygonal lines inscribed into K such that at $m \to \infty$ $\lambda(L_m) \to 0$. In this case $s(L_m) \to s(K)$, $\kappa(L_m) \to \kappa(K)$. For every m let us define a sphere which contains the polygonal line L_m and such that a straight line passing through the centre of the sphere cannot contain a whole link of the polygonal line L_m, while the diameter Δ_m of the sphere S_m tends to $\Delta(K)$ at $m \to \infty$. At every m we have: $s(L_m) \leqslant \Delta_m\varphi[\kappa(K_m)]$.

Going over to a limit at $m \to \infty$ and taking into account the continuity of the function $\varphi(x)$, we get: $s(K) \leqslant \Delta(K)\varphi[\kappa(K)]$, which completely proves the theorem.

5.9. Convergence with a Turn

5.9.1. Let us say that the curves K_m, $m = 1, 2, \ldots$, converge to the curve K with a turn, if $\kappa(K) < \infty$, the curves K_m converge to the curve K, and the turns of the curves K_m converge to $\kappa(K)$. The basic result of this paragraph is expressed in the following theorem.

THEOREM 5.9.1. *If the curve K has no points of return and the curves K_m, $m = 1, 2, \ldots$, converge to the curve K with a turn, then the indicatrices of the tangents of the curves K_m converge to that of the curve K. In more exact words, the curves K, K_m and their indicatrices of the tangents allow such correlated parametrizations $(X(u), t(u))$, $(X_m(u), t_m(u))$, $0 \leqslant u \leqslant 1$, that at*

$m \rightarrow \infty$ *the functions* $X_m(u)$ *and* $t_m(u)$ *converge uniformly to the functions* $X(u)$ *and* $t(u)$, *respectively in the segment* $[0, 1]$.

Proof of the theorem is based on a number of lemmas, which will be proved in succession.

LEMMA 5.9.1. *Let the curves* K_m *converge with a turn to the curve* K. *In this case, if* X *and* Y *are smooth points of the curve* K, X_m *and* Y_m *are the points of the curve* K_m, *such that at* $m \rightarrow \infty$ *the arcs* $[X_mY_m]$ *converge to the arc* $[XY]$, *then the curves* $[X_mY_m]$ *converge to the curve* $[XY]$ *with a turn.*

Proof. Let A, B, A_m and B_m be the terminal ends of the curves K and K_m. Let us choose a subsequence K_{m_i}, such that

$$\lim_{i \to \infty} \kappa(X_{m_i}Y_{m_i}) = \overline{\lim_{m \to \infty}} \kappa(X_mY_m).$$

In view of the smoothness of the points X and Y, from Theorems 5.1.1 and 5.1.3 we get:

$$\kappa(AB) = \kappa(AX) + \kappa(XY) + \kappa(YB)$$
$$\leqslant \lim_{i \to \infty} \kappa(A_{m_i}X_{m_i}) + \lim_{i \to \infty} \kappa(X_{m_i}Y_{m_i}) + \lim_{i \to \infty} \kappa(Y_{m_i}B_{m_i})$$
$$\leqslant \lim_{i \to \infty} [\kappa(A_{m_i}X_{m_i}) + \kappa(X_{m_i}Y_{m_i}) + \kappa(Y_{m_i}B_{m_i})]$$
$$\leqslant \lim_{i \to \infty} \kappa(A_{m_i}B_{m_i}) = \kappa(AB).$$

From this we conclude that in the relations presented the sign of the equality should be throughout and, hence,

$$\kappa(XY) = \lim_{i \to \infty} \kappa(X_{m_i}Y_{m_i}) = \overline{\lim_{m \to \infty}} \kappa(X_mY_m).$$

Since, in conformity with Theorem 5.1.1, $\kappa(XY) \leqslant \underline{\lim}_{m \to \infty} \kappa(X_mY_m)$, then the limit $\lim_{m \to \infty} \kappa(X_mY_m)$ does exist and is equal to $\kappa(XY)$, which is the required proof.

LEMMA 5.9.2. *Let the curves* K_m *converge with a turn to the curve* K, *and let* X *be a smooth point of the curve* K; *let* A, B, A_m, B_m *be the terminal ends of the curves* K, K_m; *let* X_m *be the points of the curves* K_m, *such that at* $m \rightarrow \infty$ *the arcs* $[A_mX_m]$ *and* $[X_mB_m]$ *converge to the arcs* AX *and* XB; *let* t *be a tangent of the curve* K *at the point* X; *and let* t_m *be an arbitrary intermediate tangent of the curve* K_m *at the point* X_m. *In this case at* $m \rightarrow \infty$ *a directed straight line* t_m *converges to the directed straight line* t.

Proof. Let t'_m be a unit vector the right-hand tangent at the point X_m, t''_m be a unit vector of the left-hand tangent at X_m, let t be a tangential unit vector at the point X of the curve K. It will be sufficient to show that at $m \rightarrow \infty$ $t'_m \rightarrow t$, $t''_m \rightarrow t$. Let us limit ourselves to the case of the tangents t'_m, since for t''_m the proof is analogous. Let us arbitrarily choose ε, $0 < \varepsilon < \pi$, and find on K a smooth point Y lying to the right from X, $Y > X$, such that $\kappa(XY) < \varepsilon$.

It should be remarked that for any curve in a space the angle between the tangent $t(X)$ and the secant $l(XY)$, which pass through one and the same point of the curve, is not greater than the turn of the arc $[XY]$, which can easily be proved by tending the turn of the inscribed polygonal line XZY to a limit at $Z \to X$.

Let us find on the curve K_m a point Y_m, such that at $m \to \infty$ the arc $[X_m Y_m]$ converges to the curve $[XY]$ with a turn. Since $\kappa(XY) < \varepsilon$, then there can be found m_0, such that at $m > m_0$ we will also have $\kappa(X_m Y_m) < \varepsilon$. At such m the angle between the tangent t'_m and the secant $l_m = l(X_m Y_m)$ will also be less than ε. At $m \to \infty$ the straight line $l_m = l(X_m Y_m)$ converges to $l = l(XY)$ and hence the angle between them tends to zero. Let us find $m_1 \geqslant m_0$, such that at $m > m_1$ $(l', \widehat{\ } l_m) < \varepsilon$. In this case, if $m > m_1$, then we have:

$$(t'_m, \widehat{\ } t) \leqslant (t'_m, \widehat{\ } l_m) + (l_m, \widehat{\ } l) + (l', \widehat{\ } t) < \varepsilon + \varepsilon + \varepsilon = 3\varepsilon.$$

In view of the arbitrariness of $\varepsilon > 0$, the lemma is proved.

LEMMA 5.9.3. *Let K be a rectifiable spherical curve and let L be a spherical polygonal line inscribed into it, having with K general terminal points and such that the lengths of the arcs into which the vertices of the polygonal line L subdivide the curve K, are less than a certain $\gamma < \pi$. In this case the distance between the curves K and L is not greater than*

$$\pi \sqrt{\frac{s(K) - s(L)}{2 \cos \frac{\gamma}{2}}}.$$

Proof. Let $A_0, A_1, A_2, \ldots, A_m$ be the successive vertices of the polygonal line L. Let us consider the arc $A_i A_{i+1}$ of the curve K and a link of the polygonal line L connecting the points A_i and A_{i+1}. As a parameter on the curve K, let us introduce the length of an arc read off from the beginning. Let $A_i = X(s_i)$, $A_{i+1} = X(s_{i+1})$, $s_i < s_{i+1}$. Let $Y(s)$, $0 \leqslant s \leqslant s(K)$ be the parametrization of the polygonal line L, such that $A_i = Y(s_i)$ and on each of the links of the polygonal line the parameter s is a linear function of the length of the arc read off from the origin of the link. Let us prove that at all s the distance between the points $X(s)$ and $Y(s)$ is not greater than $\pi \sqrt{[s(K) - s(L)]/(2 \cos \gamma/2)}$. This obviously will be the proof of the Lemma.

Let $s_i < s < s_{i+1}$. Let us set $X(s) = X$, $Y(s) = Y$, $|A_i Y| = \alpha$, $|Y A_{i+1}| = \beta$, $s_{i+1} - s = \beta_1$, $s - s_i = \alpha_1$, $|X A_i| = \alpha'$, $|X A_{i+1}| = \beta'$. Since by the condition $\beta_1/\alpha_1 = \beta/\alpha$ and $\beta_1 + \alpha_1 \geqslant \beta + \alpha$, then $\beta_1 \geqslant \beta$ and $\alpha_1 \geqslant \alpha$, and then $\alpha_1 \geqslant \alpha'$ and $\beta_1 \geqslant \beta'$. Since one of the arcs α_1, β_1 is not greater than $\gamma/2$, for instance, $\beta_1 \leqslant \gamma/2$ and $\beta \leqslant \beta_1$, then from the triangle $XA_{i+1}Y$ we find $XY \leqslant \gamma < \pi$, so that all the sides of the spherical triangles XA_iY, $XA_{i+1}Y$ are less than π.

The angle Y of the spherical triangle $A_1 XY$ will be denoted by φ. Let us also set $XY = \rho$. In view of the cosine formula of spherical geometry, we have:

$$\cos \alpha_1 \leqslant \cos \alpha' = \cos \rho \cos \alpha + \sin \rho \sin \alpha \cos \varphi,$$

$$\cos \beta_1 \leqslant \cos \beta' = \cos \rho \cos \beta - \sin \rho \sin \beta \cos \varphi.$$

Multiplying the first inequality by $\sin \beta$ and the second one by $\sin \alpha$ and summing, we get:

$$\cos \rho \geqslant \frac{\cos \alpha_1 \sin \beta + \cos \beta_1 \sin \alpha}{\sin (\alpha + \beta)}.$$

In line with the Lagrange theorem on a finite increment we can write:

$$\cos \alpha_1 = \cos \alpha - (\alpha_1 - \alpha) \sin \theta_1, \quad \cos \beta_1 = \cos \beta - (\beta_1 - \beta) \sin \theta_2$$

where $\alpha \leqslant \theta \leqslant \alpha_1, \beta \leqslant \theta_2 \leqslant \beta_1$.
From this we get:

$$\cos \rho \geqslant 1 - \frac{(\alpha_1 - \alpha) \sin \theta_1 \sin \beta + (\beta_1 - \beta) \sin \theta_2 \sin \alpha}{\sin (\alpha + \beta)}$$

$$\geqslant 1 - \frac{[\alpha_1 + \beta_1 - (\alpha + \beta)] (\sin \alpha + \sin \beta)}{\sin (\alpha + \beta)}.$$

Furthermore, we obtain

$$\frac{\sin \alpha + \sin \beta}{\sin (\alpha + \beta)} = \frac{\cos \frac{\alpha - \beta}{2}}{\cos \frac{\alpha + \beta}{2}} \leqslant \frac{1}{\cos \frac{\gamma}{2}},$$

$$\alpha_1 + \beta_1 - (\alpha + \beta) \leqslant s(K) - s(L),$$

which yields

$$\cos \rho \geqslant 1 - \frac{s(K) - s(L)}{\cos \frac{\gamma}{2}}.$$

It should be remarked that if $\cos \rho \geqslant 1 - \alpha$ and $0 \leqslant \rho \leqslant \pi$, then $\rho \leqslant \pi \sqrt{\alpha/2}$. Indeed, we have $1 - \cos \rho \leqslant \alpha$, or $2 \sin^2 \rho/2 \leqslant \alpha$, and, hence, $\sin \rho/2 \leqslant \sqrt{\alpha/2}$. At $0 \leqslant x \leqslant \pi/2$, as is known, the following inequality holds: $2x/\pi \leqslant \sin x$. Therefore, $\rho/2 \leqslant \sqrt{\alpha/2}$, which is the required proof. Thus, we get

$$\rho \leqslant \pi \sqrt{\frac{s(K) - s(L)}{2 \cos \frac{\gamma}{2}}}$$

and the lemma is proved.

Proof of Theorem 5.9.1. According to the condition, the curve K has no points of return. Let $\alpha_0 < \pi$ be the greatest among the turns at the angular points of the curve K and let $\alpha = (\alpha_0 + \pi)/2$. Let us choose $\varepsilon > 0$ and, using it, find $\delta > 0$, such that at $x < \delta$

$$\pi \sqrt{\frac{x}{\cos \frac{\alpha}{2}}} < \varepsilon.$$

Let us find on the curve K a sequence of smooth points $Y_0 < Y_1 < \cdots < Y_n$, such that the turns of the arcs into which they subdivide the curve is less than α and, if t_i is a unit vector of the tangent at the point Y_i, then

$$\sum_{i=0}^{m-1} (t_1, \widehat{} t_{i+1}) > \kappa(K) - \delta.$$

The possibility of such a choice is ensured by Theorem 5.1.3.

Plotting the vectors t_i from the centre of the unit sphere Ω^{n-1} and successively connecting their ends with the arcs of large circumferences, we obtain a spherical polygonal line Q inscribed into the indicatrix of the tangents of the curve K. In view of Theorem 5.2.2, the vertices Q subdivide S into arcs of the length less than α, and $s(S) - s(Q) < \delta$. In line with Lemma 5.9.3, we have

$$\phi(S, Q) \leqslant \pi \sqrt{\frac{s(S)-s(Q)}{2 \cos \frac{\alpha}{2}}} < \varepsilon.$$

Let us now find on the curve K_m a sequence of the points $Y_0^m < Y_1^m < \cdots < Y_n^m$, such that at $m \to \infty$ the arcs $[Y_i^m Y_{i+1}^m]$ converge to the arcs $[Y_i Y_{i+1}]$ of the curve K. Let t_i^m be a unit vector of the right-hand tangent at the point Y_i^m of the curve K_m. According to Lemma 5.9.2, $t_i^m \to t_i$ at $m \to \infty$. Let us plot the vectors t_i^m from the centre of the unit sphere Ω^{n-1}. Connecting their ends successively with the arcs of a large circle, we obtain a spherical polygonal line Q_m, inscribed into the indicatrix of the tangents S_m of the curve K_m. At $m \to \infty$, by the condition, $s(S_m) \to s(S)$, and the vertices of the polygonal line Q_m converge to the corresponding vertices of the polygonal line Q. In view of Lemma 5.9.1 and Theorem 5.2.2, at $m \to \infty$ the lengths of the arcs into which the vertices of Q_m divide S_m, tend to the lengths of the corresponding arcs of S. Therefore, there is such an $m_0 > 0$ that at $m > m_0$ the vertices of Q_m divide S_m into the arcs which are less than α, the difference $s(S_m) - s(Q_m)$ is less than δ, and the distance between the polygonal lines Q_m and Q is less than ε. In this case, by Lemma 5.9.3, we get

$$\rho(S_m, Q_m) \leqslant \pi \sqrt{[s(S_m) - s(Q_m)]/2 \cos \frac{\alpha}{2}} < \varepsilon$$

and

$$\rho(S_m, S) \leqslant \rho(S_m, Q_m) + \rho(Q_m, Q) + \rho(S, S) < 3\varepsilon.$$

Due to the arbitrariness of $\varepsilon > 0$, the theorem is proved.

5.10. Turn of a Plane Curve

5.10.1. In classical differential geometry a curvature as a function of a point of a plane curve is a complete system of the curve's invariants. It would therefore be natural to assume that in our more general theory a complete system of the invariants of a plane curve is set by defining a turn as a function of the arc of a curve. In this case the notion of a turn should be somewhat modified by introducing the notion of a turn with a sign or, as it will be henceforth called, a rotation of a plane curve.

Plane curves of finite turn were first considered by Radon [23] who called them curves with limited rotation in connection with certain problems of the theory of potentials.

The notations introduced below are especially designed for the case of angles on an oriented plane. The angles between vectors in a space will be denoted as above.

Let us consider curves in E^2, under the assumption that a certain orientation of E^2 is given.

Let a and b be two non-zero vectors on the plane, and let φ be an angle at which the vector a must be turned to become coincident in direction with the vector b.

Let us set $(a\frown b) = \pm \varphi$, where the positive sign is taken when the rotation considered takes place in the positive direction of the plane, and the negative sign in the opposite case. The quantity $(a\frown b)$ will be called an angle between a and b. The angle $(a\frown b)$ is defined only to the accuracy of an integer multiple to 2π, in which case $(a\frown b) = -(b\frown a)$. The value of the angle $(a\frown b)$ lying in the semi-open segment $(-\pi, \pi]$ will be called its main value and denoted by $[a\frown b]$. The following equality is valid:

$$(a\frown b) + (b\frown c) = (a\frown c). \tag{29}$$

Let $t(u)$, $a \leqslant u \leqslant b$, be a continuous vector-function, the values of which are vectors in E^2, such that $t(u) \neq 0$ for all u. In this case there exists, and only one, continuous function $\varphi : [a, b] \to \mathbb{R}$, such that $\varphi(a) = 0$ and $\varphi(u) = (t(a)\frown t(u))$ for all $u \in [a, b]$. Such a function φ can be constructed by dividing the segment $[a, b]$ into sufficiently small partial segments by the points $u_0 = a < u_1 < \cdots < u_m = b$ and setting

$$\varphi(u) = \sum_{k=1}^{i} [t(u_{k-1})\frown t(u_k)] + [t(u_i)\frown t(u)]$$

at $u_i \leqslant u \leqslant u_{i+1}$. If φ_1 and φ_2 are two functions, such that $\varphi_1(u) = (t(a)\frown t(u))$ and $\varphi_2(u) = (t(a)\frown t(u))$ for all $u \in [a, b]$, then the difference $\varphi_1 - \varphi_2$ assumes only the values equal to an integer multiple to 2π. Therefore, if φ_1 and φ_2 are continuous and $\varphi_1(a) = \varphi_2(a) = 0$, then $\varphi_1(u) - \varphi_2(u) \equiv 0$ and φ_1 and φ_2 coincide. This shows the uniqueness of the function φ which possesses all the required properties.

Let K be an arbitrary one-sidedly smooth curve on a plane, X be an internal point of the curve K. Let us set

$$\tilde{\kappa}(X) = [t_l(X)\frown t_r(X)].$$

Let us call the quantity $\tilde{\kappa}(X)$ a rotation of the curve K at the point X. It should in particular be noted that a rotation at the point of return of the curve K proves to be equal to π.

Let K be a one-sidedly smooth curve in E^2, and let $X \mapsto \tau(X)$ be a tangent correspondence of the curve K. For any point X $\tau(X)$ is a shortest arc of the circumference Ω^1 with the ends at the points $t_l(X)$ and $t_r(X)$. In the case when X is a point of return, the arc $\tau(X)$ is ambiguously determined. Let us eliminate the ambiguity arising by the following assumption. If $t_l(X) = -t_r(X)$ then out of two arcs with the ends $t_l(X)$ and $t_r(X)$ of the circumference Ω^1 let us choose as $\tau(X)$ the one for which at the motion of the point Z along it in the direction from $t_l(X)$ to $t_r(X)$, the radius OZ rotates in E^2 in the positive direction.

Let K be a one-sidedly smooth curve on a plane, and let Q be its indicatrix of the tangents. Let Q be constructed in accordance with the assumptions made above. Let $t(u)$, $a \leqslant u \leqslant b$ be a parametrization of the curve Q, and let $\varphi(u)$ be an angular function for the vector-function t. The quantity $\varphi(b)$ is obviously independent of the choice of the parametrization of the curve K. The number $\varphi(b)$ is called a rotation of the curve K and denoted henceforth through the symbol $\tilde{\kappa}(K)$.

A rotation of a plane polygonal line is obviously equal to the sum of rotations at all vertices.

A relation between the rotation of a one-sidedly smooth curve and its turn is established in the following theorem.

THEOREM 5.10.1. *Let K be an arbitrary one-sidedly smooth plane curve, $\{X(u)$, $a \leqslant u \leqslant b\}$ be an arbitrary parametrization of the curve K, $\tilde{\kappa}(u)$ be a rotation of the arc $[X(a)X(b)]$ of the curve K. In this case a variation of the function $\tilde{\kappa}(u)$ is equal to a turn of the curve K.*

Proof. Let $\xi = (t_0, t_1, \ldots, t_m)$, where $t_0 < t_1 < \cdots < t_m$ is an arbitrary chain composed of the tangents of the curve K,

$$\kappa^*(\xi) = \sum_{i=1}^{m} |[t_{t-1} \frown t_i]|.$$

From the chain ξ let us construct a new chain ξ_1 in the following way. If t_i is an intermediate tangent at any point X of the curve K, then let us join to the chain the left-hand and the right-hand tangents at this point. Having carried out this procedure for all $i = 0, 1, 2, \ldots, m$, we get as a result a new chain which will be denoted by ξ_1. It is obvious that $\kappa^*(\xi_1) \geqslant \kappa^*(\xi)$. It should be remarked that if the chain ξ_1 contains both a left- and a right-hand tangent at any point $X \in K$, then if we exclude from ξ_1 the intermediate tangent at the point X, the value of $\kappa^*(\xi_1)$ will remain the same. Having excluded all the intermediate tangents from ξ_1, we obtain a chain ξ_2 for which $\kappa^*(\xi_2) = \kappa^*(\xi_1) \geqslant \kappa^*(\xi)$, and which consists only of one-sided tangents of the curve. Let $\xi_a = \{t'_0, t'_1, \ldots, t'_r\}$. Let us arbitrarily set $\varepsilon > 0$. Let us also assume that t'_i is a right-hand tangent at the point X of the curve K. Then at $Y \to X$ on the right $t_l(Y) \to t'_i$. We can therefore conclude that by way of substituting each right-hand tangent contained in ξ_2 by its left-hand tangent at a neighbouring point, we finally will get a chain ξ_0 for which $\kappa^*(\xi_0) > \kappa^*(\xi_2) - \varepsilon \geqslant \kappa^*(\xi) - \varepsilon$, and which is composed of the left-hand tangents. Let us assume that the chain ξ_0 is composed of left-hand tangents at the points $X(u_i)$, where $a \leqslant u_0 < u_1 < \cdots < u_r \leqslant b$ of the curve K. If we set $t(u) = t_l(X(u))$, then we have:

$$\kappa^*(\xi_0) = \sum_{i=1}^{r} |[t(u_{i-1}), t(u_i)]| \leqslant \sum_{i=1}^{r} |\tilde{\kappa}(u_i) - \tilde{\kappa}(u_{i-1})|$$
$$\leqslant \bigvee_a^b \tilde{\kappa}(u)$$

From this we get:

$$\kappa^*(\xi) \leqslant \varepsilon + \bigvee_a^b \tilde{\kappa}(u)$$

and, since $\varepsilon > 0$ and the chain ξ are arbitrarily chosen, then

$$\kappa(K) \leqslant \bigvee_a^b \tilde{\kappa}(u). \qquad (30)$$

In order to prove the opposite inequality, let us first assume that any two tangents of the curve K comprise an angle which is less than $\pi/2$. Let $a = u_0 < u_1 < \cdots < u_m = b$ be an arbitrary chain of the points of the segment $[a, b]$. For any i we have:

$$\tilde{\kappa}(u_i) - \tilde{\kappa}(u_{i-1}) = [t(u_{i-1}), t(u_i)].$$

(Here we set $t(u) = t_l[X(u)].$) From this we get:

$$\sum_{i=1}^{m} |\tilde{\kappa}(u_i) - \tilde{\kappa}(u_{i-1})| = \sum_{i=1}^{m} |[t(u_{i-1}), t(u_i)]| \leqslant \kappa(K).$$

Since the chain $u_0 < u_1 < \cdots < u_m$ of the points of the interval $[a, b]$ is arbitrarily chosen, then we have established that

$$\bigvee_a^b \tilde{\kappa}(u) \leqslant \kappa(K) \qquad (31)$$

and for the case under discussion the equality

$$\bigvee_a^b \tilde{\kappa}(u) = \kappa(K)$$

is proved.

In a general case the validity of equality (31) can easily be proved by dividing the curve K into arcs on each of which any two tangents comprise an angle less than $\pi/2$.

The theorem is proved.

In connection with the theorem proved above it should be remarked that in some papers (for instance [5, 21, 22]) considering curves in a space, the term 'variation of a turn' was used instead of the term 'turn of a curve'. This is attributed to the fact that in the monograph by A. D. Alexandrov *Internal Geometry of Convex Surfaces* [3], to which the cited papers pertain and where the fundamentals of the theory of curves on a convex surface were first presented, the term 'turn' referred to what is termed in this book a 'rotation' (for the case of plane curves).

LEMMA 5.10.1. *Let* $t_0(u)$, $a \leqslant u \leqslant b$, *and* $t_m(u)$, $a \leqslant u \leqslant b$, *be continuous vec-tor-functions with the values on the plane, and let* $|t_m(u)| > 0$ *at all* $m = 1$, $2, \ldots$ *In this case, if at* $m \to \infty$ $t_m(u) \to t_0(u)$ *uniformly in* $[a, b]$, *then the angular functions* $\varphi_m(\bar{u})$ *converge uniformly in* $[a, b]$ *to an angular func-tion for* $t_0(u)$.

Proof. Since at $m \to \infty$ $t_m(u) \to t(u)$ uniformly in $[a, b]$ and $|t_0(u)| > 0$ at $u \in [a, b]$, then at $m > m_0$ $|(t_m(u), \widehat{} t_m(u)) - (t_0(u), \widehat{} t_0(u)) - 2\pi N_1| < \varepsilon$, where N_1 is an integer. If we choose $\varepsilon < \pi$, then $|\varphi_m(u) - \varphi_0(u) - 2\pi N| < \varepsilon$ $(a \leqslant u \leqslant b)$, where N is an integer (possibly, u-dependent).

The difference $\varphi_m(u) - \varphi_0(u)$ is continuous in the segment $[a, b]$. Each of its values lies within one of the segments $(2\pi N - \pi, 2\pi N + \pi)$, where N is an inte-ger. Since $\varphi_m(a) - \varphi_0(a) = 0$, then $\varphi_m(u) - \varphi_0(u)$ are contained in the segments $(-\pi, \pi)$ and hence $|\varphi_m(u) - \varphi_0(u)| < \varepsilon$ within the whole segment $[a, b]$. The lemma is thus proved.

THEOREM 5.10.2. *Let* $K, K_1, K_2, \ldots, K_m, \ldots$ *be plane curves, in which case* $\kappa(K) < \infty$, *the curve* K *has no points of return, at* $m \to \infty$ *the curves* K_m *con-verge to the curve* K, *and* $\kappa(K_m) \to \kappa(K)$. *Then at* $m \to \infty$ $\tilde{\kappa}(K_m) \to \tilde{\kappa}(K)$.

Proof. In conformity with Theorem 5.9.1, at $m \to \infty$ the indicatrices of the tangents of the curves K_m converge to the indicatrix of the tangents of the curve K. The statement of the Theorem results directly herefrom, in line with Lemma 5.10.1.

The theorem is proved.

5.10.2. A rotation of a plane curve also allows a certain integro-geometri-cal presentation.

Let $K = [AB]$ be an arbitrary one-sidedly smooth curve in E^2. A point X, $A < X < B$, of the curve K is called a convex point of K or a point of convexity, if there exists a neighbourhood U of the point X on the curve K and a straight line l passing through the point X, such that U lies entirely to one side of l and any straight line colinear to l (the straight line l included) intersects the arc U at no more than two points.

Let X be any convex point of the curve K. Let us define a certain number $\sigma(X)$ which will be termed an index of the point X. If X is an angular point, then we set $\sigma(X) = \text{sign } \kappa_K(X)$, where $\kappa_K(X)$ is the rotation of the curve K at the point X. Hence, at an angular point $\sigma(X) = 1$, if $\kappa_K(X) > 0$, $\sigma(X) = -1$, if $\kappa_K(X) < 0$. Let X be a smooth point, $t(X)$ be a tangent at the point X, $\bar{t}(X)$ is a corresponding tangential unit vector. A certain neighbourhood U of the point X lies to one side of the curve K. Let us denote by $n(X)$ a unit vector ortho-gonal to $t(X)$ and directed to the side of the straight line $t(X)$, where the small neighbourhood U of the point X on the curve K lies. We set $\sigma(X) = 1$, if the pair $(t(X), n(X))$ is right-hand, and $\sigma(X) = -1$ if this pair is left-hand.

Let K be an arbitrary curve of a finite turn on the plane E^2. Let us choose a direction p on the plane of K, such that $\kappa(K_p) < \infty$ and p is not orthogonal to any straight arc of the curve K. Such are almost all directions p in the plane E^2. The curve K_p is a polygonal line of a finite number of links. Let $Y_1, Y_2, \ldots, Y_{m-1}$ be successive points of return of the curve K_p, let Y_0 be its beginning, and let Y_m be its end. In this case on the curve K we can find a sequence of points $X_0, X_1, \ldots, X_m$, such that the arc $[X_{i-1}X_i]$ is mapped onto the arc $[Y_{i-1}Y_i]$ in a one-to-one way. The points $X_1, X_2, \ldots, X_{m-1}$ are the points of convexity of the curve K. Let us assume

$$(\kappa)(K_p) = \sum_{i=1}^{m-1} \pi\sigma_K(X_i).$$

Therefore, on the set of one-dimensional directions Ω^1 we have a function $p \mapsto (\kappa)(K_p)$ which is defined almost everywhere in Ω^1 and

$$|(\kappa)(K_p)| \leqslant \kappa(K_p)$$

for all p for which the quantity $(\kappa)(K_p)$ is defined.

THEOREM 5.10.3. *For any curve of finite turn on the plane E^2 the function $p \mapsto (\kappa)(K_p)$ is integrable with respect to the invariant measure μ in G_1^2 and*

$$\int_{G_1^2} (\kappa) (K_p) \, d\mu(p) = \tilde{\kappa}(K).$$

Proof. Let us assume that the curve K is polygonal. A general case will, evidently, be reduced to the case when the polygonal line K consists of two links formed by the vectors a and b. In this case

$$(\kappa)(K_p) = \sigma(X)(a_p, \frown b_p)$$

where X is the only vertex of the polygonal line K. As a result of integration, we get:

$$\int_{\Omega^1} (\kappa)(K_p) \, d\mu(p) = \sigma(X)|[a_p, \frown b_p]| = [a_p, b_p] = \tilde{\kappa}(K),$$

and for this case the theorem is proved.

A general case can be derived from the one considered by way of a limiting transition. Let us first assume that the curve K has no points of return. Let us construct a sequence of polygonal lines L_m, $m = 1, 2, \ldots$, inscribed into the curve K, such that $\lambda(L_m) \to 0$ at $m \to \infty$. At every m for almost all one-dimensional directions p we have:

$$|(\kappa)(L_{m,p})| \leqslant \kappa(L_{m,p}) \leqslant \kappa(K_p)$$

so that the sequence of functions $p \mapsto (\kappa)(L_{m,p})$, $m = 1, 2, \ldots$, is majorized by the integrable function $p \mapsto \kappa(K_p)$. In order to prove the Theorem, it is therefore sufficient to prove that

$$(\kappa)(L_{m,p}) \to (\kappa)(K_p)$$

for almost all p.

Let E_1 be a set of all one-dimensional directions, each of which is orthogonal to at least one of the links of at least one of the polygonal lines L_m.

The set E_1 is not greater than countable. Let E_2 be a set of those p for which $\kappa(K_p) = \infty$. Let us set $E = E_1 \cup E_2$. It is obvious that E is a set of zero measure in G_1^2. Let us arbitrarily choose $p \notin E$. In this case $\kappa(K_p) < \infty$. The straight line p is orthogonal to no straight line containing a whole arc of the curve K, since any such straight line contains a link of the polygonal line L_m for sufficiently large m, and $p \notin E_1$. Let Y_0 be the beginning and Y_r the end of the curve K_p, and let $Y_1 < Y_2 < \cdots < Y_{r-1}$ be its points of return. Let us denote by $X_0, X_1, \ldots, X_r$ the points of the curve K which are projected into $Y_0, Y_1, \ldots, Y_r$, correspondingly. The arc $[X_{i-1}X_i]$ is projected onto the arc $[Y_{i-1}Y_i]$ in a one-to-one way. Let δ be the least of the diameters of the arcs $[X_{i-1}X_i]$. Let m_0 be such that at $m \geqslant m_0$ the vertices of the polygonal line L_m divide the curve K into arcs the diameters of which are less than $\delta/4$. In this case on each of the arcs $[X_{i-1}X_i]$ there lie at least three vertices of the polygonal line L.

Let l_i be a straight line orthogonal to p and passing through the point X_i, where $1 \leqslant i \leqslant r-1$. Let us denote by $X_i^{(m)}$ the vertex of the polygonal line L_m the nearest to X_i, lying on the arc $[X_{i-1}X_{i+1}]$. Let $X_0^{(m)}$ be the beginning and $X_r^{(m)}$ the end of the polygonal line L_m.

Let us construct an additional affinity system of coordinates with the origin at the point X_i. As an axis Ox_1, let us take a straight line l_i directed in such a way that the basis vector u of this axis coincide with the vector $t(X_i)$, if X_i is a smooth point, while in the case when X_i is an angular point, the presentation $u = \alpha t_l(X_i) + \beta t_r(X_i)$, where $\alpha \geqslant 0$, $\beta \geqslant 0$ be allowed. The axis Ox_2 is chosen in such a way that it lies in the pair of vertical angles formed by the tangents at the point X_i which does not contain l_i, while the direction of the axis Ox_2 is chosen in such a way that the basis vectors of the coordinate system form a right-hand pair of the vectors in E^2. A certain neighbourhood U_i of the point X_i on the curve in the constructed system of coordinates is defined by the equation $u_2 = f(u_1)$, $|u_1| \leqslant \delta$, where $f(0) = 0$. If $\sigma(X_i) = 1$, then $f(u_1) \geqslant 0$, while in the case when $\sigma(X_i) = -1$ $f(u_1) \leqslant 0$. At sufficiently large m the point X_i^m lies inside the arc U_i and, as is seen, $\sigma(X_i^{(m)}) = 1$ in the case when $f(u_1) \geqslant 0$ and $\sigma(X_i^{(m)}) = -1$, if $f(u_1) \leqslant 0$. We, therefore, come to the conclusion that, for sufficiently large m, $\sigma(X_i^{(m)}) = \sigma_K(X_i)$. It follows from here that, for sufficiently large m, $m \geqslant m_1 \geqslant m_0$,

$$(\kappa)(L_{m,p}) = (\kappa)(K_p).$$

We thus have established that

$$(\kappa)(L_{m,p}) \longrightarrow (\kappa)(K_p)$$

at $m \longrightarrow \infty$ and the theorem is hence proved for the case when the curve K has no points of return. An arbitrary curve of a finite turn can be divided into a finite number of arcs which have no points of return. Applying the formula proved to each of these arcs after their summation we obtain the required result.

The theorem is proved.

5.10.3. Let us consider natural parametrizations of a curve. Here we are going to prove that the length of an arc of a plane curve and its rotation, given as functions of the arc, unambiguously define the curve to the accuracy of the motion. Besides, here we are going to establish the degree of arbitrariness to which the rotation and the arc length as a parameter function themselves can be set. The results obtained in the present section are valid for arbitrary one-sidedly smooth plane curves. The limitedness of a turn is not required.

Let K be an arbitrary one-sidedly smooth plane curve, $X(u)$, $t(u)$, $a \leqslant u \leqslant b$, be the coordinated parametrizations of the curve K and its indicatrices of the tangents. Let $s(u)$ be the length of the arc $[X(a)X(u)]$ of the curve K, and let $\tilde{\kappa}(u)$ be an angular function of the vector-function $t(u)$. A pair of the functions $(s(u), \tilde{\kappa}(u))$ is called a natural parametrization of the curve K.

Let us note the following properties of natural parametrization. On the plane of the variables (s, x) let us construct a curve R with the parametric equations:

$$s = s(u), \qquad x = \tilde{\kappa}(u).$$

To each angular point of the curve K there corresponds an arc of the curve R, which is parallel to the axis Ox. Conversely, to each of such arcs there corresponds an angular point of the curve K. Oscillations of the function $\tilde{\kappa}(u)$ within this arc are contained in a semi-open interval $(-\pi, \pi]$.

The curve R is termed a natural indicatrix of the curve K. Below we shall establish that R is independent of the choice of the initial parametrizations $(X(u), t(u))$ of the curve K and its indicatrix of the tangents.

The notion of natural parametrization is an evident generalization of the notion of natural equation of a curve in differential geometry.

THEOREM 5.10.4. *Any two natural parametrizations of one and the same plane one-sidedly smooth curve K are equivalent, and each pair of the functions which is equivalent to the natural parametrization of the curve K, is its natural parametrization.*

Proof. Let $[s_1(u), \tilde{\kappa}_1(u)]$, $a \leqslant u \leqslant b$, $[s_2(v), \tilde{\kappa}_2(v)]$, $c \leqslant v \leqslant d$, be two natural parametrizations of the curve K, $X_1(u)$, $t_1(u)$, $a \leqslant u \leqslant b$, and $X_2(v)$, $t_2(v)$, $c \leqslant v \leqslant d$, be the corresponding coordinated parametrizations of the curve K and its indicatrices of the tangents. In this case, in line with what has been proved in Section 1 (Theorem 1.1.1), there exists a non-decreasing function $\psi(u)$, $a \leqslant u \leqslant b$, such that $\psi(a) = c$, $\psi(b) = d$, $X_1(u) = X_2[\psi(u)]$, $t_1(u) = t_2[\psi(u)]$ at all $u \in [a, b]$. It is easily seen also that for this function $\psi(u)$ we also have $s_1(u) = s_2[\psi(u)]$ and $\tilde{\kappa}_1(u) = \tilde{\kappa}_2[\psi(u)]$.

If $(s_1(u), \tilde{\kappa}_1(u))$, $a \leqslant u \leqslant b$, is the natural parametrization of the curve K, and a pair of functions $(s_2(v), \tilde{\kappa}(v))$, $c \leqslant v \leqslant d$, is equivalent to it, then there can be found a non-decreasing function $\psi(v)$, such that $\psi(c) = a$, $\psi(d) = b$ and $s_2(v) = s_1[\psi(v)])$, $\tilde{\kappa}_2(u) = \tilde{\kappa}_1[\psi(u)]$. Let $X_1(u)$, $t_1(u)$ be the coordinated parametrizations of the curve K and its indicatrices of the tangents corresponding to the natural parametrization $[s_1(u), \tilde{\kappa}_1(u)]$. Then, as can be easily seen, the functions $X_2(v) = X_1[\psi(v)]$ and $t_2(v) = t_1[\psi(v)]$ are continuous and hence they are a pair of coordinated parametrizations of the curve K and its indicatrix of the tangents. The natural parametrization corresponding to this pair is, obviously, $[s_2(v), \tilde{\kappa}_2(v)]$.

The theorem is thus proved.

The following property of the natural parametrization $[s(u), \tilde{\kappa}(u)]$ of a plane curve should be recalled. The function $s(u)$ is non-decreasing and can have some portions of constancy. Each such portion corresponds to an angular point of the curve, and the function $\tilde{\kappa}(u)$ is monotonous in it. In this case an increment of the function $\tilde{\kappa}(u)$ within this portion is always contained in the interval $(-\pi, \pi]$, since it is equal to the rotation at the corresponding point of the curve. The following theorem demonstrates that the natural parametrization of a curve is completely defined by these properties.

THEOREM 5.10.5. *Let* $[s(u), \tilde{\kappa}(u)]$, $a \leqslant u \leqslant b$, *be an arbitrary pair of continuous functions, where the function* $s(u)$ *is non-decreasing and* $s(u)$ *is not constant in any of the intervals of the type* $[a, a+\delta]$ *or* $[b-\delta, b]$, *where* $\delta > 0$, *while the function* $\tilde{\kappa}(u)$ *is monotonous in any portion of the function* $s(u)$ *constancy, its increment in any of such a portion always lying in the interval* $(-\pi, \pi]$ *and* $s(a) = \kappa(a) = 0$. *In this case there exists a plane curve* K, *such that the pair* $(s(u), \tilde{\kappa}(u))$ *is its natural parametrization.*

If the natural parametrization of two plane curves K_1 *and* K_2 *coincides, then the curves* K_1 *and* K_2 *can be superposed by a motion of the plane preserving the orientation.*

Proof. Let $s(u)$ and $\tilde{\kappa}(u)$, $a \leqslant u \leqslant b$, be the functions obeying the conditions of the theorem. On the plane let us set a Cartesian orthogonal system of coordinates and let $t(u)$ be a unit vector on the plane, such that $(t(u)\widehat{}e_1) = \tilde{\kappa}(u)$, where e is the basis vector of the axis Ox. This condition defines the vector-function $t(u)$ in a unique way. Let us set

$$X(u) = \int_a^u t(v)\, ds(v).$$

(The integral here is understood in the Stieltjes sense). Thus, the vector-function $X(u)$, $a \leqslant u \leqslant b$, is determined. Let K be a plane curve, the parametrization of which is the vector-function $X(u)$. Let us prove that this curve is the one sought.

Let us first establish that for any $u_1, u_2 \in [a, b]$, $u_1 < u_2$, the difference $s(u_2) - s(u_1)$ is equal to the length of the arc $[X(u_1)X(u_2)]$ of the curve K. Let us arbitrarily set $u_1, u_2 \in [a, b]$, $u_1 < u_2$. Let us arbitrarily choose a number ε, such that $0 < \varepsilon < \pi/2$ and find a sequence of the values $\alpha_0 = u_1 < \alpha_1 < \cdots < \alpha_{m-1} < \alpha_m = u_2$, such that at every $i = 1, 2, \ldots, m$ for all $t \in [\alpha_{i-1}, \alpha_i]$ the vector $t(u)$ makes with $t(\alpha_i)$ an angle less than ε. At every i we have:

$$|X(\alpha_i) - X(\alpha_{i-1})| = |\int_{\alpha_{i-1}}^{\alpha_i} t(u) \, ds(u)|$$

$$\geqslant \int_{\alpha_{i-1}}^{\alpha_i} <t(u), t(\alpha_i)> \, ds(u) \geqslant \int_{\alpha_{i-1}}^{\alpha_i} \cos \varepsilon \, ds(u)$$

$$= \cos \varepsilon \, [s(\alpha_i) - s(\alpha_{i-1})].$$

Performing a summation with respect to i, we get:

$$s([X(u_1)X(u_2)]) \geqslant \sum_{i=1}^{m} |X(\alpha_i) - X(\alpha_{i-1})|$$

$$\geqslant \cos \varepsilon \, [s(u_2) - s(u_1)].$$

Since $\varepsilon > 0$ is arbitrary, this yields that the length of the arc $[X9u_1)X(u_2)]$ of the curve K is not less than $s(u_2) - s(u_1)$. On the other hand, at every i we have:

$$|X(\alpha_i) - X(\alpha_{i-1})| = |\int_{\alpha_{i-1}}^{\alpha_i} t(u) \, ds(u)|$$

$$\leqslant \int_{\alpha_{i-1}}^{\alpha_i} |t(u)| \, ds(u) = \int_{\alpha_{i-1}}^{\alpha_i} ds(u) = s(\alpha_i - s(\alpha_{i-1}).$$

Summing with respect to i, we obtain:

$$\sum_{i=1}^{m} |X(\alpha_i) - X(\alpha_{i-1})| \leqslant s(u_2) - s(u_1).$$

Due to the arbitrariness of the sequence of points $\alpha_0 = u_1 < \alpha_1 < \cdots < a_m = u_2$, it affords

$$s([X(u_1)X(u_2)]) \leqslant s(u_2) - s(u_1).$$

Comparing the results, we come to the conclusion that

$$s([X(u_1)X(u_2)]) = s(u_2) - s(u_1).$$

Let us now prove that the curve K is one-sidedly smooth and $t(u)$ is a tangential vector of the curve K at the point $X(u)$. Let us arbitrarily choose a point X of the curve K. Let $[\alpha, \beta]$ be a portion of constancy of the function $X(u)$ corresponding to the given point X. (It is obvious that the function $X(u)$ is constant within a certain interval $[\alpha, \beta] \subset [a, b]$ if and only if the function $s(u)$ is constant in it.) In line with the condition of the theorem, either $\alpha > a$ or $\beta < b$. Let us consider the case when $\beta < b$. Let us arbitrarily set $\varepsilon > 0$ and, using it, find $\delta > 0$, such that for any $u \in [\beta, \beta+\delta]$ the inequality $|t(u) - t(\beta)| < \varepsilon/2$ holds. The function s is not constant in the interval $[\beta, \beta+\delta]$. Let us arbitrarily choose the values $u_1, u_2 \in (b-\delta, b)$, such that $u_1 < u_2$ and $s(u_2) - s(u_1) > 0$. In this case we get:

$$|X(u_2) - X(u_1) - t(\beta)[s(u_2) - s(u_1)]|$$

$$= |\int_{u_1}^{u_2} [t(u) - t(\beta)] \, ds(u)| \leqslant \int_{u_1}^{u_2} |t(u) - t(\beta)| \, ds(u)$$

$$\leqslant \frac{\varepsilon}{2} [s(u_2) - s(u_1)].$$

From this we obtain:

$$\left| \frac{X(u_2) - X(u_1)}{s(u_2) - s(u_1)} - t(\beta) \right| \leqslant \frac{\varepsilon}{2} < \varepsilon.$$

Due to the arbitrariness of $\varepsilon > 0$, it proves that the secants $l[X(u_1, X(u_2)]$ at $u_1 \to \beta + 0$, $u_2 \to \beta + 0$ converge to the straight line l with the directing vector $t(\beta)$ and, hence, $t(\beta)$ is a right-hand tangential unit vector at the point $X(t)$ of the curve K, and K has a tangent in the strong sense at this point. In an analogous way we prove (under the assumption that $\alpha > a$) that $t(\alpha)$ is a left-hand tangential unit vector at the point $X(t)$ of the curve K, in which case K has a left tangent in the strong sense at this point. At $\alpha < u < \beta$ the vector $t(u)$ makes up with $t(\alpha)$ an angle equal to $\tilde{\kappa}(u) - \tilde{\kappa}(\alpha)$ and lying between $-\pi$ and π, in which case the value of the angle equal to $-\pi$ is excluded. Therefore, for such u, $t(u)$ is an intermediate tangential unit vector at the points $X(u)$.

It obviously follows from what has been proved above that the constructed curve is the one sought.

Let us now assume that the curves K_1 and K_2 are such that their natural parametrization coincide. Let us superpose the beginnings of the curves K_1 and K_2 by way of a displacement, and their tangents at the beginning - by way of rotating around their general beginning. Let $[s(t), \tilde{\kappa}(t)]$, $a \leqslant t \leqslant b$, be a general natural parametrization of the curves K_1 and K_2; $X(s)$, $0 \leqslant s \leqslant l_1$ and $Y(s)$, $0 \leqslant s \leqslant l_2$, be the parametrizations of K_1 and K_2, respectively, where the parameter s is the length of an arc. In this case $l_1 = l_2 = s(b) = l$. Let us arbitrarily choose $s \in [0, l)$. Let u_0 be the utmost right-hand value, such that $s(u) = s$; $t_1(u)$ and $t_2(u)$ be the right-hand tangential unit vectors of the curves K_1 and K_2 at the points corresponding to the given value of u. In this case $[t_1(u), t_2(a)] = [t_2(u), t_2(a)] = \tilde{\kappa}(u)$ and, hence, the vectors $t_1(u)$ and $t_2(u)$ coincide. But, on the other hand, $t_1(u) = X_r'(s)$, $t_2(u) = Y_r'(s)$. Therefore, for all $s \in [0, l]$ $X_r'(s) = Y_r'(s)$. This affords $X(s) \equiv Y(s)$ and hence the curves K_1 and K_2 coincide.

The theorem is proved.

CHAPTER VI

Theory of a Turn for Curves on an n-Dimensional Sphere

6.1. Auxiliary Results

6.1.1. In the space E^{n+1} let us arbitrarily fix an origin O. The symbol Ω^n will henceforth denote an n-dimensional sphere in the space E^{n+1} of radius equal to 1 and the centre O,

$$\Omega^n = \{X \in E^n \,|\, |OX| = 1\}$$

An arbitrary point $x \in \Omega^n$ will be associated with the vector $Ox \in V^n$ which is a radius-vector of the point x with respect to the point O.

For $x, y \in \Omega^n$ let us set $\rho(x, y) = (x, \widehat{\ } y)$.

In line wiht the known properties of an angle, $\rho(x, y)$ is a metric on the sphere Ω^n. We have the inequalities

$$|x-y| \leqslant (x, \widehat{\ } y) \leqslant \frac{\pi}{2}|x-y|$$

which, in particular, afford that the topology defined by the metric ρ on the sphere Ω^n coincides with the natural topology induced on Ω^n by the surrounding space.

Let us consider a Grassmanian manifold G_k^{n+1} of k-dimensional directions in the space E^{n+1}. Let us identify an arbitrary element α of the manifold G_k^{n+1} with a k-dimensional plane in E^{n+1}, passing through the point O and belonging to the direction α.

Henceforth let us call a k-dimensional great sphere on Ω^n any set Γ which is a cross-section of the sphere Ω^n with an $(k+1)$-dimensional plane P passing through O of the sphere Ω^n. For $\alpha \in G_{k+1}^{n+1}$ let us denote the great sphere $\alpha \cap \Omega^n$ by the symbol $\gamma(\alpha)$. The term a 'great sphere' given with no indication of its dimensionality will henceforth denote an $(n-1)$-dimensional large sphere, unless stated otherwise.

Let $\alpha \in G_{k+1}^{n+1}$, $\beta = \nu(\alpha)$ be an $(n-k)$-dimensional direction quite orthogonal to α. In this case the $(n-k-1)$-dimensional great sphere $\beta \cap \Omega^n = \gamma(\beta)$ will be called a sphere polar to the k-dimensional sphere $\gamma(\alpha)$ and denoted by the symbol $\sigma(\alpha)$.

The cross-section of Ω^n formed by a two-dimensional plane passing through its center will be called a great circumference. Let x, y be two arbitrary points of the sphere Ω^n. In this case if $0 < \rho(x, y) < \pi$, then there exists a great circumference Γ, and only one, passing through the points x and y, i.e.,

175

the cross-section of Ω^n formed by a two-dimensional plane passing through the points x, y and O. The circumference Γ is divided into two arcs by the points x and y. The length of one of these arcs is $\rho(x, y) = (x\frown y)$. Let us henceforth call this arc the shortest line on the sphere Ω^n connecting the points x and y. If $\rho(x, y) = \pi$, then x and y are diametrically opposing points of the sphere Ω^n and in this case there exists an infinite number of circumferences passing through x and y. Each of these circumferences is divided into two arcs of the length equal to π by the points x and y. These arcs will be called the shortest arcs, in this case, connecting x and y.

Let us also define the operation of the orthogonal projection for the sphere Ω^n.

Let us arbitrarily choose a $(k+1)$-dimensional plane $\alpha \in G_{k+1}^{n+1}$. Let $\gamma(\alpha) = \alpha \cap \Omega^n$ and $\sigma(\alpha)$ be an $(n-k-1)$-dimensional large sphere polar to the k-dimensional sphere $\gamma(\alpha)$. By definition, $\sigma(\alpha) = \Omega^n \cap \nu(\alpha)$, where $\nu(\alpha)$ is an $(n-k)$-dimensional plane quite orthogonal to α. The point $x \in \Omega^n$ belongs to $\sigma(\alpha)$ if and only if the vector x is orthogonal to any vector $y \in \alpha$. It means that $\sigma(\alpha)$ is a totality of all vectors $x \in \Omega^n$ for which $\rho(x, y) = \pi/2$ for any $y \in \Omega^n$.

For $x \in E^{n+1}$ let us denote by $\pi(x)$ an orthogonal projection of the vector x onto the plane α. If $x \notin \nu(\alpha)$, then the vector $y = \pi(x)$ is non-zero and the vector $h = x - y$ is orthogonal to any vector $z \in \alpha$. For any vector $z \neq 0$ lying in the plane α we have: $(x\frown z) \geqslant (x\frown y)$. Let $x \in \Omega^n$, in which case $x \notin \sigma(\alpha)$. In this case the vector $y = \pi(x) \neq 0$ and we set

$$\pi_\alpha(x) = \frac{\pi(x)}{|\pi(x)|}.$$

In a geometrical sense, the vector $\pi_\alpha(x)$ is a point of the intersection of the intersection of the sphere Ω^n and the ray Oy. For any $z \in \gamma(\alpha)$ we have: $\rho(x, z) = (x\frown z) \geqslant (x\frown y) = x\frown \pi_\alpha(x)) = \rho(x, \pi_\alpha(x))$. Therefore, $\pi_\alpha(x)$ is the nearest point to x on the sphere $\gamma(\alpha)$.

Let Γ be an arbitrary l-dimensional large sphere on Ω^n, where $l \leqslant k$, β is the plane of the sphere Γ. In this case if Γ does not intersect $\sigma(\alpha)$, then the plane β contains no vectors orthogonal to α and, hence, at the orthogonal projection the plane β is mapped into α in a one-to-one manner. The limitation of π onto β is a linear mapping. Let x and y be two arbitrary unit vectors lying in the plane β. Let us denote by V a plane angle limited by the rays Ox and Oy. In the case when $(x\frown y) < \pi$ V is a totality of all the vectors z of the kind $z = \lambda x + \mu y$, where $\lambda \geqslant 0$, $\mu \geqslant 0$, and if $(x\frown y) = \pi$, then V is constructed in the following way: we set an arbitrary two-dimensional plane containing the vectors x and y lying anywhere in the plane β. The straight line xOy divides this two-dimensional plane into two semi-planes and V is one of these semi-planes. The set $\pi(V)$ is also an angle. Therefore, the image $\pi_\alpha([xy])$ of the shortest line $[xy]$ is the shortest line in the plane α, connecting the points $\pi_\alpha(x)$ and $\pi_\alpha(y)$.

The curve K on the sphere Ω^n is called to be a spherical polygonal line if one can find a sequence $X_0 < X_1 < \cdots < X_m$ of the points of the curve K, where X_0 is the beginning and X_m is the end of the curve K, such that each of the arcs $[X_{i-1}X_i]$ is the shortest arc of a great circumference passing through the points X_{i-1} and X_i. The points X_i, $i = 0, 1, 2, \ldots, m$, are called the vertices of the spherical polygonal line K, the arcs $[X_{i-1}X_i]$ are called its links. Any spherical polygonal line is a one-sidedly smooth curve in E^{n+1} and, moreover, a curve of a finite turn in the space E^{n+1}. Let φ_i be a turn at the vertex X_i of the spherical polygonal line K, $1 \leqslant i \leqslant m-1$. Let us set

$$\varphi_i + \varphi_2 + \cdots + \varphi_{m-1} = \kappa_g(K).$$

The quantity $\kappa_g(K)$ is called a geodesic turn of the polygonal line K. It should be remarked that since the arc $[X_{i-1}X_i]$ is an arc of a great circumference, the radius of which is 1, then the turn of this arc is equal to its length. Therefore, for any spherical polygonal line K the following equality is valid:

$$\kappa(K) = s(K) + \kappa_g(K) \tag{1}$$

Here $\kappa(K)$ is the turn of K as a curve in the space E^{n+1}.

Let K be a curve on the spere Ω^n, let L be a spherical polygonal line, let $X_0 < X_1 < \cdots < X_m$ be the successive vertices of the polygonal line L, let X_0 be its beginning, and let X_m be its end. Let us say that the polygonal line L is inscribed into K, if its vertices belong to the curve K and are located on it in the same order as on the polygonal line L, i.e., if there exists a finite sequence $X_0' < X_1' < \cdots < X_m'$ of the points of the curve K such that at every i the point X_i' coincides with the point X_i spatially.

Let $\xi = X_0 < X_1 < \cdots < X_m$ be an arbitrary chain of the points of a spherical curve K. Connecting in a successive manner the points X_0 and X_1, X_1 and X_2, ..., X_{m-1} and X_m with the shortest arcs of great circumference, we obtain a certain spherical polygonal line L inscribed into the curve K. Let us denote the biggest of the diameters of the arc $[X_{i-1}X_i]$ of the curve K by the symbol $\lambda(\xi)$, and set $\lambda(L) = \lambda(\xi)$.

LEMMA 6.1.1. *For any spherical curve K its length $s(K)$ is equal to the upper boundary of the lengths of the spherical polygonal lines inscribed into K.*

Proof. Let K be an arbitrary curve on the sphere Ω^n. Let us denote by $s_g(K)$ the least upper boundary of the lengths of the spherical polygonal lines inscribed into K. The task is to prove that $s_g(K) = s(K)$. It should be remarked that the quantity $s_g(K)$ has the following sense: it is the length of the curve K in the metric space (Ω^n, ρ).

For any points $X, Y \in \Omega^n$ we obviously have:

$$|XY| \leqslant \rho(X, Y)$$

which yields that for any chain $\xi = \{X_0 < X_1 < \cdots < X_m\}$ of the points of the curve K the following inequality holds:

$$s(\xi) = \sum_{i=1}^{m} |X_{i-1}X_i| \leqslant \sum_{i=1}^{m} \rho(X_{i-1}X_i) \leqslant s_g(K)$$

From this we get

$$s(K) = \sup_{\xi} s(\xi) \leqslant s_g(K). \tag{2}$$

Let us arbitrarily set θ such that $0 < \theta < \pi$, and let X and Y be two points on the sphere Ω^n, such that $0 < \rho(X, Y) \leqslant \theta$. Let us set $\rho(X, Y) = \alpha$. In this case $|XY| = 2 \sin \frac{1}{2}\alpha$, $\rho(X, Y) = \alpha \leqslant 2 \tan \frac{1}{2}\alpha = (2 \sin \frac{1}{2}\alpha)/\cos \frac{1}{2}a = |XY|/\cos \frac{1}{2}\alpha$. From this we get: $|XY| \geqslant \rho(X, Y) \cos \frac{1}{2}\alpha \geqslant \rho(X, Y) \cos \frac{1}{2}\theta$.

Let L be an arbitrary spherical polygonal line inscribed into the curve K. Let us construct a polygonal line L_0 inscribed into K, the vertices of which divide K into the arcs with diameters not greater than θ, and let L' be a spherical polygonal line inscribed into K such that each of the polygonal lines L and L_0 is inscribed into L. Let us denote by $Y_0, Y_1, \ldots, Y_r$ the successive vertices of the polygonal line L'. In this case at every $i = 1, 2, \ldots, r$, $\rho(Y_{i-1}, Y_i) \leqslant \theta$. We have:

$$s(L) \leqslant s(L') = \sum_{i=1}^{m} \rho(Y_{i-1}, Y_i)$$

$$\leqslant \left(\sum_{i=1}^{m} |Y_{i-1}Y_i| \right) \cos \frac{\theta}{2} \leqslant s(K) \cos \frac{\theta}{2},$$

as far as L is an arbitrary spherical polygonal line inscribed into the curve K. Therefore,

$$s_g(K) = \sup s(L) \leqslant s(K) \cos \frac{\theta}{2}.$$

Passing to a limit at $\theta \rightarrow 0$, we get

$$s_g(K) \leqslant s(K). \tag{3}$$

The result sought is obtained from inequalities (2) and (3).

6.1.2. Let us obtain here a particular integral-geometrical relation which will be used below.

Let us fix an arbitrary point N on the sphere Ω^n. For $\alpha \in G_k^{n+1}$ let $\rho(\alpha)$ be the distance from the point N to the large sphere $\gamma(\alpha)$, i.e., $\rho(\alpha) = \inf_{X \in \gamma(\alpha)} \rho(N, X)$. Let us denote through $N(\alpha)$ the nearest point to N of the sphere $\gamma(\alpha)$, i.e., $N(\alpha)$ is the point $X \in \gamma(\alpha)$ for which $\rho(N, X) = \rho(\alpha)$. Obviously, $N(\alpha) = \pi_\alpha(N)$.

LEMMA 6.1.2. *Let $f:[0, 1] \rightarrow \mathbb{R}$ be a non-negative function measurable in the Borel sense. In this case the function $\alpha \in G_k^{n+1} \mapsto f[\cos \rho(\alpha)]$ is measurable and the following equality is valid:*

$$\int_{G_k^{n+k}} f[\cos \rho(\alpha)] \, \mu_{n+1,k}(d\alpha) = C_{n,k} \int_0^1 f(x) \, x^{k-1}(1-x^2)^{\frac{n+1-k}{2}} \, dx \tag{4}$$

where $C_{n,k}$ is a constant.

REMARK. The constant $C_{n,k}$ is defined from the fact that in the case when $f(x)$

$\equiv 1$ both parts of equality (4) must turn to 1. This results in the relation:

$$1 = C_{n,k} \int_0^1 x^{k-1}(1-x^2)^{\frac{n+1-k}{2}} dx$$

$$= C_{n,k} \frac{\Gamma(\frac{k}{2}) \, \Gamma(\frac{n+1-k}{2})}{2\Gamma(\frac{n+1}{2})}$$

whence we get:

$$C_{n,k} = \frac{2\Gamma(\frac{n+1}{2})}{\Gamma(\frac{k}{2}) \, \Gamma(\frac{n+1-k}{2})}. \tag{5}$$

Proof. Let us first consider the case when the function is continuous by way of induction with respect to the number $m = n + 1 - k$.

Let $m = 1$, k is arbitrary, $n = k$. For $t \in \Omega^n$ the symbol t' will denote a point of the sphere diametrically opposing t. For $t \in \Omega^n$ let $\nu(t)$ be an n-dimensional plane passing through the point O, perpendicular to the straight line Ot. For any function $F: G_n^{n+1} \to \mathbb{R}$ we have:

$$\int F(\xi) \, \mu_{n+1,n} \, (d\xi) = \frac{1}{\omega_n} \int_{\Omega^n} F[\nu(t)] \, \sigma_n(dt). \tag{6}$$

Here the measure σ_n is a volume of the sphere Ω^n as of a Riemanian space, $\omega_n = \sigma_n(\Omega^n)$.

Let Ω_+^n be a set of points t of the sphere Ω^n, such that $\rho(N, t) \leqslant \pi/2$, Ω^{n-1} is an equatorial $(n-1)$-dimensional sphere corresponding to the pole N, $\Omega^{n-1} = \{t \in \Omega^n | r(N, t) = \pi/2\}$. For any point $t \in \Omega^n$ we have: $\nu(t) = \nu(t')$. This yields that the integral in the right-hand part of (6) is equal to

$$\frac{2}{\omega_n} \int_{\Omega_+^n} F[\nu(t)] \, \sigma_n(dt) \tag{7}$$

Let us introduce a polar system of coordinates on the sphere Ω^n. To any point $t \in \Omega^n$, such that $t \neq N$, $t \neq N'$ there corresponds a pair (φ, τ), where $0 < \varphi < \pi$, and τ is a point of the great sphere Ω^{n-1}. In this case $\varphi = \rho(N, t)$, and the point τ is constructed in the following way. The points N and N' are connected with the shortest line passing through the point t. Such a shortest line does exist and is unique, the point τ being a point of intersection of the considered curve and the sphere Ω^{n-1}. For an arbitrary function $G: \Omega^n \to \mathbb{R}$ we have:

$$\int_{\Omega_+^n} G(t) \, \sigma_n(dt) = \int_{\Omega^{n-1}} \left[\int_0^{\frac{\pi}{2}} G(\varphi, \tau) \sin^{n-1} \varphi \, d\varphi \right] \sigma_{n-1} \, (d\tau).$$

Let us consider the case of interest when $F(\alpha) = f[\cos \rho(\alpha)]$, and where f is a continuous function defined in the interval $[0, 1]$. For the point $t \in \Omega_+^n$ with the polar coordinates (φ, τ), $(t \neq N)$, we have: $\rho[\nu(t)] = \frac{1}{2}\pi - \varphi$, $\cos \rho[\nu(t)] = \sin \varphi$ and we therefore obtain

$$\int_{\Omega_+^n} F[\nu(t)] \, \sigma_n(dt) = \omega_{n-1} \int_0^{\frac{\pi}{2}} f(\sin \varphi) \, \sin^{n-1} \varphi \, d\varphi.$$

In the integral in the right-hand part let us replace the variable of integration, setting $\sin \varphi = x$. As a result,

$$\int_{G_n^{n+1}} f[\cos \rho(\alpha)] \, \mu_{n+1,n} \, (d\alpha) = C_{n+1,n} \int_0^1 f(x) \, x^{n-1} \, \frac{d\,x}{\sqrt{1-x^2}}. \qquad (8)$$

It proves the validity of equality (4) for the case when the difference $m = n + 1 - k = 1$.

Let us assume that for a certain natural m equality (4) is proved. Let us show it to be valid also for the case when $n + 1 - k = m + 1$.

Therefore, henceforth we assume that $n + 1 - k = m + 1$ and equality (4) is valid in the case when $n + 1 - k = m$.

Let us arbitrarily choose $\xi \in G_n^{n+1}$. Then, in line with Theorem 4.5.1, we have:

$$\int_{G_k^{n+1}} f[\cos \rho(\alpha)] \, \mu_{n+1,k} \, (d\alpha) =$$

$$= \int_{G_n^{n+1}} \left\{ \int_{G_k(\xi)} f[\cos \rho(\alpha)] \, \mu_{\xi,k} \, (d\alpha) \right\} \mu_{n+1,n} \, (d\xi). \qquad (9)$$

Under the sign of the integral here ξ denotes an arbitrary n-dimensional direction in E^{n+1} and $G_k(\xi)$ is the totality of all k-dimensional directions contained in ξ. For $\xi \in G_n^{n+1}$ the point $N(\xi) = \gamma(\xi)$ is defined and let $\rho_\xi(\alpha)$ be the distance (measured on the sphere Ω^n) from the point $N(\xi)$ to the sphere $\gamma(\alpha)$ (i.e., $\rho_\xi(\alpha) = \inf_{x \in \gamma(\alpha)} \rho[N(\xi), X]$. The equality

$$\cos \rho(\alpha) = \cos \rho(\xi) \cos \rho_\xi(\alpha) \qquad (10)$$

is valid. Indeed, let A be the nearest point to N of the plane ξ, and let B be the nearest point to A of the plane $\alpha \subset \xi$. In this case: $\angle NOA = \rho(\xi)$, $\angle AOB = \rho_\xi(\alpha)$. The vector $\overline{AN}$ is orthogonal to the plane ξ and, hence, to the plane $\alpha \subset \xi$. The vector $\overline{BA}$ is orthogonal to α. Therefore, the vector $\overline{BN} = \overline{BA} + \overline{AN}$ is orthogonal to the plane α and, hence, B is the nearest point to N of the plane α. It yields $\angle BON = \rho(\alpha)$, which fact makes it possible to conclude that $|OB| = \cos \rho(\alpha)$. On the other hand, we have: $|OB| = |OA| \cos \rho_\xi(\alpha) = \cos \rho(\xi) \cos \rho_\xi(\alpha)$ and equality (10) is proved. Relation (9) can therefore be rewritten in the following way:

$$\int_{G_k^{n+1}} f[\cos \rho(\alpha)] \, \mu_{n+1,k} \, (d\alpha) =$$

$$= \int_{G_n^{n+1}} \left\{ \int_{G_k(\xi)} f[\cos \rho(\xi) \cos \rho_\xi(\alpha)] \, \mu_{\xi,k} \, (d\alpha) \right\} \mu_{n+1,n} \, (d\xi). \qquad (11)$$

The internal integral can be easily defined on the basis of the induction assumption. The dimensionality of the plane ξ is n, and the difference $n - k = m$. Replacing $n + 1$ by n in equality (4) and $f(x)$ by $f(x \cos \rho(\xi))$, we find that internal integral in the right-hand part of (11) is

$$C \int_0^1 f(x \cos \rho(\xi)) \, x^{k-1}(1-x^2)^{\frac{n-k}{2}-1} \, dx = \Phi \, (\cos \rho(\xi))$$

where C is a constant. Here use has been made of the supposition that for given m the equality sought is valid. It affords

$$\int_{G_k^{n+1}} f[\cos \rho(\alpha)] \, \mu_{n+1,k} \, (d\alpha) = = \int_{G_n^{n+1}} \Phi(\cos \rho(\xi)) \, \mu_{n+1,k} \, (d\xi).$$

In line with the above proof, the latter integral is

$$C \int_0^1 \Phi(\lambda) \, \lambda^{n-1} \, \frac{d\lambda}{\sqrt{1-\lambda^2}}.$$

If we take into account the expression for the function Φ, then we get:

$$\int_{G_k^{n+1}} f[\cos \rho(\alpha)] \, \mu_{n+1,k} \, (d\alpha) =$$

$$= C \int_0^1 \left[\int_0^1 f(x\lambda) \, x^{k-1}(1-x^2)^{\frac{n-k}{2}-1} \, dx \right] \frac{\lambda^{n-1} \, d\lambda}{\sqrt{1-\lambda^2}}. \tag{12}$$

The exact values of the constant multipliers in intermediate calculations, due to the remark preceding the proof, are not necessary and, therefore, all the multipliers will be denoted by the same letter C.

Let us transform the integral in the right-hand part of (12). First of all, setting $x\lambda = y$, we get:

$$\int_0^1 \left\{ \int_0^1 f(x\lambda) \, x^{k-1}(1-x^2)^{\frac{n-k}{2}-1} \, dx \right\} \frac{\lambda^{n-1} \, d\lambda}{\sqrt{1-\lambda^2}} =$$

$$= \int_0^1 \left\{ \int_0^\lambda f(y) \, y^{k-1}(\lambda^2-y^2)^{\frac{n-k}{2}-1} \, dy \right\} \frac{\lambda \, d\lambda}{\sqrt{1-\lambda^2}}. \tag{13}$$

The latter integral can be transformed by way of changing the order of integration. Thus we get the integral:

$$\int_0^1 f(y) \, y^{1-k} \left[\int_\lambda^1 (\lambda^2-y^2)^{\frac{n-k}{2}-1} \, \frac{\lambda \, d\lambda}{\sqrt{1-\lambda^2}} \right] dy. \tag{14}$$

In the internal integral let us substitute the integration variable t, setting $1 - \lambda^2 = (1 - y^2)t^2$. At λ changing from y to 1, t changes from 1 to 0 and, after obvious transformations, we get:

$$\int_\lambda^1 (\lambda^2-y^2)^{\frac{n-k}{2}-1} \, \frac{\lambda \, d\lambda}{\sqrt{1-\lambda^2}} = (1-y^2)^{\frac{n+1-k}{2}-1} \int_0^1 (1-t^2)^{\frac{n-k}{2}-1} \, dt. \tag{15}$$

Comparing expressions (13), (14) and (15), we conclude that

$$\int_0^1 \left\{ \int_0^1 f(x\lambda) \, x^{k-1}(1-x^2)^{\frac{n-k}{2}-1} \, dx \right\} \frac{\lambda^{n-1} \, d\lambda}{\sqrt{1-\lambda^2}}$$

$$= C \int_0^1 f(x) \, x^{k-1}(1-x^2)^{\frac{n+1-k}{2}-1} \, dx.$$

and, therefore, in conformity with (12), the required equality (4) is proved.

In the above considerations the function f is assumed to be continuous. A general case can be easily reduced to the one discussed by a limiting transition, which the reader can do for himself.

It should be remarked that the calculations carried out above are valid under essentially much weaker assumptions as regards the function f.

The lemma is proved.

COROLLARY. *Let N be an arbitrary point on the sphere Ω^n, $r \in (0, \pi/2)$, and let $H(N, r)$ be a set of all planes $\alpha \in G_k^{n+1}$, intersecting the geodesic ball $\overline{B}(N, r) = \{t \in \Omega^n | \rho(t, N) \leqslant r\}$. The set $H(N, r)$ is closed, its measure $\mu_{n+1,k}$ $[H(N, r)]$ is independent of N and*

$$\mu_{n+1,k}\,[H(N, r)] = Cr^{n+1-k} + o(r^{n+1-k}) \tag{16}$$

at $r \longrightarrow 0$ where $C = \mathrm{const} > 0$ and for all r

$$\mu_{n+1,k}\,[H(N, r)] \leqslant C_1 r^{n+1-k}, \tag{17}$$

where $C_1 = \mathrm{const} < \infty$.

Proof. The fact that the set $H(N, r)$ is closed is obvious. Let us introduce the function $f : [0, 1] \longrightarrow \mathbb{R}$, setting $f(x) = 1$ at $\cos r \leqslant x \leqslant 1$ and $f(x) = 0$ at $0 \leqslant x \leqslant \cos r$. In this case $f[\cos \rho(\alpha)] = 1$ if $\rho(\alpha) \leqslant r$ and $f[\cos \rho(\alpha)] = 0$ at $\rho(\alpha) > r$, i.e., $F(\alpha) = f[\cos \rho(\alpha)] = 1$ at $\alpha \in H(N, r)$ and $F(\alpha) = 0$ at $\alpha \notin H(n, r)$. It yields

$$\mu_{n+1,k}\,[H(N, r)] = \int_{G_k^{n+1}} f[\cos \rho(\alpha)]\, \mu_{n+1,k}\,(d\alpha).$$

In conformity with the lemma we conclude:

$$\mu_{n+1,k}\,[H(N, r)] = C_{n,k} \int_{\cos r}^1 x^{k-1}(1-x^2)^{\frac{n+1-k}{2}}\, dx.$$

We, therefore, can see that $\mu_{n+1,k}[H(N, r)]$ is N-independent. In the latter integral let us replace the integration variable, setting $x = \sqrt{1-y^2}$. As a result, we get

$$\mu_{n+1,k}\,[H(N, r)] = C_{n,k} \int_0^{\sin r} (1-y^2)^{\frac{k-3}{2}}\, y^{n-k}\, dy.$$

Applying L'Hôpital rule, we find that at $r \longrightarrow 0$ the relation

$$\frac{\mu_{n+1,k}[H(N, r)]}{r^{n-k+1}}$$

tends to a finite limit equal to

$$\frac{C_{n,k}}{n-k+1} = \frac{2\Gamma(\frac{n+1}{2})}{\Gamma(\frac{k}{2})\,\Gamma(\frac{n+1-k}{2})(n-k+1)} = \frac{\Gamma(\frac{n+1}{2})}{\Gamma(\frac{k}{2})\,\Gamma(\frac{n+1-k}{2} + 1)}$$

whence, obviously, we get relation (14). At $r \longrightarrow 0$ the relation

$$\frac{\mu_{n+1,k}[H(N, r)]}{r^{n-k+1}}$$

has a finite limit and is, consequently, limited, which results in the second statement of the corollary.

6.1.3. Let us define a certain characteristic of the thinness of a set in a metric space.

Let us arbitrarily set a metric space M and let ρ be its metric. A closed sphere with centre $x \in M$ and radius $r > 0$ in the space M is a set $\overline{B}(x, r)$ of all points $y \in M$, such that $\rho(y, x) \leqslant r$. Let E be an arbitrary set of points of the space M. It is assumed that E is a set of the k-dimensional Hausdorff measure, which is equal to zero and where $k > 0$, if for any $\varepsilon > 0$ we can give a sequence of closed spheres $\overline{B}_m = B(x_m, r_m)$, $m = 1, 2, \ldots$, such that $E \subset \bigcup_{m=1}^{\infty} \overline{B}_m$ and the inequality $\sum_{m=1}^{\infty} r_m^k < \varepsilon$ holds.

LEMMA 6.1.3. *Let E be a set of the points of the sphere Ω^n. In this case, if E is a set of the $(n+1-k)$-dimensional Hausdorff measure, which is equal to zero, where $1 \leqslant k \leqslant n$ and k is an integer, then for nearly all $\xi \in G_k^{n+1}$ the large circumference $\gamma(\xi)$ contains no points of the set E.*

Proof. Let $X \in \Omega^n$, $r > 0$ and $\overline{B}(X, r) = \{Y \in \Omega^n | \rho(X, Y) \leqslant r\}$. Let us denote through $H_k(X, r)$ a set of all k-dimensional directions $\xi \in G_k^{n+1}$, for which the great sphere $\gamma(\xi)$ contains the points of the sphere $\overline{B}(X, r)$. In this case, in line with inequality (17), we have:

$$\mu_{n+1,k} [H_k(X, r)] \leqslant Cr^{n+1-k}$$

where $C = \text{const} > 0$. Let H be a set of all $\xi \in G_k^{n+1}$, such that $\gamma(\xi) \cap E$ is not empty. Let us arbitrarily set $\varepsilon > 0$, and assume $\varepsilon_1 = \varepsilon/C$. Since E, by the condition, is a set of the $(n+1-k)$-dimensional Hausdorff measure which is equal to zero, then we can find a sequence $(\overline{B}_m = \overline{B}(X_m, r_m)$, $m = 1, 2, \ldots)$, of closed spheres covering E and such that $\sum_{m=1}^{\infty} r_m^{n+1-k} < \varepsilon_1$. Let us set $\Delta_m = H_k(X_m, r_m)$. If $\xi \in H$, then $\gamma(\xi) \cap E \neq \varnothing$ and hence there exists an m, such that the sphere $\overline{B}(X_m, r_m)$ contains the points of the sphere $\gamma(\xi)$, i.e., $\xi \in \Delta_m$. This affords $H \subset \bigcup_{m=1}^{\infty} \Delta_m$ and hence

$$\mu_{n+1,k}(H) \leqslant \sum_{m=1}^{\infty} \mu_{n+1,k}(\Delta_m) \leqslant \sum_{m=1}^{\infty} Cr_m^{n+1-k} < C\varepsilon_1 = \varepsilon.$$

Since $\varepsilon > 0$ was chosen arbitrarily, then $\mu_{n+1,k}(H) = 0$ and the lemma is thus proved.

COROLLARY. *For any rectifiable curve K on the sphere Ω^n the set of those $\xi \in G_{n-1}^{n+1}$ for which the large sphere $\gamma(\xi)$ intersects K is a set of the zero measure.*

Proof. At $k = n - 1$ we have: $n + 1 - k = 2$. In order to prove the corollary it is sufficient to establish that if the curve K is rectifiable, then its carrier $|K|$ is a set of a two-dimensional Hausdorff measure equal to zero. Let $l = s(K)$, $x(s)$, $0 \leqslant s \leqslant l$, be a parametrization of the curve K, where the parameter s is an arc length. Let us arbitrarily set a natural m and let $\overline{B}_k = \overline{B}[x(k/m), \overline{m}]$, $k = 1, 2, \ldots, m$. The closed spheres $\overline{B}_k$ obviously cover the set

$|K|$ and the sum of the squares of their radii is $m(l^2/m^2) = l^2/m$. At $m \to \infty$, $l^2/m \to 0$ and it is thus proved that $|K|$ is a set of a zero two-dimensional Hausdorff measure.

6.2. Integro–Geometrical Theorem on Angles and its Corollaries

6.2.1. Let us first prove some integro-geometrical relations which are of major importance for further considerations.

Let us begin with stating a certain general theorem of an integro-geometrical character as regards the notion of an angle between k-dimensional planes.

Let P be a k-dimensional subspace of the space V^n, $n \geqslant k$. A reference frame of the subspace P will be any ordered system $u = \{u_1, u_2, \ldots, u_k\}$ of k linearly independent vectors lying in the plane P. If $u = \{u_1, u_2, \ldots, u_k\}$ is an arbitrary reference frame of the subspace P, then any vector $x \in P$ can be, and uniquely, presented as $x = \lambda_1 u_1 + \lambda_2 u_2 + \cdots + \lambda_k u_k$, where $\lambda_1, \lambda_2, \ldots, \lambda_k$ are real numbers. Let $u = (u_1, u_2, \ldots, u_k)$ and $v = (v_1, v_2, \ldots, v_k)$ be two arbitrary reference frames of the subspace P. In this case we have:

$$v_i = \sum_{j=1}^{k} \lambda_{ij} u_j, \quad i = 1, 2, \ldots, k,$$

where λ_{ij} are real numbers. The reference frames u and v are called equally oriented if the determinant

$$\det \|\lambda_{ij}\|, \quad i, j = 1, 2, \ldots, k,$$

is positive, and oppositely oriented if this determinant is negative. The totality of all reference frames of the subspace P falls into two classes, so that the reference frames belonging to one class are orientated in the same way, while those taken from different classes are oppositely orientated. We say that a certain orientation P is given or, in other words, that P is orientated, if the reference frames belonging to one of these classes are termed right-handed, while those belonging to another class are termed left-handed. Any k-dimensional subspace P exactly defines two differently oriented subspaces. The reference frames which are right-hand for one of those subspaces are left-hand for another and *vice versa*.

Let us now define the notion of an angle between orientated k-dimensional subspaces of V^n. An angle is a distance of a certain kind between k-dimensional subspaces and it can be defined in various ways. Let us give the definition which in a geometrical sense seems most natural.

Let P and Q be orientated k-dimensional subspaces of V^n, π_Q be a mapping of the orthogonal projection of V^n onto Q. If the plane P contains at least one vector $u \neq 0$ which is orthogonal to Q, then $\pi_Q(P)$ is a proper subspace of Q. In this case we assume the angle between P and Q to equal $\pi/2$.

Let us assume that P contains no vectors which are orthogonal to the subspace Q. Then π_Q maps P onto Q in a one-to-one manner. For $x \in P$ let us set $H(x) = \langle \pi_Q(x), \pi_Q(x) \rangle$. The function H is a quadratic form in the subspace P.

At $x \neq 0$ $\pi_Q(x) \neq 0$ and, hence $H(x) = |\pi_Q(x)|^2 > 0$, i.e, the quadratic form $H(x)$ is determined positively. Let $u_1, u_2, \ldots, u_k$ be a system of k proper vectors of quadratic form H, let $\lambda_1^2, \lambda_2^2, \ldots, \lambda_k^2$ be the corresponding eigennumbers to them of quadratic form H. We have: $H(x) = <x, \pi_Q^* \pi_Q x.$, where π_Q^* is a linear mapping conjugate to π_Q, and for any i $\pi_Q^* \pi_Q u_i = \lambda_i^2 u_i$. From this we conclude that

$$\lambda_i^2 = <\pi_Q u_i, \pi_Q u_i> = |\pi_Q(u_i)|^2$$

and at $i \neq j$ $<\pi_Q u_i, \pi_Q u_j> = <\pi_Q^* \pi_Q u_i, u_j> = \lambda_i^2 <u_i, u_j> = 0$. The vectors $v_i = \pi_Q(u_i)$ are, therefore, mutually orthogonal and $|v_i| = \lambda_i$. Let us assume that $(u_1, u_2, \ldots, u_k)$ is a right-hand reference frame in the subspace P. Since for any vector $x \in R$ $|\pi_Q(x)| \leqslant |x|$, in which case $|\pi_Q(x)| = |x|$ if and only if $x = \pi_Q(x)$, then $0 < \lambda_i \leqslant 1$ at all $i = 1, 2, \ldots, k$. Let us now define the number $\varphi \in [0, \pi]$, setting:

$$\cos \varphi = \sigma \lambda_1 \lambda_2 \ldots \lambda_k$$

where $\sigma = 1$, if $(v_1, v_2, \ldots, v_k)$ is the right-hand reference frame in the plane Q, or $\sigma = -1$, if it is the left-hand one. In line with what has been said above, we always have: $0 < \lambda_1 \lambda_2 \ldots \lambda_k \leqslant 1$. It should be remarked that if $\cos \varphi = \pm 1$, then $\lambda_1 \lambda_2 \ldots \lambda_k = 1$ and, hence, $\lambda_1 = \lambda_2 = \cdots = \lambda_k = 1$. In this case, therefore, the vectors $u_i = v_i$ at every $i = 1, 2, \ldots, k$. Hence we conclude that if $\varphi = 0$, then the subspaces P and Q coincide with the orientation. In the case when $\varphi = \pi$, so that $\cos \varphi = -1$, then the reference point $(u_1, u_2, \ldots, u_k)$ which is right-hand in P, proves to be left-hand in Q, so that the subspaces P and Q differ in their orientation. The number φ defined in the above way will be denoted by the symbol $(P, \frown Q)$ and termed an angle between the oriented subspaces P and Q.

Let the reader himself prove the fact that the function $(P, Q) \mapsto (P, \frown Q)$ is a metric on the set of all oriented k-dimensional subspaces V^n.

Let us consider a special case when there exists a $(k-1)$-dimensional subspace $R \subset P \cap Q$. Let us set an arbitrary system of $k-1$ mutually orthogonal unit vectors $u_1, u_2, \ldots, u_k$ lying in the plane R and let p and q be unit vectors lying in P and Q, respectively, which are orthogonal to R and such that $(u_1, \ldots, u_k, p)$ is the right-hand reference frame in P, and $(u_1, \ldots, u_{k-1}, q)$ is the right-hand reference frame of the subspace Q. The vectors $u_1, \ldots, u_{k-1}, p$ form in this case a set of the proper vectors of the quadratic form $H(x) = |\pi_Q x|^2$. To the vector p there corresponds an eigennumber λ_k. We have: $\pi_Q(p) = \sigma \lambda_k q$, where $\sigma = 1$, if the vectors $\pi_Q(p)$ and q are directed in the same way and, hence, the reference frame $(u_1, \ldots, u_{k-1}, \pi_Q(p)) = (\pi_Q(u_1), \ldots, \pi_Q(u_{k-1}), \pi_Q(p))$ is right-hand, and $\sigma = -1$ if the vectors $\pi_Q(p)$ and q are oppositely directed; in the latter case the reference frame $(u_1, \ldots, u_{k-1}, \pi_Q(p))$ being left-hand. On the other hand, by definition, $\cos \varphi = \sigma \lambda_k$, where $\varphi = (P, \frown Q)$. The vector $h = p - \pi_Q(p)$ is orthogonal to q and, hence, $<p, q> = <h, + <\pi_Q(p), q> = <\pi_Q(p), q> = \sigma \lambda_k$. From this we conclude that $\sigma \lambda_k = \cos (p, \frown q)$

and hence an angle between the oriented planes P and Q is, in the case considered, equal to that between the vectors p and q.

Let P be an arbitrary oriented k-dimensional subspace V^n and $\alpha \in G_l^m$, where $l \geqslant k$. Let us assume that P does not intersect the $(m-l)$-dimensional subspace $\nu(\alpha)$. In this case the mapping η of the orthogonal projecting of the space V^n onto the subspace α gives a one-to-one mapping of P onto a certain subspace $P_\alpha \subset \alpha$. The space P_α is assumed orientated in the following way. If $(u_1, u_2, \ldots, u_k)$ is a right-hand reference frame of the plane P, then $(\eta(u_1), \eta(u_2), \ldots, \eta(u_k))$ is a right-hand reference frame of P_α. The oriented plane P_α defined in such a way will be termed an orthogonal projection of the orientated subspace P onto the l-dimensional direction $\alpha \in G_l^m$.

6.2.2. LEMMA 6.2.1. *Let $k > 0$, $m > 0$ be integers and let $n = k + m$. Let us arbitrarily set $Q_0 \in G_k^n$ and let A be a totality of all $P \in G_m^n$, for which the intersection $P \cap Q_0$ contains the points other than the point O. In this case the set A is closed and its measure $\mu_{n,m}(A)$ equals zero.*

Proof. In the space V^n let us introduce a Cartesian orthogonal system of coordinates with the basis $e_1, e_2, \ldots, e_n$ in such a way that Q_0 coincides with the plane stretched onto the vectors $e_1, e_2, \ldots, e_k$. For a set of indices $J = (j_1, j_2, \ldots, j_m)$, $1 \leqslant j_1 \leqslant n$, $1 \leqslant j_2 \leqslant n, \ldots, 1 \leqslant j_m \leqslant n$, where $j_1, j_2, \ldots, j_m$ are mutually different, let us denote through $E_j = E_{j_1 j_2 \ldots j_m}$ an m-dimensional subspace V^n stretched onto the vectors $e_{j_1}, e_{j_2}, \ldots, e_{j_m}$ and let U_J be a set of all $P \in G_m^n$ which are projected onto E_J in a one-to-one manner. As has been shown in Section 4.1.6, the sets U_J cover G_m^n, so that $P \in G_m^n$ belongs to U_J at least at one J. Let $I = (i_1, i_2, \ldots, i_k)$ be a set of numbers obtained when we delete $j_1, j_2, \ldots, j_m$ in the sequence $(1, 2, \ldots, m)$. Any plane $P \in U_J$ is defined by the equation

$$y = Hz$$

where $y = (x_{i_1}, x_{i_2}, \ldots, x_{i_k})$, $z = (x_{j_1}, x_{j_2}, \ldots, x_{j_m})$, H is an $k \times m$ matrix the elements of which are the coordinates of P in a canonical coordinate system in G_m^n which corresponds to the plane E_J.

For an arbitrary Borel set $E \subset G_m^n$ we have:

$$\mu_{n,m} (E \cap U_J) = \int_{\varphi_J \mathbb{R}^k} g(H) \, dh_{11} \ldots dh_{1m} \ldots dh_{k,1} \ldots dh_{k,m}$$

where the mapping φ_J is a canonical system of coordinates in U_J, and $g(H)$ is a a continuous and positive throughout function in $\mathbb{R}^{km}$. In order to prove that the set $E \subset G_m^n$ is a set of the zero measure it is sufficient to state that $\varphi_J(E \cap U_J)$ is a set of the zero measure in $\mathbb{R}^{mk}$ for any set of indices $J = (j_1, j_2, \ldots, j_m)$.

Let us now consider the set of interest A. The closeness of A is obvious from the definition. The task is to prove that for any $J = (j_1, j_2, \ldots, j_m)$ $\varphi_J(A \cap U_J)$ is a set of zero measure in the space $\mathbb{R}^{mk}$. Let the plane $P \in U_J$. In this case P is defined by the system of equations

$$x_{j_s} - \sum_{r=1}^{m} h_{sr} x_{i_r} = 0, \quad s = 1, 2, \ldots, k. \tag{17}$$

The plane Q_0 is defined by the system of equations

$$x_{k+r} = 0, \quad r = 1, 2, \ldots, m \tag{18}$$

The plane P belongs to $A \cap U_J$ if and only if the system of $n = k + m$ equations (17) and (18) has a non-trivial solution. Let us assume that P obeys this condition. In this case the set of indices $I = (i_1, i_2, \ldots, i_m)$ does not coincide with the set $(k+1, k+2, \ldots, k+m)$, since in the opposite case equalities (17) and (18) afford $x = 0$. Let $i_{l+1}, i_{l+2}, \ldots, i_m$ be those of the indices $i_1, i_2, \ldots, i_m$ which are included between $k + 1$ and $k + m$, and let $j_1, j_2, \ldots, j_l$ be those of the j_s values which lie between $k + 1$ and $k + m$. In this case the system of equations (17) and (18) is equivalent to the system:

$$\sum_{r=1}^{l} h_{sr} x_{i_r} = 0, \quad s = 1, 2, \ldots, l$$

This system has a non-trivial solution if and only if

$$0 = \det(h_{sr}); \quad s = 1, 2, \ldots, l; \quad r = 1, \ldots, l. \tag{19}$$

Equality (19) is the sought equation of the set $\varphi_J(A \cap U_J)$ in the space $\mathbb{R}^{km}$. Therefore, $\varphi_J(A \cap U_J)$ is a set of points $H \in \mathbb{R}^{km}$, on which a certain polinomial which is not identically equal to zero, turns to zero. Hence $\varphi_J(A \cap U_J)$ is a set of the zero measure and the lemma is thus proved.

COROLLARY. *Let there be given a k-dimensional subspace P of the space V^{n+1} and let $l \geqslant k$. In this case the set B of the $Q \in G_l^{n+1}$, such that the plane P contains non-zero vectors orthogonal to Q is the set of zero measure in G_l^{n+1}.*

Proof. Let arbitrarily set an l-dimensional subspace $P_0 \supset P$. In this case if P contains the vectors orthogonal to the subspace Q, then the intersection of P_0 and the subspace $\nu(Q)$, which quite orthogonal to Q, contains the vectors other than zero. The dimensionality of $\nu(Q)$ is $n + 1 - l$. In conformity with the lemma, the measure of the set of those $P \in G_{n+1-l}^{n+1}$ for which $P \cap P_0$ is non-trivial is equal to zero and, hence, the measure of the set of those $Q \in G_l^{n+1}$, for which $\nu(Q) \cap P_0$ contains the vectors other than zero, also equals zero.

The closure of B is obvious. The corollary is proved.

6.2.3. THEOREM 6.2.1. *Let P and Q be orientated k-dimensional subspaces V^{n+1}, such that the intersection $P \cap Q$ contains a $(k-1)$-dimensional plane. For an arbitrary $\alpha \in G_l^{n+1}$, where $l \geqslant k$, let P_α and Q_α be projections of P and Q into the subspace α. In this case for almost all $\alpha \in G_l^{n+1}$ P_α and Q_α are k-dimensional subspaces α, the function*

$$\alpha \mapsto (P_\alpha, \widehat{} Q_\alpha)$$

is measurable and the following equality is valid:

$$\int_{G_l^{n+1}} (P_\alpha \widehat{} Q_\alpha)\, \mu_{n+1,l}\, (d\alpha) = (P \widehat{} Q). \tag{20}$$

Proof. Let U be a set of those $\alpha \in G_l^{n+1}$ for which the intersection $\nu(\alpha) \cap (P \cup Q)$ does not contain non-zero vectors. In line with the corollary to Lemma 6.2.1, U is an open set in G_l^{n+1} and $\mu_{n+1,l}(U) = \mu_{n+1,l}(G_l^{n+1}) = 1$. The planes P_α and Q_α are continuously α-dependent since the mapping η_α of the orthogonal projections into α is a continuous function of α. Hence, $\alpha \mapsto (P_\alpha \widehat{} Q_\alpha)$ is a continuous function. Therefore, this function is measurable and, since it is limited, $0 \leqslant (P_\alpha \widehat{} Q_\alpha) \leqslant \pi$ for all $\alpha \in G_l^{n+1}$, then obviously it is integrable with respect to G_l^{n+1}.

Let us first prove that the value of

$$\theta(P, Q) = \int_{G_l^{n+1}} (P_\alpha \widehat{} Q_\alpha)\, \mu_{n+1,l}\, (d\alpha)$$

depends only on that of $\varphi = (P \widehat{} Q)$. Indeed, let P_1 and Q_1, P_2 and Q_2 be two pairs of oriented k-dimensional subspaces V^{n+1}, such that the intersections $P_1 \cap Q_1$ and $P_2 \cap Q_2$ contain $(k-1)$-dimensional subspaces and $(P_1 \widehat{} Q_1) = (P_2 \widehat{} Q_2)$. In this case the pairs (P_1, Q_1) and (P_2, Q_2) can be superposed by orthogonal transformations of V^{n+1}. Indeed, let S_1 and S_2 be $(k+1)$-dimensional subspaces of V^{n+1}, such that $P_1 \cup Q_1 \subset S_1$ and $P_2 \cup Q_2 \subset S_2$, and R_1 and R_2 be $(k-1)$-dimensional subspaces contained in $P_1 \cap Q_1$ and $P_2 \cap Q_2$, respectively. Let us first, by means of orthogonal transformations of V^{n+1}, superpose S_2 with S_1. Then by another orthogonal transformation which maps the plane S_1 onto itself, let us superpose R_2 with R_1. The next step is to superpose P_2 with P_1 by rotating them around R_1 in S_2. Due to the equality $(P_1 \widehat{} Q_1) = (P_2 \widehat{} Q_2)$, Q_2 will be automatically superposed with Q_1. It should be remarked that if $\varphi \in \mathbb{O}^{n+1}$, then, in accordance with equality (8) in Section 4.2.1 for any $x \in V^{n+1}$ we have:

$$\eta_{\varphi(\alpha)}(x) = \varphi[\eta_\alpha(\varphi^{-1}(x))].$$

This affords

$$P_{\varphi(\alpha)} = \eta_{\varphi(\alpha)}(P) = \varphi[\eta_\alpha(\varphi^{-1})(P))].$$

for any plane $P \in G_l^{n+1}$. Let $\varphi \in \mathbb{O}^{n+1}$ be such that $P_1 = \varphi(P_2)$, $Q_1 = \varphi(Q_2)$. Then we get

$$(P_{1,\,\varphi(\alpha)} \widehat{} Q_{i,\varphi(\alpha)}) = (\varphi^{-1}(P_1)_\alpha \widehat{} \varphi^{-1}(Q_1)_\alpha) = (P_{2,\alpha} \widehat{} Q_{2,\alpha}).$$

Herefrom we have:

$$\int_{G_l^{n+1}} (P_{2,\alpha} \widehat{} Q_{2,\alpha})\, \mu_{n+1,l}\, (d\alpha) = \int_{G_l^{n+1}} (P_{1,\varphi(\alpha)} \widehat{} Q_{1,\varphi(\alpha)})\, \mu_{n+1}\, (d\alpha)$$

$$= \int_{G_l^{n+1}} (P_{1,\alpha} \widehat{} Q_{1,\alpha})\, \mu_{n+1,l}\, (d\alpha),$$

i.e., $\theta(P_1, Q_1) = \theta(P_2, Q_2)$. Therefore, if the pairs of k-dimensional subspaces (P_1, Q_1) and $P_2, Q_2)$ are such that $(P_1 \widehat{} Q_1) = (P_2 \widehat{} Q_2)$ and $P_1 \cap Q_1$ and $P_2 \cap Q_2$ contain $(k-1)$-dimensional subspaces, then $\theta(P_1, Q_1) = \theta(P_2, Q_2)$. It affords $\theta(P, Q) = f(\varphi)$, where $\varphi = (P \widehat{} Q)$.

The function f is defined and non-negative for all $\varphi \in [0, \pi]$. It is easily seen that $f(0) = 0$ and $f(\pi) = \pi$. Let us prove that f obeys the functional Cauchy equation. Let us arbitrarily set φ_1, φ_2, such that $\varphi_1 > 0$, $\varphi_2 > 0$ and $\varphi_1 + \varphi_2 \leqslant \pi$. In V^{n+1} let us set an orthonormal basis $e_1, \ldots, e_{k-1}, e_k, e_{k+1}$, $\ldots, e_{n+1}$. Let us also set $p = e_k \cos \varphi_1 - e_{k+1} \sin \varphi_1$, $q = e_k$, $r = e_k \cos \varphi_2 + e_{k+1} \sin \varphi_2$. The vectors p, q, r are unit vectors and $(p \overset{\frown}{,} q) = \varphi_1$, $(q \overset{\frown}{,} r) = \varphi_2$, $(p \overset{\frown}{,} r) = \varphi_1 + \varphi_2$. Let P, Q and R are oriented k-dimensional subspaces V^{n+1}, such that $(e_1, \ldots, e_{k-1}, p)$ is a right-hand reference frame of P, $(e_1, \ldots, e_{k-1}, q)$ is the right-hand reference frame of Q, and, finally, $(e_1, \ldots, e_{k-1}, r)$ is the right-hand reference frame of the subspace R. All the subspaces P, Q and R intersect through one $(k-1)$-dimensional plane and $(P \overset{\frown}{,} R) = \varphi_1 + \varphi_2$, $(P \overset{\frown}{,} Q) = \varphi_1$, $(Q \overset{\frown}{,} R) = \varphi_2$.

Let us first consider the case when $l = k$. Let $\alpha \in G_l^{n+1}$ be such that not a single plane P, Q and R contains non-zero vectors orthogonal to α. Let us prove that in this case we always have:

$$(P_\alpha \overset{\frown}{,} R_\alpha) = (P_\alpha \overset{\frown}{,} Q_\alpha) + (Q_\alpha \overset{\frown}{,} R_\alpha) \tag{21}$$

Each of the angles contained in this equality can be equal either to 0 or to π. Let us assume that $(P_\alpha \overset{\frown}{,} R_\alpha) = \pi$. Then P_α and R_α differ only in their orientation and Q_α coincides either with P_α or with R_α. In the former case $(P_\alpha \overset{\frown}{,} Q_\alpha) = 0$, $(Q_\alpha \overset{\frown}{,} R_\alpha) = \pi$, in the latter case $(P_\alpha \overset{\frown}{,} Q_\alpha) = \pi$, $(Q_\alpha \overset{\frown}{,} R_\alpha) = 0$ and equality (21) is fulfilled in both cases. Let us now assume that $(P_\alpha \overset{\frown}{,} R_\alpha) = 0$. It means that the oriented subspaces P_α and R_α coincide. Let us prove that $Q_\alpha = P_\alpha = R_\alpha$ also. It should be remarked that $(P_\alpha \overset{\frown}{,} R_\alpha) < \pi$, since in the opposite case, obviously, we would have $(P_\alpha \overset{\frown}{,} R_\alpha) = \pi$. We have: $p \sin \varphi_2 + r \sin \varphi_1 = e_k \sin (\varphi_1 + \varphi_2)$, and hence $q = e_k = \alpha p + \beta r$, where $\alpha > 0$, $\beta > 0$. Let e_i', $i = 1, 2, \ldots, k-1$, p', q', r' be the orthogonal projections of the vectors e_i, p, q and r onto the plane α, respectively. In this case $q' = \alpha p' + \beta r'$. The reference frames $(e_1', \ldots, e_{k-1}', p')$ and $(e_1', \ldots, e_{k-1}', r')$ are oriented in the same way as $P_\alpha = R_\alpha$, and it follows from the above said that the reference frame $(e_1', \ldots, e_{k-1}', q')$ is also oriented in the same way as the first two ones, i.e., $Q_\alpha = P_\alpha = R_\alpha$. Equality (21) therefore, is valid in this case as well. Hence, equality (21) is fulfilled for nearly all $\alpha \in G_l^{n+1}$. Integrating it, we obtain

$$\theta(P, R) = \theta(P, Q) + \theta(Q, R)$$

and, hence,

$$f(\varphi_1 + \varphi_2) = f(\varphi_1) + f(\varphi_2). \tag{22}$$

The function f, as has already been noted, is non-negative, $f(0) = 0$, $f(\pi) = \pi$ and for any φ_1, φ_2, such that $\varphi_1 > 0$, $\varphi_2 > 0$, $\varphi_1 + \varphi_2 \leqslant \pi$, equality (22) holds. Therefore $f(\varphi) = \varphi$ for all $\varphi \in [0, \pi]$ and, hence, the theorem is proved for the case $k = l$.

The case when $k < l$ is reduced to the one considered above in Theorem 4.5.1.

Let $\alpha \in G_l^{n+1}$, $\beta \in G_k^{n+1}$, $\beta \subset \alpha$. Let us arbitrarily choose orientated k-dimensional subspaces P and Q, obeying the conditions of the theorem. In this case the projections $P_\alpha = \eta_\alpha(P)$ and $Q_\alpha = \eta_\alpha(Q)$, $P_\beta = \eta_\beta(P)$, $Q_\beta = \eta_\beta(Q)$ are defined. In conformity with the known properties of the projections we have: $P_\beta = \eta_\beta(P_\alpha)$, $Q_\beta = \eta_\beta(Q_\alpha)$. In line with Theorem 4.5.1, we get:

$$(P \frown Q) = \int_{G_k^{n+1}} (P_\beta \frown Q_\beta) \, \mu_{n+1,k} \, (d\beta) =$$

$$= \int_{G_l^{n+1}} \left[\int_{G_{k(\alpha)}} (P_\beta \frown Q_\beta) \, \mu_{\alpha,k} \, (d\beta) \right] \mu_{n+1,l} \, (d\alpha).$$

Since $P_\beta = \eta_\beta(P_\alpha)$, $Q_\beta = \eta_\beta(Q_\alpha)$, then the internal integral, according to the above proved, is $(P_\alpha \frown Q_\alpha)$, which obviously yields (21). The theorem is proved.

6.2.4. Let us prove some integro-geometrical theorems on the length and geodesic turn of a spherical polygonal line, which are of major importance in further considerations.

THEOREM 6.2.2. *For any spherical polygonal line valid are the equalities*:

$$s(L) = \int_{G_2^{n+1}} s[\pi_\alpha(L)] \, \mu_{n+1,2} \, (d\alpha), \tag{23}$$

$$\kappa_g(L) = \int_{G_2^{n+1}} \kappa_g[\pi_\alpha(L)] \, \mu_{n+1,2} \, (d\alpha). \tag{24}$$

REMARK. The curve $\pi_\alpha(L)$ is defined, as has been shown above, for almost all two-dimensional directions α and, consequently, the functions standing in the integrals in (23) and (24) are defined on the set G_2^{n+1} for almost all α.

Proof. Let $x_0 < x_1 < \cdots < x_m$ be subsequent vertices of the spherical polygonal line L. Let P_i be a two-dimensional plane of the shortest line $[x_{i-1} x_i]$, $i = 1, 2, \ldots, m$. Let us orient the plane P_i, assuming the pair $(x_{i-1}, t_r(x_{i-1}))$ to be the right-hand reference frame. In this case, obviously, the pair $(x_i, t_l(x_i))$ will also be a right-hand reference frame in the plane P_i. At every $i = 1, 2, \ldots, m-1$ a turn at the point x_i of the polygonal line L equals the angle $(P_i \frown P_{i+1})$. We therefore obtain

$$s(L) = \sum_{i=1}^{m} (x_{i-1} \frown x_i)$$

and

$$\kappa_g(L) = \sum_{i=1}^{m-1} (P_{i-1} \frown P_i).$$

Let $\alpha \in G_2^{n+1}$ be such that a polar sphere $\sigma(\alpha)$ does not intersect one of the planes P_i, $i = 1, 2, \ldots, m$. According to Lemma 6.2.1, such are almost all $\alpha \in G_2^{n+1}$. Let $y_i = \pi_\alpha(x_i)$. We have: $y_i = \eta_\alpha(x_i)/|\eta_\alpha(x_i)|$, where η_α is the operation of an orthogonal projection onto α. This therefore affords:

$$(y_{i-1} \frown y_i) = (\eta_\alpha(x_{i-1}) \frown \eta_\alpha(x_i))$$

and hence

$$\kappa_g(\pi_\alpha(L)) = \sum_{i=1}^{m} (\eta_\alpha(x_{i-1}) \widehat{} \eta_\alpha(x_i)). \tag{25}$$

In conformity with Theorem 6.2.1, we have:

$$\int_{G_l^{n+1}} \left[(\eta_\alpha(x_{i-1}) \widehat{} \eta_\alpha(x_i)) \right] \mu_{n+1,2} (d\alpha) = (x_{i-1} \widehat{} x_i).$$

Integrating equality (25) term by term, we get relation (23).

Let us now consider the plane Q_i of the link $[y_{i-1}, y_i]$ of the polygonal line $\pi_\alpha(L)$. Any pair $u, v \in [x_{i-1}x_i]$, such that $x_{i-1} < u < v < x_i$, in the sense of the order on the polygonal line L, forms a right-hand reference frame on the plane P_i. (We must here allow for the case when x_{i-1} and x_i are diametrically opposed points of Ω^n, which prevents us from simply taking the reference frame (x_{i-1}, x_i).) The vectors $\pi_\alpha(u)$ and $\pi_\alpha(v)$ taken in the given order form a right-hand reference frame of the plane Q_i. It makes it possible to conclude that $Q_i = \eta_\alpha(P_i)$. Therefore,

$$\kappa_g(\pi_\alpha(L)) = \sum_{i=1}^{m-1} [\eta_\alpha(P_i) \widehat{} \eta_\alpha(P_{i+1})].$$

Integrating this equality term by term, and allowing for Theorem 6.2.1, we get:

$$\int_{G_l^{n+1}} \kappa_g[\pi_\alpha(L)] \, \mu_{n+1,2} (d\alpha) = \sum_{i=1}^{m-1} \int_{G_l^{n+1}} (\eta_\alpha(P_i) \widehat{} \eta_\alpha(P_{i+1})) \, \mu_{n+1,2} (d\alpha)$$

$$= \sum_{i=1}^{m-1} (P_i \widehat{} P_{i+1}) = \kappa_g(L).$$

The theorem is proved.

6.3. Definition and Basic Properties of Spherical Curves of a Finite Geodesic Turn

6.3.1. Let K be a curve on the sphere Ω^n. A geodesic turn of the curve K is the quantity

$$\kappa_g(K) = \inf \varlimsup_{m \to \infty} \kappa_g(L_m)$$

where $L_1, L_2, \ldots, L_m, \ldots$ is an arbitrary sequence of spherical polygonal lines inscribed into the curve K, such that $\lambda(L_m) \to 0$ at $m \to \infty$, and the greatest lower boundary is taken on the set of all such sequences. For any spherical curve K, as can be easily proved, there exists a sequence (L_m), $m = 1, 2, \ldots$, of polygonal lines inscribed into it, such that $\lambda(L_m) \to 0$ and $\kappa_g(L_m) \to \kappa_g(K)$ at $m \to \infty$.

The reader can easily prove that in the case of a spherical polygonal line this definition of an geodesic turn coincides with the initial one, which follows from the theorem proved below.

The term 'geodesic' will henceforth be omitted each time it results in no ambiguity. A turn of a spherical curve in the sense of Chapter V will be termed, unlike a geodesic turn, a spatial turn.

Let L be an arbitrary spherical polygonal line. In this case, generally speaking, it is not true that for any polygonal line M inscribed into L the inequality $\kappa_g(M) \leqslant \kappa_g(L)$ holds, which accounts for the reason why we do not act here by analogy with the case considered in Chapter V.

THEOREM 6.3.1. *If a geodesic turn of the spherical curve K is finite, then its spatial turn is also finite.*

Proof. Let δ be a spherical diameter of the curve K, i.e., $\delta = \sup_{X,Y \in K} \rho(X, Y)$. Let us construct a sequence (L_m), $m = 1, 2, \ldots$, of the spherical polygonal lines inscribed into the curve K such that $\lambda(L_m) \longrightarrow 0$ and $\kappa_g(L_m) \longrightarrow \kappa_g(L_m)$ at $m \longrightarrow \infty$. At every m $\kappa(L_m) = s(L_m) + \kappa_g(L_m)$. The spherical and, hence, the spatial diameter of each of the polygonal lines L_m is not greater than δ. In line with Theorem 5.6.1, at every m we have the estimate $s(L_m) \leqslant \delta C_n[\kappa(l_m) + \pi]$, where C_n is a constant, $C_n > 0$. Let us set $\delta_0 = \frac{1}{2}C_n$ and assume that $\delta \leqslant \delta_0$. In this case we get:

$$\kappa(L_m) = s(L_m) + \kappa_g(L_m) \leqslant \delta C_n[\kappa(L_m) + \pi] + $$
$$+\kappa_g(L_m) \leqslant \tfrac{1}{2}\kappa(L_m) + \frac{\pi}{2} + \kappa_g(L_m).$$

From this we have:

$$\tfrac{1}{2}\kappa(L_m) \leqslant \frac{\pi}{2} + \kappa_g(L_m)$$

and, finally,

$$\kappa(L_m) \leqslant \pi + 2\kappa_g(L_m).$$

The sequence $(\kappa_g(L_m))$, $m = 1, 2, \ldots$, is bounded, since $\lim \kappa_g(L_m) = \kappa_g(K)$ is finite. This results in the boundedness of the sequence $(\kappa(L_m))$, $m = 1, 2, \ldots$, and, hence, $\kappa(K) \leqslant \underline{\lim}_{m \to \infty} \kappa(L_m)$ is finite. We have therefore established that $\kappa(K) < \infty$ if the spherical diameter δ of the curve K is not greater than δ_0. Let K be an arbitrary spherical curve. In this case K can be divided into a finite number of arcs the spherical diameter of each of which is not greater than δ_0. If a geodesic turn of the spherical curve is finite, then a geodesic turn of any of its arcs is also finite. In particular, geodesic turns of the arcs of the constructed division are finite. This means that the spatial turn of each of these arcs is finite and, hence $\kappa(K) < \infty$. The theorem is proved.

COROLLARY. *If for the spherical curve K $\kappa_g(K) < \infty$, then the curve K is one-sidedly smooth.*

6.3.2. A spherical curve is termed linear if it is contained in one great circumference.

LEMMA 6.3.1. *If K is a linear spherical curve then for any sequence of spherical polygonal lines $L_1, L_2, \ldots, L_m, \ldots$, inscribed into the curve K and such that $\lambda(L_m) \longrightarrow 0$ at $m \longrightarrow \infty$ then the relation $\kappa_g(L_m) \longrightarrow \kappa_g(K)$ at $m \longrightarrow \infty$ is valid.*

Proof. If $\kappa_g(K) = \infty$, then, by the definition of a geodesic turn, $\kappa_g(L_m) \longrightarrow \infty$

at $m \rightarrow \infty$. Let us assume that $\kappa_g(K) < \infty$. Then, in line with Theorem 6.3.1, the curve K is one-sidedly smooth and hence it is a particular spherical polygonal line; let the vertices of this polygonal line in successive order be $A_0, A_1, \ldots, A_n$. Some of these vertices are the points of return of the curve K. Let n' be their number. Let us denote by δ the least of the lengths of the arcs $[A_i A_{i+1}]$, $i = 0, 1, \ldots, n-1$. Let us find m_0, such that at $m > m_0$

$$\lambda(L_m) < \min\left(\frac{\delta}{2}, \pi\right).$$

At $m > m_0$ on each arc $[A_i A_{i+1}]$, $i = 0, 1, \ldots, n-1$, at least two vertices of the polygonal line L_m are located. let A_i'' be the utmost left-hand, and A_{i+1}' the utmost right-hand vertex of the polygonal line L_m which lie on the arc $[A_i A_{i+1}]$. The turns of the polygonal line L_m at the vertices lying between A_i'' and A_{i+1}' are obviously equal to zero. Let us consider the points A_{i+1}' and A_{i+1}''. Obviously, $A_{i+1}' \leqslant A_{i+1} \leqslant A_{i+1}''$. If a turn of the curve K at the point A_i equals zero, then the L_m turns at the points A_{i+1}' and A_{i+1}'' are also equal to zero. In the case when A_{i+1} is a point of return, however, it is one (and only one) of the points A_{i+1}' and A_{i+1}'' that is a point of return of L_m. Therefore, to each point of return of the curve K there corresponds a point of return of L_m. The curve L_m has no other points of return than those obtained in the way discussed above and, hence, a turn of L_m is equal to $\pi n' = \kappa_g(K)$. Since to prove it we only require that $m > m_0$, then the lemma is proved.

LEMMA 6.3.2. *Let K be an arbitrary linear spherical curve. In this case an absolute turn of any spherical polygonal line L which is inscribed into K and lies in the same great circumference as does K, is not greater than $2[s(K) + \kappa_g(K)]$.*

Proof. In the case when $\kappa_g(K) < \infty$, the lemma is obvious. Let us assume that $\kappa_g(K) < \infty$. In this case the curve K is a spherical polygonal line. Let $A_0, A_1, \ldots, A_m$ be vertices of the polygonal line L. Let us consider an arbitrary vertex A_i. Let $[A_{i-1} A_{i+1}]$ be an arc of the curve K between the vertices $[A_{i-1}, A_{i+1}]$ of the polygonal line L. If $\kappa_{g,L}(A_i) = \pi$, then there can be two possibilities: either $|\kappa_g|(A_{i-1} A_{i+1}) > 0$, and hence $|\kappa_g|(A_{i-1} A_{i+1}) \geqslant \pi$, which yields $\kappa_{g,L}(A_i) \leqslant |\kappa_g|(A_{i-1} A_{i+1})$; or $|\kappa_g|(A_{i-1} A_{i+1}) = 0$, in which case the e-quality $\kappa_g(A_i) = \pi$ is valid, as can be seen, only if $s(A_{i-1} A_{i-1}) \geqslant \pi$, which affords $\kappa_g(A_i) \leqslant s(A_{i-1} A_{i+1})$. In both cases

$$\kappa_{g,L}(A_i) \leqslant s(A_{i-1} A_{i+1}) + |\kappa_g|(A_{i-1} A_{i+1})$$

when $\kappa_{g,L}(A_i) = 0$, the latter inequality is obvious. By summing we obtain the inequality given in the lemma.

6.3.3. LEMMA 6.3.3. *Let E be a set of one-dimensional directions $\lambda \in G_1^{n+1}$, such that its measure $\mu_{n+1,1}(E)$ is zero. In this case there can be found a system of $n+1$ one-dimensional directions $\lambda_0, \lambda_1, \ldots, \lambda_n$, such that for any i, j, $i \neq j$ the straight lines λ_i and λ_j are orthogonal and $\lambda_i \notin E$ for all $i = 0, 1, 2, \ldots, n$.*

This is an evident corollary of Lemma 4.9.2.

6.3.4. THEOREM 6.3.2. *For any spherical curve K and for any sequence (L_m),* $m = 1, 2, \ldots$, *of the spherical polygonal lines inscribed into K, such that* $\lambda(L_m) \longrightarrow 0$ *at $m \longrightarrow \infty$ the limit $\lim_{m \to \infty} \kappa_g(L_m)$ exists and equals $\kappa_g(K)$. If a two-dimensional Hausdorff measure of the curve K is zero, then for almost all* $\alpha \in G_2^{n+1}$ *the quantity $\kappa_g(K_\alpha)$ is defined, the function $\alpha \mapsto \kappa_g(K_\alpha)$ is measurable and*

$$\kappa_g(K) = \int_{G_2^{n+1}} \kappa_g(K_\alpha) \, \mu_{n+1,2} \, (d\alpha).$$

Proof. If $\kappa_g(K) = \infty$, then from the definition of $\kappa_g(K)$ it follows that for any sequence (L_m) of the polygonal lines inscribed into K and such that $\lambda(L_m) \longrightarrow 0$ at $m \longrightarrow \infty$, $\kappa_g(L_m) \longrightarrow \infty$.

If $\kappa_g(K) < \infty$, then, according to the corollary of Theorem 6.3.1, the curve K is rectifiable and hence its two-dimensional Hausdorff measure equals zero. Therefore, in this case $\sigma(\alpha)$ does not intersect K for almost all $\alpha \in G_2^{n+1}$ and the curve $K_\alpha = \pi_\alpha(K)$ is defined for almost all such α.

Let us assume that K is a curve on the sphere Ω^n, such that a two-dimensional Hausdorff measure of the set $|K|$ equals zero. Let (L_m), $m = 1, 2, \ldots$, be an arbitrary sequence of spherical polygonal lines inscribed into the curve K, such that $\lambda(L_m) \longrightarrow 0$ at $m \longrightarrow \infty$. Let us set

$$A = |K| \cup \bigcup_{m=1}^{\infty} |L_m|.$$

A two-dimensional Hausdorff measure of the set A equals zero. Let U be a set of those $\alpha \in G_2^{n+1}$, for which the $(n-2)$-dimensional large sphere $\sigma(\alpha)$, which is polar to the large circumference $\gamma(\alpha)$, contains no points of the set A. The set U is measurable and $\mu_{n+1,2}(U) = 1 = \mu_{n+1,2}(G_2^{n+1})$. Let us arbitrarily choose $\alpha \in U$. For this α the curve K_α and the spherical polygonal line $L_{m,\alpha}$ inscribed into it are defined at every m. Let $\lambda_m(\alpha)$ be the greatest of the diameters of the arcs into which the curve K_α is divided by the vertices of the polygonal line $L_{m,\alpha}$. In this case $\lambda_m(\alpha) \longrightarrow 0$ at $m \longrightarrow \infty$. In line with Lemma 6.3.1, it affords $\kappa_g(L_{m,\alpha}) \longrightarrow \kappa_g(K_\alpha)$ for any $\alpha \in U$. The function $\alpha \mapsto \kappa_g(L_{m,\alpha})$ is measurable and hence the function $\alpha \mapsto \kappa_g(K_\alpha)$ is also measurable. According to Fatou's theorem, we have:

$$\int_{G_2^{n+1}} \kappa_g(K_\alpha) \, (d\alpha) \leqslant \varliminf_{m \to \infty} \int_{G_2^{n+1}} \kappa_g(L_{m,\alpha}) \, \mu_{n+1,2} \, (d\alpha) = \varliminf_{m \to \infty} \kappa_g(L_m). (26)$$

If

$$\int_{G_2^{n+1}} \kappa_g(K_\alpha) \, \mu_{n+1,2} \, (d\alpha) = \infty$$

then

$$\lim_{m \to \infty} \kappa_g(L_m) = \infty.$$

and, since the sequence (L_m) of the polygonal lines inscribed into K and such that $\lambda(L_m) \to 0$ at $m \to \infty$, is taken arbitrarily, then $\kappa_g(K) = \infty$.

Let us assume that a two-dimensional Hausdorff measure of the set $|K|$ equals zero and

$$\int_{G_2^{n+1}} \kappa_g(K_\alpha) \, \mu_{n+1,2} \, (d\alpha) < \infty \qquad (27)$$

In this case the curve K_α is one-sidedly smooth and hence rectifiable for almost all $\alpha \in G_2^{n+1}$. Let us first prove that the curve K is rectifiable, having preliminarily established that if for almost all n-dimensional directions α the projection $\pi_\alpha(K)$ of the curve K onto the great sphere $\gamma(\alpha)$ is rectifiable, then K is a rectifiable curve. Let us suppose that K obeys this condition. Let E' be a set of those $\alpha \in G_n^{n+1}$, for which the curve $\pi_\alpha(K)$ is not rectifiable, let E be a set of all one-dimensional directions λ orthogonal to the planes $\alpha \in E'$. In line with Lemma 6.3.3, there can be found a system of $(n+1)$ one-dimensional directions $\lambda_0, \lambda_1, \ldots, \lambda_n$, such that none of them belongs to $E \cup |K|$ and λ_i is orthogonal to λ_j at $i \neq j$, $i, j = 0, 1, \ldots, n$. Let α_i be an n-dimensional direction in E^{n+1} orthogonal to λ_i. It is obvious that α_i contains all λ_j, where $i \neq j$. Let us denote the curve K_i through $\pi_{\alpha_i}(K)$. As far as $\lambda_i \notin |K|$, then the curve K_i is defined and, since $\lambda_i \notin E$, then the curve K_i is rectifiable. Let X and Y be two arbitrary points of the sphere Ω^n. Let us denote by X_i and Y_i the points $\pi_{\alpha_i}(X)$ and $\pi_{\alpha_i}(Y)$, respectively. In accordance with Lemma 4.9.1, we have:

$$\rho(X, Y) \leqslant \frac{1}{\sqrt{n-1}} \sum_{i=0}^{n} \rho(X_i, Y). \qquad (28)$$

Let us arbitrarily set a sequence $X_0 < X_1 < \cdots < X_m$ of the points of the curve K. Let $X_{k,i} = \pi_{\alpha_i}(X_k)$. In line with (28), we get:

$$\sum_{k=1}^{m} \rho(X_{k-1}, X_k) \leqslant \frac{1}{\sqrt{n-1}} \sum_{i=0}^{n} \sum_{k=1}^{m} \rho(X_{k-1,i}, X_{k,i}) \leqslant \frac{1}{\sqrt{n-1}} \sum_{i=0}^{n} s(K_i) < \infty$$

and, since the sequence of the points $X_0 < X_1 < \cdots X_m$ of the curve K was chosen arbitrarily, it affords

$$s(K) \leqslant \frac{1}{\sqrt{n-1}} \sum_{i=0}^{n} s(K_i) < \infty$$

and hence the curve K is rectifiable.

Let us prove that for almost all $\alpha \in G_k^{n+1}$, where $2 \leqslant k \leqslant n$, the projection $\pi_\alpha(K)$ of the curve K onto α is a rectifiable curve. It is valid for $k = 2$. Let us assume that for a certain $k < n$ this statement is true. Let E be a set of those $\alpha \in G_k^{n+1}$, for which the curve $\pi_\alpha(K)$ is either not defined or is not rectifiable. Under the supposition we have $\mu_{n+1,k}(E) = 0$. As follows from Theorem 4.5.1, for nearly all $\lambda \in G_{k+1}^{n+1}$ the set $G_k(\lambda) \cap E$ is a set of zero measure in the space $G_k(\lambda)$. Let $\lambda \in G_{k+1}^{n+1}$ be such as has just been mentioned. Let us arbitrarily choose $\alpha \in G_k^{n+1}$, such that $\lambda \supset \alpha$. Then $\pi_\alpha(K) = \pi_\alpha(\pi_\lambda(K))$. For almost all $\alpha \in G_k(\lambda)$ the curve $\pi_\alpha(K)$ is rectifiable. Substituting n for k in the above considerations we find that the curve $\pi_\lambda(K)$ is rectifiable. Therefore we have established that $\pi_\lambda(K)$ is rectifiable for almost all $\alpha \in G_{k+1}^{n+1}$. By way of induction it follows from the above proof that for almost all $\alpha \in G_n^{n+1}$ the curve K_α is defined and rectifiable and thus we have proved that, if condition (27) is fulfilled for the curve K, then it is rectifiable.

Now we can easily finish proving the theorem. Let the curve K be such that a two-dimensional Hausdorff measure of the set $|K|$ equals zero and inequality (27) holds. Let us arbitrarily set a sequence (L_m), $m = 1, 2, \ldots$, of the spherical polygonal lines inscribed into the curve K and such that $\lambda(L_m) \to 0$ at $m \to \infty$. In this case for almost all $\alpha \in G_2^{n+1}$ we have:

$$\kappa_g(L_{m,\alpha}) \to \kappa_g(K_\alpha)$$

and, in line with Lemma 6.3.2:

$$\kappa_g(L_{m,\alpha}) \leqslant 2[\kappa_g(K_\alpha) + s(K_\alpha)].$$

The right-hand part of the latter inequality is a function integrable with respect to α and, hence, according to the Lebesgue theorem on a limiting transition,

$$\lim_{m \to \infty} \kappa_g(L_m) = \lim_{m \to \infty} \int_{G_2^{n+1}} \kappa_g(L_{m,\alpha}) \, \mu_{n+1,2}\,(d\alpha) = \int_{G_2^{n+1}} \kappa_g(K_\alpha) \, \mu_{n+1,2}\,(d\alpha).$$

The sequence of the spherical polygonal lines (L_m) inscribed into K, such that $\lambda(L_m) \to 0$ at $m \to \infty$ is taken arbitrarily and, hence, it proves that for any such sequence the limit $\lim_{m \to \infty} \kappa_g(L_m)$ does exist and equals the integral

$$\int_{G_2^{n+1}} \kappa_g(K_\alpha) \, \mu_{n+1,2}\,(d\alpha).$$

Therefore, this integral equals $\kappa_g(K)$ and for any sequence of the inscribed into K polygonal lines (L_m), such that $\lambda(L_m) \to 0$ at $m \to \infty$ we have

$$\kappa_g(L_m) \to \kappa_g(K).$$

The theorem is thus proved.

THEOREM 6.3.3. *Let $K = [AB]$ be a spherical curve and let X be an arbitrary internal point in it. In this case if $\kappa_g(AB) < \infty$, then*

$$\kappa_g(AB) = \kappa_g(AX) + \kappa_g(X) + \kappa_g(XB).$$

If, inversely, the geodesic turns of the arcs $[AX]$ and $[XB]$ are finite, then $\kappa_g(AB) < \infty$.

Proof. Let $\kappa_g(K) < \infty$. Let us construct a sequence of the polygonal lines $L_1, L_2, \ldots, L_m, \ldots$ inscribed into the curve K and having the point X as its vertex, in which case $\lambda(L_m) \to 0$ at $m \to \infty$. The polygonal line L_m is divided by the point X into two arcs: L_m' and L_m''. Let α_m be a turn of the polygonal line L_m at the vertex X. Obviously,

$$\kappa_g(L_m) = \kappa_g(L_m') + \alpha_m + \kappa_g(L_m'').$$

At $m \to \infty$, in line with Theorem 6.3.2, $\kappa_g(L_m') \to \kappa_g(AX)$ and $\kappa_g(L_m'') \to \kappa_g(XB)$. As far as K is a one-sidedly smooth curve, then $\alpha_m \to \kappa_g(X)$. Taking into account the fact that $\kappa_g(L_m) \to \kappa_g(K)$, we come to the conclusion that

$$\kappa_g(K) = \kappa_g(AX) + \kappa_g(X) + \kappa_g(XB).$$

Now, if $\kappa_g(AX) < \infty$, $\kappa_g(XB) < \infty$, then the arcs $[AX]$ and $[XB]$ have finite spatial turn; in particular, there exists a left- and a right-hand tangents at the point X. It obviously affords that $\kappa(K) < \infty$ and then $\kappa_g(K) < \infty$.

THEOREM 6.3.4. *Let K be a spherical curve of a finite geodesic turn, X be an arbitrary point of the curve. Then, if $Y \to X$ along the curve, then $\kappa_g(XY) \to 0$.*

Proof. According to Theorem 6.3.2, we have:

$$\kappa_g(K) = \int_{G_2^{n+1}} \kappa_g(K_\alpha)\, \mu_{n+1,2}\,(d\alpha)$$

and, for almost all α, $\kappa_g(K_\alpha) < \infty$. At such α the curve K_α is a spherical polygonal line with a finite number of links. Let, for the sake of definitness, $Y \to X$ from the left. If $[YX]_\alpha$ is a projection of the arc $[YX]$ onto $\gamma(\alpha)$, then $\kappa_g([YX]) \leqslant \kappa_g(K_\alpha)$, $\kappa_g([YX]_\alpha) \to 0$, since at Y, which is sufficiently close to X, the arc $[YX]_\alpha$ is an arc of a great circumference. Integrating, we obtain that

$$\kappa_g(YX) = \int_{G_2^{n+1}} \kappa_g([YX]_\alpha)\, \mu_{n+1,2}\,(d\alpha) \to 0$$

which is the required proof.

6.4. Definition of a Geodesic Turn by Means of Tangents

6.4.1. By analogy with the case of general spatial curves, an absolute turn of a spherical curve can be defined by means of its tangents. In order to formulate a definition that may possibly be applicable to a more general class of spherical curves, the notion of a geodesic contingency of a curve is used here.

A right- (left-)hand contingency at the point X of a spherical curve K is a totality $q_r(X)$ $(q_l(X))$ of all oriented great circumferences which are the limits of all possible sequences of the intersecting circumferences $C(XX_n)$ when the points X_n converge to the point X along the curve from the right (left).

A set of all the contingencies gets, naturally, ordered if each of the contingencies at the point X is assumed to precede each contingency at the point Y, if $X < Y$, and if, for any point X, $q_l(X) < q_r(X)$. A finite sequence ξ of the spherical contingencies of the curve K located in the order of increasing is termed a chain of the K contingencies.

Let C_1 and C_2 be two oriented great circumferences of the sphere Ω^n, P_1 and P_2 be the planes of the circumferences C_1 and C_2, respectively, oriented in such a way that at the motion of the point x along C_i in the positive direction the vector $\overline{Ox}$ also turns in the positive direction in the plane P_i.

We assume:

$$\theta(C_1, C_2) = \theta(P_1, P_2) = \int_{G_2^{n+1}} (P_{1,\alpha}\overset{\frown}{,}P_{2,\alpha})\, \mu_{n+1,2}\,(d\alpha).$$

The quantity $\theta(C_1, C_2)$ will be termed an integro-geometrical angle between the great circumferences C_1 and C_2. The term 'integro-geometrical' will henceforth be omitted in cases when no ambiguity results. For any three large circumferences C_1, C_2 and C_3 the following inequality is valid: $\theta(C_1, C_2) \leqslant \theta(C_1, C_2) + \theta(C_2, C_3)$. Indeed, in this case for any $\alpha \in G_2^{n+1}$ which is orthogonal to none of the planes P_1, P_2, P_3, we have: $(P_{1,\alpha} \widehat{} P_{3,\alpha}) \leqslant (P_{1,\alpha} \widehat{} P_{2,\alpha}) + (P_{2,\alpha} \widehat{} P_{3,\alpha})$. The required result is obtained by integrating this inequality term by term.

Let $\xi = \{q_1 < \cdots < q_m\}$ be an arbitrary chain of contingencies of the curve K, C_i be an arbitrary great circumference belonging to the contingency q_i. Let us set

$$\kappa_g^*(\xi) = \sup_{C_i \in q_i} \sum_{i=1}^{m-1} \theta(C_i, C_{i+1}),$$

where the upper boundary is shared by a set of all possible sequences of great circumferences $C_1, C_2, \ldots, C_m$, where $C_i \in q_i$ at every i.

The exact upper boundary of the quantity $\kappa_g^*(\xi)$ on a set of all the chains of the contingencies of the curve K will be denoted through $\mu(K)$. The basic result of this paragraph is expressed in the following theorem.

THEOREM 6.4.1. *For any spherical curve the value $\mu(K)$ is equal to its geodesic turn.*

Let us first dwell in detail on the case of one-sidedly smooth spherical curves. Each of the contingencies $q_l(X)$ and $q_r(X)$ in this case consists of a single great circumference - a tangent of the curve K. If $\xi = \{t_1, t_2, \ldots, t_m\}$ is an arbitrary chain of contingencies of a curve, then $\kappa_g^*(\xi)$ is equal to the sum $\sum_{i=1}^{m-1} \theta(t_i, t_{i+1})$. The points of a curve, the tangents at which comprise the the chain ξ, divide the curve K into arcs. The largest of the diameters of these arcs will be termed a modulus of the chain x and denoted by $\lambda(\xi)$. The inequality for a triangle for the angles between large circumferences (i.e., the relation $\theta(C_1, C_3) \leqslant \theta(C_1, C_2) + \theta(C_2, C_3)$) yields that if the chain ξ' is obtained from ξ by adding new elements, or, in other words, if $\xi' \supset \xi$, then $\kappa_g^*(\xi') \geqslant \kappa_g^*(\xi)$. Let us prove that in the case under discussion there exists a sequence of chains of the tangents $\xi_1, \xi_2, \ldots, \xi_n, \ldots$, such that at $n \to \infty$ $\lambda(\xi_n) \to 0$ and $\kappa_g^*(\xi_n) \to \mu(K)$. Indeed, let there first be a sequence of chains $\eta_1, \eta_2, \ldots, \eta_n, \ldots$, such that, at every n, $\kappa_g^*(\eta_n) > \mu(K) - 1/n$. Let us add to the chain η_n new tangents in such a way that a chain ξ_n with $\lambda(\xi_n) < 1/n$ would result. As far as $\kappa_g^*(\xi_n) \geqslant \kappa_g^*(\eta_n) > \mu(K) - 1/n$, then $\kappa_g^*(\xi_n) \to \mu(K)$ at $n \to \infty$ and, hence, the sequence $\xi_1, \xi_2, \ldots, \xi_n, \ldots$ is the one sought.

The following lemmas are proved in complete analogy with the corresponding lemmas of Chapter V.

LEMMA 6.4.1. *If $\mu(K) < \infty$, then the spherical curve K is one-sidedly smooth.*

LEMMA 6.4.2. *For any spherical curve the inequality $\kappa_g(K) \geqslant \mu(K)$ is valid.*

Out of these lemmas only Lemma 6.4.2 might require some clarification. When proving this lemma, one should act in the following way. By analogy with what has been done when proving the lemma, it is established that for any $\varepsilon > 0$ and for any chain ξ there can be found a spherical polygonal line L inscribed into K, such that $\kappa_g(L) > \kappa_g^*(\xi) - \varepsilon$, $\lambda(L) \leqslant \lambda(\xi)$. Therefore, for any $\delta > 0$ $\sup_{\lambda(L) \leqslant \delta} \kappa_g(L) \geqslant \sup_{\lambda(\xi) \leqslant \delta} \kappa_g^*(\xi) - \varepsilon$, which yields

$$\kappa_g(K) = \lim_{\delta \to 0} \left(\sup_{\lambda(L) \leqslant \delta} \kappa_g(L) \right) \geqslant \mu(K)$$

and Lemma 6.4.2 is thus proved.

To complete the proof it is necessary to establish that for any curve K the following inequality holds

$$\kappa_g(K) \leqslant \mu(K).$$

For the case of curves in E^n (Chapter V) in order to prove an analogous statement use was made of the considerations which cannot be employed in the case under discussion. Let us here use considerations based on the integro-geometrical representation of a geodesic turn proved in Section 6.3.

It should be remarked that if the curve K lies in one great circumference, then $\kappa_g(K) = \mu(K)$. Indeed, if the curve K is one-sidedly smooth, then it is a spherical polygonal line and in this case the equality $\kappa_g(K) = \mu(K)$ is obvious. In the case when the curve K is not one-sidedly smooth, $\kappa_g(K) = \infty$, and, in line with Lemma 6.4.1, also $\mu(K) = \infty$, we have $\kappa_g(K) = \mu(K)$, too.

We shall also need the following supposition.

LEMMA 6.4.3. *If for a spherical curve K, $\mu(K) < \infty$, then for almost all $\alpha \in G_2^{n+1}$ the curve K_α is defined and $\kappa_g(K_\alpha) < \infty$.*

Before proving the last lemma, let us demonstrate in what way Theorem 6.4.1 can be deduced from it and from Lemmas 6.4.1 and 6.4.2.

At $\mu(K) = \infty$, we have, according to Lemma 6.4.2, $\kappa_g(K) = \infty$, therefore in this case the equality $\mu(K) = |\kappa_g|(K)$ is valid.

Let $\mu(K) < \infty$. Then, by Lemma 6.4.1, the curve K is one-sidedly smooth and, hence, rectifiable, and, in line with Theorem 6.3.2

$$\kappa_g(K) = \int_{G_2^{n+1}} \kappa_g(K_\alpha) \, \mu_{n+1,2} \, (d\alpha).$$

Let us also prove that

$$\mu(K) = \int_{G_2^{n+1}} \kappa(K_\alpha) \, \mu_{n+1,2} \, (d\alpha).$$

Let us choose an arbitrary sequence $\xi_1, \xi_2, \ldots, \xi_m, \ldots$ of the chains of the tangents of the curve K, such that $\lambda(\xi_m) \to 0$ at $m \to \infty$. Let E_0 be a set of those $\alpha \in G_2^{n+1}$ for which the polar sphere $\sigma(\alpha)$ intersects the curve K, E_m be a totality of those $\alpha \in G_2^{n+1}$, for which $\sigma(\alpha)$ intersects at least one of the tangent great circumferences, which form the chain ξ_m, $E = \bigcup_{m=0}^{\infty} E_m$. In this case E is a set of zero measure in G_2^{n+1}.

Let $\alpha \notin E$. Projections of the tangents of the chain ξ_m form a certain chain

$\xi_{m,\alpha}$ of the tangents of the curve K_α. It is seen that

$$\kappa_g^*(\xi_m) = \int_{G_2^{n+1}} \kappa_g^*(\xi_{m,\alpha}) \, \mu_{n+1,2} \, (d\alpha).$$

Furthermore, we have $\kappa_g^*(\xi_{m,\alpha}) \leqslant \kappa_g(K_\alpha)$ and, if $\kappa_g(K_\alpha) < \infty$, then at $m \to \infty$ $\kappa_g^*(\xi_{m,\alpha}) \to \kappa_g(K_\alpha)$. If $\kappa_g(K_\alpha) = \infty$, then, generally speaking, $\kappa_g^*(\xi_{m,\alpha})$ may not converge to $\kappa_g(K_\alpha)$, as can be shown by examples. But, according to Lemma 6.4.3 a set of such α has the measure equal to zero and, hence $\kappa_g^*(\xi_{m,\alpha}) \to \kappa_g(K_\alpha)$ for almost all α. From this we get

$$\lim_{m \to \infty} \kappa_g^*(\xi_m) = \lim_{m \to \infty} \int_{G_2^{n+1}} \kappa_g^*(\xi_{m,\alpha}) \, \mu_{n+1,2} \, (d\alpha) =$$

$$= \int_{G_2^{n+1}} \kappa_g(K_\alpha) \, \mu_{n+1,2} \, (d\alpha).$$

The sequence (ξ_m) can be chosen in such a way that at $m \to \infty$ $\kappa_g^*(\xi_m) \to \mu(K)$ and the theorem is therefore proved.

<u>Proof of Lemma 6.4.3.</u> Since $\mu(K) < \infty$, then, by Lemma 6.4.1 the curve K is one-sidedly smooth and, in particular, K is rectifiable. Therefore, the curve K_α is defined for almost all $\alpha \in G_2^{n+1}$.

Let A_k, where k is a natural number, be a totality of all α, for which the curve K_α is defined and $\kappa_g(K_\alpha) > k$. The set A_k is measurable, since the function $\kappa_g(K_\alpha) = \lim_{m \to \infty} \kappa_g(L_{m,\alpha})$ is measurable, where L_m are inscribed into K polygonal lines, such that $\lambda(L_m) \to 0$ at $m \to \infty$. Obviously, $A_1 \supset A_2 \supset \cdots \supset A_k \supset \cdots$ and the intersection M of all the sets A_k is a totality of all $\alpha \in G_2^{n+1}$, for which $\kappa_g(K_\alpha) = \infty$.

Let us prove that, at every k, $\mu_{n+1,2}(A_k) \leqslant \mu(K)/k$. Let us exclude from consideration a set $\mathcal{E}$ of those $\alpha \in G_2^{n+1}$, for which the polar sphere $\sigma(\alpha)$ intersects a large circumference containing a whole arc of the curve K. Obviously, $\mu_{n+1,2}(\mathcal{E}) = 0$. Let us assume $A_k' = A_k \backslash \mathcal{E}$. Let F be an arbitrary closed set contained in A_k'.

Let us arbitrarily choose $\alpha \in F$. In this case, as has been noted above,

$$\mu(K_\alpha) = \kappa_g(K_\alpha) > k.$$

Therefore, on the curve K_α there can be found a sequence of the points $y_1, y_2, \ldots, y_m$, such that out of the contingencies K_α at the points y_i we can choose great circumferences $l_1, l_2, \ldots, l_m$, for which $\sum_{i=1}^{m-1} (l_i, \widehat{} l_{i+1}) > k$.

In any neighbourhood of the point y_i there is a point $z \in K_\alpha$, for which $l(y_i z) = l_i$. For every i let us choose such a point $z = z_i$, which is sufficiently close to y_i. The arcs $[y_i z_i]$ can be considered pair by pair having no common internal points, the length of each of them being less than π. If $[y_i' z_i']$ is an arc of the curve K, a projection of which onto $\gamma(\alpha)$ is the arc $[y_i z_i]$, then on the arc $[y_i' z_i']$ there can be found such a point t_i at which one of the tangents λ_i possesses the following properties: (1) λ_i is not orthogonal to α and (2) when projecting onto $\gamma(\alpha)$ the direction λ_i goes over to the direction l_i.

The sequence $\lambda_1, \lambda_2, \ldots, \lambda_m$ forms a chain ξ. Relative to continuity there can be found a vicinity V of the two-dimensional direction α, such that for any $\beta \in V$:

$$\sum_{i=1}^{m-1} (\lambda_{i,\beta}, \widehat{\quad} \lambda_{i-1,\beta}) = \sum_{i=1}^{m-1} (\lambda_{i,\alpha}, \widehat{\quad} \lambda_{i-1,\alpha}) > k.$$

(Here $\lambda_{i,\beta}$ and $\lambda_{i,\alpha}$ are the projections of the large circumference λ_i onto $\gamma(\beta)$ or $\gamma(\alpha)$, respectively.)

Therefore, for any point $\alpha \in F$ we have its vicinity V in the manifold G_2^{n+1} and the chain ξ of the tangents of the curve K, such that, for any $\beta \in V$ $\kappa_g^*(\xi_\beta) > k$. The set F is closed and hence compact. According to the Borel lemma, there exists a finite system of neighbourhoods $V_1, V_2, \ldots, V_r$, such that $F \subset \bigcup_{j=1}^{r} V_j$, and at every $j = 1, 2, \ldots, r$, there exists a chain ξ of the tangents of the curve K, for which $\kappa_g^*(\xi_{j,\beta}) > k$ for any $b \in V_j$. Let ξ be a chain obtained by uniting the chains ξ_j. Each $\alpha \in F$ belongs to at least one of the neighbourhoods V_j and, hence,

$$\kappa_g^*(\xi_\alpha) \geqslant \kappa_g^*(\xi_{j,\alpha}) > k.$$

From this we conclude:

$$\mu(K) \geqslant \kappa_g^*(\xi_\alpha) = \int_{G_2^{n+1}} \kappa_g^*(\xi_{j,\alpha}) \, \mu_{n+1,2} \, (d\alpha) \geqslant k\mu_{n+1,2} \, (F)$$

and, hence

$$\mu_{n+1,2} \, (F) \leqslant \frac{\mu(K)}{k}.$$

Since F is an arbitrarily closed set included in A_k', what we have proved above yields

$$\mu_{n+1,2}(A_k) = \mu_{n+1,2}(A_k') = \sup_{F \subset A_k'} \mu_{n+1,2}(F) \leqslant \frac{\mu(K)}{k},$$

which is the required proof.

In particular, we obtain $\mu_{n+1,2}(A_k) \to 0$ at $k \to \infty$ and, hence, the measure of the set M of those α for which $\kappa_g(K_\alpha) = \infty$ equals zero. The lemma is proved.

In the process of proving Theorem 6.4.1 we have, in essence, established the following assumption.

THEOREM 6.4.2. *For an arbitrary one-sidedly smooth spherical curve K under the condition $\lambda(\xi) \to 0$ a limit of the sums $\kappa_g^*(\xi)$, where ξ is a chain of the tangents of the curve, exists and is equal to $\kappa_g(K)$.*

6.4.2. Let us prove that for a spherical curve of a finite geodesic turn the quantity $\kappa_g(K)$ can be defined with the help of sequences of the secants of the double chains of this curve.

Let us arbitrarily set a spherical curve K and let $\xi = \{X_1, Y_1; X_2, Y_2; \ldots; X_p, Y_p\}$ be an arbitrary double chain of the curve K. Let us denote through $l_i = l(X_i Y_i)$ the circumference of a large circle connecting the points X_i and Y_i. If such a circumference is unambiguously defined, then as $l_i(X_i Y_i)$ we can arbitrarily take a great circumference passing through the points X_i and Y_i. Let us orientate the circumference l_i in such a way that at the motion in the positive direction along the shortest arc $[X_iY_i]$ the point X_i is in front of the point Y_i. Let $\lambda(\xi)$ and $\nu(\xi)$ be a modulus and a multiplicity of the double chain, respectively. Let us set

$$\kappa_g(\xi) = \sum_{i=1}^{p-1} \theta(l_i, l_{i+1}).$$

The quantity $\kappa_g(\xi)$ is termed a spherical turn of the double chain ξ. Let us prove the following theorem, which is analogous to Theorem 5.7.1.

THEOREM 6.4.3. *Let K be an arbitrary spherical curve with a finite absolute turn. Then for any sequence of the double chains $\xi_1, \xi_2, \ldots, \xi_m, \ldots$ of the curve K, such that $\lambda(\xi_m) \to 0$ at $m \to \infty$ and $\nu(\xi_m) < N < \infty$ at all m, $\lim_{m \to \infty} |\kappa_g|(\xi_m)$ exists and is equal to the geodesic turn of the curve K.*

The proof is based on the following Lemmas 6.4.4 and 6.4.5.

LEMMA 6.4.4. *Let K be an arbitrary spherical polygonal line lying in one large circumference, and let $\xi_1, \xi_2, \ldots, \xi_m, \ldots$ be an arbitrary sequence of double chains of the polygonal line K, such that $\lambda(\xi_m) \to 0$ at $m \to \infty$. In this case $\kappa_g(\xi_m) \to |\kappa_g|(K)$ at $m \to \infty$.*

The lemma is proved in nearly the same way as Lemma 5.7.1. Namely, in either case we establish that at $\lambda(\xi) < \min \{\delta_0, \pi/2\}$ the equality $|\kappa_g|(\xi_m) = |\kappa_g|(K)$ holds, where δ_0 is the least of the lengths of the links of the polygonal line K.

It is necessary to require $\lambda(\xi_m) < \pi/2$ for the secant $l(X_iY_i)$ of the chain ξ_m, for which the both points X_i and Y_i lie in one link of the polygonal line K, to have the direction coinciding with that of the link.

LEMMA 6.4.5. *Let K be an arbitrary spherical polygonal line included in a large circumference C. In this case for any double chain of the curve K the following inequality is valid:*

$$\kappa_g(\xi) \leqslant [\nu(\xi) + 1][\kappa_g(K) + s(K)].$$

(*It is assumed here that all the secants of the double chain are also included in the same large circumference C.*)

Proof. Let $\xi = \{X_1, Y_1; X_2, Y_2; \ldots; X_p, Y_p\}$ be an arbitrary double chain of the curve K. Let us consider the secants $l_i = l(X_iY_i)$ and $l_{i+1} = (X_{i+1}Y_{i+1})$. If $(l_i, l_{i+1}) = \pi$, then either $\kappa_g(X_iY_{i+1}) > 0$ and, hence, $\kappa_g(X_iY_{i+1}) \geqslant \pi$, or $\kappa_g(X_iY_{i+1}) = 0$. In the latter case there must be $s(X_iY_{i+1}) \geqslant \pi$, since at $s(X_iY_{i+1}) < \pi$ and $\kappa_g(X_iY_{i+1}) = 0$ all the intersecting arcs $[X_iY_{i+1}]$ have the same direction. Therefore, we have:

$$(l_i, l_{i+1}) \leqslant \kappa_g(X_iY_{i+1}) + s(X_iY_{i+1}).$$

Summing up with respect to i, we get:

$$\kappa_g(\xi) \leqslant \sum_{i=1}^{p-1} \kappa_g(X_i Y_{i+1}) + \sum_{i=1}^{p-1} s(X_i Y_{i+1})$$

$$\leqslant [\nu(\xi) + 1][\kappa_g(K) + s(K)].$$

The latter inequality is proved using the same considerations as when proving Lemma 5.7.2.

With the help of Lemmas 6.4.4 and 6.4.5, Theorem 6.4.3 can be proved by the considerations absolutely analogous to those cited above when proving Theorem 5.7.1.

6.5. Curves on a Two–Dimensional Sphere

6.5.1. Let us consider the case of curves on a sphere in a three-dimensional Euclidean space E^3.

The space E^3 is considered oriented. All the tangential planes of the sphere Ω^2 will be henceforth assumed oriented in the following way. Let P be a tangential plane of the sphere Ω^2 at the point X. A pair of vectors (a, b) lying in the plane P is considered right-hand if $(\overline{OX}, a, b)$ is a right-hand trio in E^3.

If a and b are two non-zero vectors lying in the tangential plane P of the sphere Ω^2 at the point X, then $[a\frown b]$ denotes an angle between them taken with allowances made for the orientation, i.e. $[a\frown b]$ is a number φ lying in the segment $(-\pi, \pi]$ and such that $|\varphi|$ is an angle in the sense of former definitions, $\varphi > 0$ if the pair (a, b) is right-handed, $\varphi < 0$ if this pair is left-handed.

Let K be a one-sidedly smooth spherical curve on Ω^2, K be an internal point of the curve K. The angle $(\kappa_g)(X) = [t_l(X)\frown t_r(X)]$ is termed a rotation at the point X of the curve K. Obviously, $|(\kappa_g)(X)| = \kappa(X)$ and $(\kappa_g)(X) = 0$ if and only if X is an angular point of the curve K.

Let L be a spherical polygonal line on the sphere Ω^2. A sum of rotations at all its vertices is termed a geodesic rotation of the polygonal line L and is denoted herefrom through the symbol $(\kappa_g)(L)$. (The term 'geodesic' will be henceforth omitted each time the omition results in no ambiguity.)

Let K be a simple closed curve on the sphere Ω^2. The set $\Omega^2/|K|$ falls into two connected components D_1 and D_2. When by-passing the curve K in the positive direction, one of them appears to lie to the right of K (let it be D_1), and the other to the left. Let us say that D_1 is the right-hand side of the curve K, and D_2 is the left-hand one.

LEMMA 6.5.1. *Let L be a simple closed spherical polygonal line. In this case the following inequality is valid:*

$$\kappa_g(L) + \omega(D) = 2\pi,$$

where D is the right-hand side of the polygonal line L, $\omega(D)$ is the area of the domain of D.

Proof. Let $A_0, A_i, \ldots, A_m = A_0$ be the vertices of the polygonal line L numbered in the order of their location on L. Through α_i let us denote an angle of the polygon D at the vertex A_i. A geodesic turn of the polygonal line L at the point A_i is equal to $\pi - \alpha_i$ and hence

$$\kappa_g(L) = \sum_{i=0}^{m-1} (\pi - \alpha_i) = m\pi - \sum_{i=0}^{m-1} \alpha_i. \tag{29}$$

On the other hand, the following formula for the domain of a spherical polygon D is known:

$$\omega(D) = \sum_{i=0}^{m-1} \alpha_i - (m-2)\pi. \tag{30}$$

Summing the above two equalities, we obtain the required result.

THEOREM 6.5.1. *Let K be an arbitrary one-sidedly smooth spherical curve having no points of return. In this case at $\lambda(L) \to 0$ the geodesic turns of the spherical polygonal lines inscribed into K converge to a certain finite limit.*

Proof. (a) Let us first assume that the curve K lies in a semi-sphere, on the boundary of which there can be found a point S, such that any large circumference passing through the point intersects the curve K at no more than one point. Let us assume besides that, when moving along K in the positive direction, the point S appears to be lying to the right. The latter assumption can obviously always be true by way of substituting, when necessary, the point S by a diametrically opposing point.

Let A and B be the beginning and the end of the curve K, respectively. Let us connect the points A and B with S with the shortest arcs of the great circumference. The arcs $[SA]$ and $[SB]$ and the curve K all together limit a certain spherical domain R. Let us denote by α and β the angles of this domain at the points A and B, and by σ the angle of the domain R at the point S. (Namely, α is the value of the angle between the great circumference SA and a tangent at the point A of the curve K, which contains all the points of the domain R. The angle β is analogously determined.)

Let $L_1, L_2, \ldots, L_n, \ldots$ be an arbitrary sequence of polygonal lines inscribed into the curve K, such that at $n \to \infty$ $\lambda(L_n) \to 0$. Let us denote by A_n the beginning, and by B_n the end of the polygonal line L_n. The points A_n and B_n will be connected with the point S by the shortest arc of the great circumference. The arcs $[SA_n]$ and $[SB_n]$ and the polygonal line L_n limit a certain spherical polygonal R_n. The angles at its vertices A_n, B_n and S will be denoted through α_n, β_n and σ_n, respectively.

It is obvious that at $n \to \infty$ $\sigma_n \hookrightarrow \sigma$. As far as the curve K is one-sidedly smooth, at $n \to \infty$ we also have $\alpha_n \to \alpha$ and $\beta_n \to \beta$.

The polygonal line L_n and the arcs $[SA_n]$ and $[SB_n]$ comprise a simple closed polygonal line M_n. Let us orient it in such a way that the arc L_n included in M_n had its original direction. In this case the domain R_n is the right-hand side of the polygonal line M_n and, on the basis of the Lemma, we have:

$$\omega(R_n) + \kappa_g(M_n) = 2\pi.$$

On the other hand, we, obviously, have:

$$\kappa_g(M_n) = \pi - \alpha_n + \pi - \beta_n + \pi - \sigma_n + \kappa_g(L_n).$$

And, herefrom,

$$\kappa_g(L_n) = \alpha_n + \beta_n + \sigma_n - \pi - \omega(R_n).$$

At $n \to \infty$ $\omega(R_n) \to \omega(R)$. In order to prove it, let us introduce a polar system of coordinates (ρ, φ) with the origin S. Let

$$\rho = \rho(\varphi), \qquad\qquad \varphi_1 \leqslant \varphi \leqslant \varphi_2;$$
$$\rho = \rho_n(\varphi), \qquad\qquad \varphi_1^{(n)} \leqslant \varphi \leqslant \varphi_2^{(n)},$$

be the equations of the curve K and the polygonal line L, respectively, in these coordinates. In this case at $n \to \infty$ $\rho_n(\varphi) \to \rho(\varphi)$ uniformly, $\varphi_i^{(n)} \to \varphi_i$, $i = 1, 2$. For the areas, obviously, we have:

$$\omega(R) = \int_{\varphi_1}^{\varphi_2} [1 - \cos \rho(\varphi)] \, d\varphi, \qquad \omega(R_n) = \int_{\varphi_1^{(n)}}^{\varphi_2^{(n)}} [1 - \cos \rho_n(\varphi)] \, d\varphi.$$

From this, on the basis of the known theorems on integral convergence, we conclude that at $n \to \infty$ $\omega(R_n) \to \omega(R)$.

It follows from the above proved that at $n \to \infty$ $\kappa_g(L_n)$ converges to a limit equal to $\alpha + \beta + \sigma - \pi - \omega(R)$. Therefore, for the particular case (a) the theorem is proved.

(b) Let us consider a general case. Let us demonstrate that the curve K can be subdivided into the arcs $[A_0 A_i]$, $[A_1 A_2]$, ..., $[A_{m-1} A_m]$, where A_0 is the origin, A_m is the end of the curve K, in such a way that each of the points A_i, $i = 1, 2, \ldots, m-1$, is smooth, and each of the arcs $[A_i A_{i+1}]$ obeys all the conditions of the case (a).

Since, by condition, the curve is one-sidedly smooth and has no points of return, each of its points has a neighbourhood obeying the mentioned conditions. Covering the curve, by the Borel lemma, with a finite number of such neighbourhoods, we obtain the required subdivision.

Let L_1, L_2, ..., L_n, ... be an arbitrary sequence of spherical polygonal lines inscribed into the curve K, such that $\lambda(L_n) \to 0$ at $n \to \infty$. Let us denote through δ the largest of the diameters of the arcs $[A_i A_{i+1}]$. Let m be such that, at $m > m_0$, $\lambda(L_m) < \delta/3$. At $m > m_0$ on each of the arcs $[A_i A_{i+1}]$ there are at least two vertices of the polygonal line L_m. Let $A_{i,n}''$ be the utmost left-hand, and $A_{i,n}'$ the utmost right-hand of the vertices of the polygonal line L_n, lying on tha arc $[A_i A_{i+1}]$, $i = 0, 1, 2, \ldots, m-1$. The points $A_{i,n}'$ and $A_{i,n}''$ either coincide or are the ends of one and the same link of the polygonal line L_n. The arc $[A_{i,n}'' A_{i+1,n}']$ of the polygonal line L_n will be denoted by $L_n^{(i)}$. We obviously have:

$$\kappa_g(L) = \sum_{i=1}^{m-1} \kappa_g(L_n^{(i)}) + \sum_{i=1}^{m-1} [\kappa_g(A'_{i,n}) + \kappa_g(A''_{i,n})].$$

At $n \to \infty$, on the basis of what has been said above, each of the addends $\kappa_g(L_n^{(i)})$ converges to a certain limit. The points $A'_{i,n}$ and $A''_{i,n}$ converge to the point A_i at $n \to \infty$. The point A_i, by the condition, is smooth. As far as the curve K is one-sidedly smooth, turns of the polygonal line L_n at the points $A'_{i,n}$ and $A''_{i,n}$ tend to zero at $n \to \infty$, i.e., at $n \to \infty$ $\kappa_g(A'_{i,n}) + \kappa_g(A''_{i,n}) \to 0$. Therefore $\kappa_g(L_n)$ is a sum of addends, each of which has a finite limit. From this we can conclude that the quantity $\kappa_g(L_n)$ itself converges to a certain finite limit at $n \to \infty$. The theorem is thus proved.

A geodesic rotation of a one-sidedly smooth spherical curve K having no points of return, is a limit $\kappa_g(K)$ of geodesic turns of the spherical polygonal lines L inscribed into the curve K at $\lambda(L) \to 0$. Let us assume that a one-sidedly smooth curve K, set on a sphere Ω^2, has points of return, the number of which is finite. Let $X_1 < X_2 < \cdots < X_r$ be the points of return of the curve K, let X_0 be its origin, and X_{r+1} its end. In this case we have:

$$(\kappa_g)(K) = \sum_{i=1}^{r+1} (\kappa_g)([X_{i-1}X_i]) + r\pi.$$

(Each of the addends in the right-hand part is defined since, by the condition, the curves $[X_{i-1}X_i]$ have no points of return.)

THEOREM 6.5.2. *Let K be an arbitrary one-sidedly smooth spherical curve, X be an arbitrary internal point of the curve K, A be its origin and B be its end. Then the following formula is valid:*

$$(\kappa_g)(AB) = (\kappa_g)(AX) + (\kappa_g)(X) + (\kappa_g(XB).$$

Proof. Let us first consider the case when the curve K has no points of return. Let us construct an arbitrary sequence of the polygonal lines $L_1, L_2,$ $\ldots, L_m, \ldots$, inscribed into the curve K, and such that $\lambda(L_m) \to 0$ at $m \to \infty$ and the point X is a vertex of each of these polygonal lines. The point X subdivides the polygonal line L_m into two polygonal lines L'_m and L''_m, inscribed into the arcs $[AX]$ and $[XB]$, respectively. At all m we have:

$$(\kappa_g)(L_m) = (\kappa_g)(L'_m) + (\kappa_g)(L'_m) + (\kappa_{g,L_m})(X).$$

On the basis of Theorem 6.5.1 and the definition of the quantity (κ_g), at $m \to \infty$ we have:

$$(\kappa_g)(L_m) \to (\kappa_g)(K) = (\kappa_g)(AB),$$

$$(\kappa_g)(L'_m) \to (\kappa_g)(AX),$$

$$(\kappa_g)(L''_m) \to (\kappa_g)(XB).$$

As far as the curve K is one-sidedly smooth, and the point X is not a point of return of the curve K, then at $m \to \infty$:

$$(\kappa_{g,L_m})(X) \rightarrow (\kappa_{g,K})(X),$$

where $(\kappa_{g,K})(X)$ is a turn of the curve K at the point X. Therefore, passing to a limit, we see that the theorem is valid for the case when a curve has no points of return.

The theorem is obviously valid in the general case as well.

CHAPTER VII

Osculating Planes and Class of Curves with an Osculating Plane in the Strong Sense

7.1. Notion of an Osculating Plane

7.1.1. Let us begin by making certain remarks concerning the notion of orientation for the case of two-dimensional planes in E^n.

Let X, Y, X be three arbitrary points in space, given in a certain order. The totality (X, Y, Z) will be termed a trio of points in the space. The trio (X, Y, Z) is called non-degenerate if the points X, Y and Z do not lie in one straight line.

Let (X, Y, Z) be an arbitrary non-degenerate trio of points on a certain oriented two-dimensional plane P. The trio (X, Y, Z) will be considered to be positively oriented if the vectors $(\overline{XY}, \overline{YX})$ form a right-hand reference frame in the plane P, and negatively oriented if the reference frame $(\overline{XY}, \overline{YZ})$ is left-hand.

Let us assume that there is a Cartesian orthogonal system of coordinates (x, y) introduced in the plane in such a way that the basis vectors e_x and e_y of the coordinate axes form a right-hand reference frame (e_x, e_y). Let (x_1, y_1), (x_2, y_2), (x_3, y_3) be the coordinates of the points (X, Y, Z) in this system. In this case the trio (X, Y, Z) will be positively oriented if the determinant

$$\begin{vmatrix} x_1, & y_1, & 1 \\ x_2, & y_2, & 1 \\ x_3, & y_3, & 1 \end{vmatrix} > 0$$

and negatively oriented if this determinant is negative. Let P and Q be two oriented two-dimensional planes in the space E^n, not orthogonal relative to each other. Let (a, b) be an arbitrary right-hand reference frame in the plane P and let a', b' be orthogonal projections of the vectors a and b onto the plane Q. Let us say that the projection P onto Q coincides with Q if (a', b') is a right-hand reference frame in the plane Q; if (a', b') is a left-hand reference frame in the plane Q, let us say that the projection of P onto Q is contrary oriented in Q, or, in other words, P is projected onto the plane $-Q$. It is obvious that the projection of P onto Q coincides with Q if and only if $(P, \widehat{\ } Q) < \pi/2$.

Let K be an arbitrary curve in the space E^n, not lying in one straight line, and let X, Y and Z be three arbitrary points of this curve not lying in one straight line. Let us number these points in the order of their location on the curve. Let (x_1, x_2, x_3) be a trio formed by the points X, Y, Z which are located in the order of their numeration. Let us draw through the points (X, Y, Z) a two-dimensional plane and orient it in such a way that the trio (x_1, x_2, x_3) is positively oriented. Let us denote this oriented plane by $P(X, Y, Z)$ and term it a secant plane of the curve K, passing through the points X, Y and Z.

In the case when X, Y and Z are three points of the curve lying in one straight line, let us assume that no secant plane $P(X, Y, Z)$ exists.

Now let K be an arbitrary curve and let X be its point. An oriented plane $P_f^K(X)$ $(P_l^K(X))$ is termed a right-hand (left-hand) osculating plane of the curve at the point X if for any $\varepsilon > 0$ one can find a right-hand (left-hand) semi-neighbourhood U of the point X not lying in one straight line and such that all the secant planes $P(X, Y, Z)$ where $Y \in U$, $Z \in U$, form with $P_r^K(X)$ $(P_l^K(X))$ angles less than ε. The index K denoting the curve, an osculating plane of which is the plane under discussion, will henceforth be omitted any time it results in no ambiguity.

It should be remarked that the above definition of a right-hand osculating plane is also applicable in the case when a certain right-hand semi-neighbourhood of a point of a curve lies in one straight line, the only requirement being that the whole of the arc $[XB]$, where B is the end of the curve K, does not lie in one straight line. An analogous remark is valid for a left-hand osculating plane as well.

Let X be an arbitrary point of the curve K. Let us say that the plane $P_m^K(X)^*$ passing through the point X is a middle osculating plane of the curve K at the point X, if the secant planes $P(Y, X, Z)$ converge to $P_m^K(X)$ when $Y \longrightarrow X$ from the left, and $Z \longrightarrow X$ from the right.

It should be remarked that the definition of an osculating plane of a curve is also meaningful in the case when the curve K lies in one two-dimensional plane. In this case any secant plane of the curve K either coincides with the plane of the curve itself or differs from it in its orientation. An angle between two secant planes of the curve K can in this case be equal to either 0 or π. Therefore, the curve K lying in the plane P has a left-hand (right-hand) osculating plane $P_l(X)$ $(P_r(X))$ at the point X if and only if there exists a left-hand (right-hand) semi-neighbourhood U of the point X, not lying in one straight line, such that all the trios of the points (X, Y, Z), where $Y \in U$, $Z \in U$, numbered in the order of their location on K, have the same orientation.

LEMMA 7.1.1. *If the curve K has a left-hand (right-hand) osculating plane at the point X, then the curve K has a left-hand (right-hand) tangent at this point.*

---

* Index 'm' from the word 'middle'.

Proof. Let the curve K have a right-hand osculating plane $P_r(X)$ at a point X. Let us first assume that a certain right-hand semi-neighbourhood U_0 of the point X lies in one straight line. Let l be this straight line. In order to prove that U_0 lies in the straight line l to one side of the point X, let us choose $\varepsilon > 0$, $\varepsilon < \pi/2$ and find a right-hand semi-neighbourhood U_ε of the point X not lying in the straight line l and such that all the secant planes $P(X, Y, Z)$, where $Y \in U_\varepsilon$, $Z \in U_\varepsilon$, comprise with $P_r(X)$ an angle less than ε. Obviously, the arc $U_\varepsilon > U_0$. Let us take an arbitrary point Z_0 in the arc U_ε not lying in the straight line l. Let us assume that the arc U_0 contains two points Y_1 and Y_2, lying in l to different sides of the point X. In this case the secant planes $P(X, Y_1, Z_0)$ and $P(X, Y_2, Z_0)$ differ from one another only in their orientation and therefore the angle between them equals π. This, however, contradicts the fact that an angle between any two secant planes of the type $P(X, Y, Z)$ of the arc U_ε is less than $2\varepsilon < \pi$. The obtained contradiction proves that U_0 lies to one side of the point X. Therefore, obviously, a right-hand tangent at the point X does exist.

Let us now assume that no right-hand semi-neighbourhood of the point X on the curve K lies in one straight line.

Let us first consider the case when the curve K lies in one two-dimensional plane.

Any secant plane of a plane curve coincides, to the accuracy of the orientation, with the plane of the curve itself. The definition of an osculating plane tells us that there is always such a right-hand semi-neighbourhood U of the points X, that all the non-degenerate trios (X, Y, Z), $X < Y < Z$, $Y \in U$, $Z \in U$, are oriented in the same way. This affords that at $X < Y < Y_0$ a ray XY lies to one side of the straight line XY_0, and when the point Y moves along U monotonically in the direction to the point X, the ray XY rotates around the point X in one direction. Therefore, the secant $l(XY)$ has a limit on the right at $Y \to X$, i.e., the curve has a right-hand tangent $t_r(X)$.

Let us now consider a general case. Let us find a right-hand (left-hand) semi-neighbourhood U of the point X, such that for any $Y, Z \in U$ the secant plane $P(X, Y, Z)$ makes with the plane $P = P_r(X)$ $(P = P_l(X))$ an angle less than ε, where $0 < \varepsilon < \pi/2$, and consider an orthogonal projection of the curve K onto the plane P. Let U' be a projection of the arc U. In this case, for any Y', $Z' \in U'$, the secant plane of the curve K', obviously, coincides with P. It means that the curve K' has a right-hand (left-hand) osculating plane at the point X. Therefore, in line with what has been proved above, at the point X', K' has a right-hand (left-hand) tangent. Any secant $l(X, Y)$, where $Y \in U$, forms with P an angle less than ε, and, since $\varepsilon > 0$ is arbitrary, the tangent $t_r(X)$ $(t_l(X))$ does exist. The lemma is proved.

7.1.2. Let us now consider the case of an arbitrary curve. Let us project it onto an osculating plane $P_r(X)$ $(P_l(X))$. At the point X the projection K has a left-hand (right-hand) tangent. This tangent, as can be seen, is also a right-hand (left-hand) tangent to the curve K, and the theorem is thus proved.

Let K be an arbitrary curve and let X be any point on it. An oriented plane P is termed a right-hand (left-hand) osculating plane, in the strong sense, of the curve K at the point X, if for any $\varepsilon > 0$ we can find a right-hand (left-hand) semi-neighbourhood of the point X not lying in one straight line and such that all its secant planes form with P angles less than ε.

Obviously, if at the point X the curve has a right-hand or a left-hand osculating plane in the strong sense, then it has a corresponding common osculating plane at this point, and these planes coinciding.

LEMMA 7.1.2. *If the curve K in E^n has a right-hand (left-hand) osculating plane in the strong sense at the point X, and this plane is P, then for any two-dimensional plane Q, not orthogonal to P, the projection K_1 of the curve K onto the plane Q at the point X_1, which corresponds to X, has a right-hand (left-hand) osculating plane in the strong sense. In this case $P_r^{K_1}(X)$ coincides with the projection of the oriented plane $P_r^K(X)$ onto Q. (An analogous statement is valid for the case of a left-hand osculating plane).*

Proof. Let us assume that the curve K has a right-hand osculating plane in the strong sense at the point X. Let Q be an arbitrary plane not orthogonal to $P_r^K(X) = P$. Let us orient Q in such a way that the angle $(Q,\widehat{\ }P_r)$ were less than $\pi/2$. Let us set $\varepsilon = \pi/2 - (Q,\widehat{\ }P_2)$ and find a right-hand (left-hand) semi-neighbourhood U_ε of the point X, such that all the secant planes of this semi-neighbourhood make with P_2 an angle less than ε. Each of such planes forms with Q an angle less than $\pi/2$. Therefore, each positively oriented triangle in an arbitrary secant plane of the arc U_ε is transformed, at an orthogonal projection onto Q, into a positively oriented triangle in the plane Q. Therefore, all the secant planes of the arc U_ε' - the projections U_ε onto Q - coincide with the plane Q in the orientation. Hence, Q is a right-hand osculating plane of the arc U_ε', which is the required proof. In the case of a left-hand osculating plane the considerations are analogous.

7.2. Osculating Plane of a Plane Curve

7.2.1. All the curves considered in this paragraph are assumed to be lying in a certain oriented two-dimensional plane P.

A plane one-sidedly smooth curve K is termed convex in the small if it has no points of return and the rotations of its any two arcs have the same signs, provided none of them turns to zero. As follows from the definition, the curve K with the parametrization $\{x(t), a \leqslant t \leqslant b\}$ is convex in the small if and only if the function $\tilde{\kappa}(t)$ (the rotation of the arc $[x(a)x(t)]$) is monotonic.

For a plane one-sidedly smooth curve K to be convex in the small, it is necessary and sufficient that it has no points of return and the following equality be fulfilled:

$$|\tilde{\kappa}(K)| = \kappa(K).$$

LEMMA 7.2.1. *If a plane curve K is convex in the small and $\kappa(K) < \pi$, then all non-degenerate trios (X, Y, Z) of the curve K, where $X < Y < Z$, are positively oriented, if $\tilde{\kappa}(K) > 0$, and negatively oriented if $\tilde{\kappa}(K) < 0$.*

Proof. Let us introduce on the plane P a Cartesian orthogonal system of coordinates in such a way that the basis vectors of this system formed a right-hand pair. If a and b are two vectors on the plane, then let

$$a \times b = \begin{vmatrix} a_x & a_y \\ b_x & b_y \end{vmatrix} = |a| \cdot |b| \cdot \sin(a, b),$$

where (a_x, a_y), (b_x, b_y) are the coordinates of the vectors a and b, respectively. The pair (a, b) is right-hand if $a \times b > 0$, and left-hand if $a \times b < 0$. At $a \times b = 0$ the vectors a and b are colinear.

Let $X(s)$, $0 < s < l$, be the parametrization of the curve K, where the parameter s is the length of an arc. Let $t_l(s)$ and $t_r(s)$ be unit vectors of a left- and a right-hand tangents of the curve K at the point $X(s)$. The angle $(t_r(s_1), \widehat{\ } t_l(s_2))$, where $s_1 < s_2$ is equal to the rotation of the arc $[X(s_1) X(s_2)]$ of the curve K, and the product $t_l(s_1) \times t_r(s_2)$ is non-negative, if $\tilde{\kappa}(K) \geqslant 0$ and, non-positive, if $\tilde{\kappa}(K) \leqslant 0$.

Let $X_1 < X_2 < X_3$ be an arbitrary non-degenerate trio of the points of the curve K, $X_1 = X(s_1)$, $X_2 = X(s_2)$, $X_3 = X(s_3)$. In this case we have obviously:

$$\overline{X_1 X_2} \times \overline{X_2 X_3} = \int_{s_1}^{s_2} \left[\int_{s_2}^{s_3} t_r(u) \times t_l(v) \, dv \right] du.$$

In the integral in the right-hand part of the equation $u \leqslant v$, and, hence, the subintegral function is non-negative if $\tilde{\kappa}(K) \geqslant 0$, and non-positive if $\tilde{\kappa}(K) \leqslant 0$. From this we get:

$$\operatorname{sgn} (\overline{X_1 X_2} \times \overline{X_2 X_3}) = \operatorname{sgn} \tilde{\kappa}(K).$$

The lemma is proved.

A plane curve K is termed convex if it is a simple arc lying on the boundary of a plane convex domain.

LEMMA 7.2.2. *Let K be a plane curve not lying in one straight line. Then, if all non-degenerate trios (X, Y, Z) of the points of the curve K, where $X < Y < Z$, are oriented in the same way, then the curve K is convex.*

Proof. Let us assume that all non-degenerate trios (X, Y, Z), $X < Y < Z$, are points of the curve K oriented positively. This can be attained by way of changing the orientation of the plane P of the curve K. Let A be the origin, and B the end of the curve K. All secant planes of the curve K coincide with the plane P, which affords that at its every point X at $A < X \leqslant B$ there exists a left-hand osculating plane in the strong sense, while at $A \leqslant X < B$ there is a right-hand osculating plane in the strong sense. Therefore, the curve has a tangent at every point X of the curve K.

Let X and Y be two arbitrary points of K, such that $X < Y$ and X does not coincide spatially with Y, and let $l = l(X, Y)$ be a secant. Let Z be another arbitrary point of K. If $X < Y < Z$, then the points X, Y and Z either lie on the straight line l, or do not, the trio (X, Y, Z) being oriented positively in the latter case. This affords that Z lies to the left of the oriented straight line l. If $Z < X < Y$, then we obviously come to the conclusion that Z either lies in the straight line or to the left of l. It follows from what has been said above that the arcs $[AX]$ and $[YB]$ of the curve K are located to the left of the oriented straight line l. Fixing one of the points X and Y and tending the other one to it in the limit, we find that the curve K lies to one side, namely, to the left of any of its tangents.

Let G be a convex envelope $|K|$. In this case G is a limited plane convex domain and its boundary is a simple closed curve Γ. Since through every point of the curve K there passes a straight line, such that K is contained in one of two closed semi-planes into which this straight line divides the plane of the curve, then K lies on the boundary G.

The curve K may have no points of return. Indeed, if X were a point of return of K, then the curve K would lie to the left of each of the straight lines $t_l(X)$ and $t_r(X)$. Since these straight lines are directed in opposite directions, it is possible only if K is contained in each of the given straight lines, which contradicts the condition.

Let $X(t)$, $a \leqslant t \leqslant b$, be an arbitrary normal parametrization of the curve K. Let us define $\tau \in [a, b]$, setting $\tau = a$, if no right-hand semi-neighbourhood of the point $A = X(a)$ lies in one straight line. If a certain right-hand semi-neighbourhood of the point A is rectilinear, then let τ be the least value of t, such that the arc $[X(a)X(t)]$ lies in one straight line, $\tau < t < b$. The arc $[X(a)X(\tau)]$ is simple since, in the opposite case, it would have points of return. Let us demonstrate that at $a < t < b$ $X(t) \neq A$. If $t \leqslant \tau$, it follows from the fact that the arc $[X(a)X\tau]$ is simple. Let us assume that there is a $t = t_0$ such that $\tau < t_0 < b$ and $X(t_0) = A$. Let $l = t_r(A)$. For all $t \in [a, b]$ the point $X(t)$ belongs to a left-hand semi-plane of the straight line l. Let us set a sequence of values (t_m), such that $X_m = X(t_m) \notin l$ at all m, $\tau < t_m < t_0$ and $t_m \to \tau$ at $m \to \infty$. Let l_m be a secant $l(A, X_m)$. The arc $[X(t_0)B]$ is contained in an angle formed by the intersection of the left-hand semi-planes of the straight lines l and l_m. On the other hand, it lies to the left of the secant $l(X(t_m), X(t_0))$. The latter consists of the same points as l_m does, but is directed oppositely to l_m. Therefore, the arc $[X(t_0)B]$ lies to the right of the straight line l. It is possible only in the case when the arc $[X(t_0)B]$ lies on the ray $AX(\tau_m)$. The rays $AX(\tau_m)$, corresponding to different m, have the same only common point - the point A - and hence the arc $[X(t_0)B]$ of the curve K degenerates into the point A, which contradicts the fact that the parametrization $X(t)$ is normal.

For $a < t < b$ let us set $\varphi(t) = (t_r(A), \widehat{AX(t)})$. It is obvious that $\varphi(t) \to 0$ at $t \to a$, and we set $\varphi(a) = 0$. At $a < t_1 < t_2 < b$ the point $X(t_2)$ lies to the left of the secant $l(A, X(t_1))$ and to the left of the straight line $l(A)$.

Therefore, $\pi \geqslant \varphi(t_2) \geqslant \varphi(t_1)$, so that the function φ is non-decreasing. We set $\varphi(b) = \lim_{t \to b} \varphi(t)$. Let $a < t_1 < t_2 < b$. If $\varphi(t_1) < \varphi(t_2)$, then $X(t_1) \neq X(t_2)$. In the case when $\varphi(t_1) = \varphi(t_2)$ the arc $[X(t_1)X(t_2]$ of the curve K lies in one straight line and is therefore simple, since in the opposite case it would have a point of return. Therefore, in this case the points $X(t_1)$ and $X(t_2)$ are different.

We have thus obtained that the curve K lies on the boundary of a plane convex domain and any its two points, the terminal points, perhaps, excluded, do not coincide as far as their location on the plane. The lemma is proved.

From 7.2.1 and 7.2.2 the following suppositions can be derived.

COROLLARY 1. *Any plane convex in the small curve, a turn of which is less than π, is convex.*

COROLLARY 2. *If a plane curve K has at its point X a left-hand (right-hand) osculating plane in the strong sense, then a certain left-hand (right-hand) semi-plane of the point X on the curve K is a convex curve.*

Indeed, in the case under discussion the existence of a left-hand (right-hand) osculating plane in the strong sense denotes that all non-degenerate trios of a certain left-hand (right-hand) semi-neighbourhood of the point X on the curve K have the same orientation, and the result obviously follows from Lemma 7.2.2.

7.3. Properties of Curves with an Osculating Plane in the Strong Sense

7.3.1. The results of Section 7.2 and Lemma 7.2.2 make it possible to formulate the following theorem.

THEOREM 7.3.1. *If the curve K has at the point X a left-hand (right-hand) osculating plane in the strong sense, then a turn of a certain left-hand (right-hand) semi-neighbourhood of the point X is finite.*

Proof is carried out by way of induction with respect to n. Let $n = 2$. Then for a certain left-hand (right-hand) semi-neighbourhood U of the point X all trios (Y, Z, T) of the points of the curve K, where $Y < Z < T$, are oriented in the same way. Therefore, this semi-neighbourhood is convex and hence its turn is finite. Let us assume that $n \geqslant 3$ and that for the curves in an $(n-1)$-dimensional Euclidean space the theorem is valid. Let $H_1, H_2, \ldots, H_n$ be n mutually orthogonal hyperplanes, none of which is orthogonal to $P_r(X)$ ($P_l(X)$), let K_i be an orthogonal projection of K onto H_i, and let X_i be a point of the curve K_i, corresponding to the point X. In this case the curve K_i has at the point X_i a right-hand (left-hand) osculating plane in the strong sense. From this, due to the inductive assumption, we conclude that there is a right-hand (left-hand) semi-neighbourhood U of the point X on the curve K, such that its projections onto the plane H_i are curves of a finite turn. An angle between any two vectors, in line with Lemma 4.9.1, is not greater than the sum of the angles of the projections of these vectors onto the planes H_i divided by $\sqrt{n-2}$. Therefore, a turn of an arbitrary polygonal line and, hence, also of an arbitrary curve does not exceed the sum of turns of its projections divided by $\sqrt{n-2}$. It follows that $\kappa(U) < \infty$, which is the required proof.

THEOREM 7.3.2. *Let X be a point of the curve K, a certain right-hand (left-hand) semi-neighbourhood U of which is one-sidedly smooth and does not lie in one straight line. For a right-hand (left-hand) osculating plane in the strong sense to exist at the point X, it is necessary and sufficient that an indicatrix of the tangents of the arc U have a right-hand (left-hand) tangential large circumference in the strong sense at the point $t(X)$. In this case the (oriented) plane of this tangential circumference coincides with the osculating plane.*

Proof. Let t_1 and t_2 be non-colinear tangential unit vectors of the curve at the points of a certain right-hand semi-neighbourhood U of the point X, in which case $t_1 < t_2$. Let us plot them from a certain point O and draw a plane through them. This plane will be oriented in such a way that the direction of rotation from t_1 to t_2 would be positive, and will be denoted through $P(t_1, t_2)$. Instead of Theorem 7.3.2 let us prove the following equivalent supposition.

For a right-hand osculating plane in the strong sense to exist it is necessary and sufficient that there existed a plane P with the following properties: whatever $\varepsilon > 0$, we can find such a non-rectilinear right-hand semi-neighbourhood U of the point X that for any two non-colinear tangential unit vectors $t_1 < t_2$ at the points of this semi-neighbourhood $(P, \hat{\ } P(t_1, t_2)) < \varepsilon$; in this case P coincides with $P_r(X)$.

For a left-hand osculating plane the considerations are completely analogous.

Let us first establish the sufficiency. As a parameter, let us choose on the curve K a length s read off from the point X. Let $t(s)$ be a left-hand tangential unit vector of the curve at the point $X(s)$, and $X_1 = X(s_1)$, $X_2 = X(s_2)$, $X_3 = X(s_3)$ be three consequent points of the curve. In this case:

$$\overline{X_1 X_2} = \int_{s_1}^{s_2} t(s)\mathrm{d}s; \quad \overline{X_2 X_3} = \int_{s_2}^{s_3} t(s)\mathrm{d}s.$$

A normal to the plane $P(X_1, X_2, X_3)$ is directed in the same way as the vector

$$\overline{X_1 X_2} \times \overline{X_2 X_3} = \int_{s_1}^{s_2} t(s)\mathrm{d}s \times \int_{s_2}^{s_3} t(s)\mathrm{d}s = \int_{s_1}^{s_2}\left[\int_{s_2}^{s_3} t(s) \times t(\sigma)\mathrm{d}\sigma\right]\mathrm{d}s.$$

If the vectors $t(s) \times t(\sigma)$ form with the normal n of the plane P angles less than $\varepsilon < \pi/2$, then it follows from the above integral relation that the vectors $\overline{X_1 X_2} \times \overline{X_2 X_3}$ forms with n an angle not greater than ε. Indeed, when proving Lemma 3.3.1 we have already used the fact that if $(a, \hat{\ } e) < \varepsilon < \pi/2$ and $(b, \hat{\ } e) < \varepsilon < \pi/2$, and $c = a + b$, then $(c, \hat{\ } e) < \varepsilon$. By induction, the corresponding statement can be easily proved for a sum of any number of vectors and, going over to a limit, we can spread it to integrals. The sufficiency of our condition is established.

Let us prove its necessity. Let a curve have at the point X a right-hand osculating plane in the strong sense $P_r(X)$. Let us assume that for a certain ε $= \varepsilon_0 > 0$ at any non-rectilinear right-hand semi-neighbourhood of the point X there can be found a pair of tangential unit vectors t_1 and t_2, $t_1 < t_2$, for which $(P(t_1, t_2), \hat{\ } P_r(X)) \geqslant \varepsilon$. Let us choose a decreasing sequence of consequent right-hand semi-neighbourhoods $U_1 \supset U_2 \supset \cdots \supset U_n \supset \cdots$ of the point X, the intersection of which is linear and empty; and a sequence of pairs of tangential unit vectors $t_1^m < t_2^m$ at the points of these neighbourhoods, such that the planes $P(t_1^m, t_2^m)$ converge to a certain plane P_0, for which $(P_0, \hat{\ } P_r(X)) \geqslant \varepsilon$. Furthermore, let Q be a plane forming with $P_r(X)$ an angle less than $\pi/2$, and which P_0 an angle greater than $\pi/2$ (such planes, obviously, do exist). In this case at sufficiently large m a projection of the plane $P(t_1^m, t_2^m)$ onto the plane Q is oppositely oriented to the plane Q. On the other hand, a projection of the plane $P_r(X)$ onto Q coincides with Q, and a projection of a certain non-rectilinear right-hand semi-neighbourhood of the point X onto Q is a convex curve (Lemmas 7.2.1 and 7.2.2). Projections of the vectors t_1^m and t_2^m are tangential unit vectors of this curve, and, according to Lemma 7.2.1 at sufficiently large m a projection of the plane $P(t_1^m, t_2^m)$ onto the plane Q is oriented in the same way as Q. Therefore we have obtained a contradiction and the theorem is thus proved.

CHAPTER VIII

Torsion of a Curve in a Three-Dimensional Euclidean Space

8.1. Torsion of a Plane Curve

8.1.1. Studying a turn of a curve employing the integro-geometrical relations obtained above, required some preliminary considerations of the notion of a turn of a curve lying in one straight line. In an analogous way, studying a torsion of a spatial curve is based on considerations referring to plane curves.

A plane curve K is termed a curve of a finite complete torsion (hereafter referred to as an f.c.t. curve) provided it can be divided into a finite number of curves that are convex in the small, in which case the curve K is non-degenerate and has no points of return.

The definition, in particular, affords that any plane polygonal line having no points of return is an f.c.t. curve.

Let K be an arbitrary locally convex curve on a plane, not lying in one straight line. The curve K will be termed positively oriented if its rotation is positive; and negatively oriented if its rotation is negative.

For an oriented plane P the symbol $-P$ will hereafter denote a plane obtained by changing P orientation. Let K be a locally convex curve lying in an oriented plane. Let us set $P(K) = P$ if the curve K is positively oriented, and $P(K) = -P$ if K is negatively oriented.

Let K be an arbitrary plane f.c.t. curve. Let us define for the curve K a certain division into arcs convex in the small. Let $A = A_0$ be the origin of the curve K, and B its end. Let us denote through A_1 the utmost right-hand point of the curve K, such that the arc $[A_1A_2]$ is convex in the small; if $A_2 \neq B$, then let A_3 be the utmost right-hand point of the curve K, such that the arc $[A_2A_3]$ is convex in the small, etc.. As a result, we get a certain sequence of the points of the curve K: $A_0, A_1, ..., A_r, ...$. Since, by the condition, K is an f.c.t. curve, this sequence breaks up after a finite number of steps and its last point A_r coincides with the point B, which is the end of the curve K. The points $A_0, A_1, ..., A_r, ...$ subdivide the curve K into arcs, each of which is convex in the small. Let us call the subdivision of the curve K obtained in such a way canonical.

Let K be an f.c.t. curve, $A_0, A_1, ..., A_r, ...$ be the points realizing a ca-

217

nonical subdivision of the curve K. Below we shall build a certain canonical set of osculating planes of the curve K at the points A_0, A_1, ..., A_r, ...; the planes of this set will be called marked. In the set of marked planes let us introduce the ratio of the order, denoted hereafter by the symbol $<$.

Let $0 < j < r$. Let us study the structure of the curve K in the vicinity of the point A_j. Several cases are possible here, and we are going to consider them one by one.

I. Let us assume that the arcs $[A_{j-1}A_j]$ and $[A_jA_{j+1}]$ of the curve K are both non-rectilinear and differ in their orientation. Let us set: $P_l(A_j) = P([A_{j-1} A_j])$, $P_r(A_j) = P([A_jA_{j+1}])$. Let us also set $P_l(A_j) < P_r(A_j)$.

II. The arcs $[A_{j-1}A_j]$ and $[A_jA_{j+1}]$ both are non-rectilinear and their orientations coincide. In this case the point A_j of the curve K is by all means an angular point, the mean osculating plane $P_m(A_j)$ being different from each of the planes $P_l(A_j)$ and $P_r(A_j)$. Indeed, in the opposite case the arcs $[A_{j-1}A_j]$ and $[A_jA_{j+1}]$ together would make one locally convex curve, which contradicts the definition of A_j, as the utmost right-hand point X of the curve K, such that the arc $[A_{j-1}X]$ is locally convex. In this case the marked planes at the point A_j will be planes $P_l(A_j)$, $P_m(A_j)$ and $P_r(A_j)$. The ratio of the order for these planes will be defined, setting

$$P_l(A_j) < P_m(A_j) < P_r(A_j).$$

III. One of the arcs $[A_{j-1}A_j]$ and $[A_jA_{j+1}]$ is rectilinear. In this case the other arc does not lie in one straight line, since in the opposite case the arcs $[A_{j-1}A_j]$ and $[A_jA_{j+1}]$ together would comprise a locally convex curve. Let us denote by P' a plane coinciding with P provided that that of the arcs $[A_{j-1} A_j]$ and $[A_jA_{j+1}]$ which is non-rectilinear, is positively oriented, and equal to $-P$, if the arc in question is negatively oriented. The point A_j is angular, in which case the plane $P_m(A_j)$ is oriented oppositely to P'. Indeed, if it were not the case, then the arcs $[A_{j-1}A_j]$ and $[A_jA_{j+1}]$ together would make up a locally convex curve, which is impossible. In this case the marked planes at the point A_j will be the planes P' and $P_m(A_j)$. Let us bring them in order in the following way. If the arc $[A_{j-1}A_j]$ is non-rectilinear, then let us set $P' < P_m(A_j)$. (In this case we will also write: $P' = P_l(A_j)$.) If non-linear is the arc $[A_jA_{j+1}]$, then we set $P_m(A_j) < P'$. In this case we will denote P' through $P_r(A_j)$.

Let P be a marked plane at the point A_j, Q be a marked plane at the point A_k. Let us set $P < Q$ if $j < k$, $P > Q$, if $j > k$.

A set of marked osculating planes is thus defined for any plane f.c.t. curve K. It is an ordered set. In this case, if the curve K is locally convex, then a set of marked osculating planes is empty for it.

As follows from the definition of the notion of a marked osculating plane, at each of the points A_j, $j = 0, 1, 2, \ldots, s$, realizing a canonical subdivision of the curve K at $0 < j < s$ there are at least two oppositely oriented marked planes.

For each f.c.t. curve K let us define a certain value of $\tau(K)$, which will be called a complete torsion of a curve.

If the curve K is locally convex, then let us set $\tau(K) = 0$.

Let us assume that the curve K is not locally convex. Let P_0, P_1, P_2, ..., P_s be a set of all its marked osculating planes numbered in the order of their sequence. A complete torsion of the curve K is, in this case, the quantity:

$$\tau(K) = \sum_{i=1}^{s} (P_{i-1} \frown P_i)$$

in which case an angle between oriented planes, i.e., between their positive normals is equal here to either zero or π. Obviously, $\tau(K) = \pi m$, where $m \geqslant 0$ is a non-negative integer.

If the curve K of a finite complete torsion is not locally convex, then a set of its marked osculating planes is not empty. In this case, as is seen from the definition of the notion of a marked plane, among the marked osculating planes there are at least two oppositely oriented. Therefore, in this case $\tau(K) > 0$. We, thus, come to the conclusion that a complete torsion of a plane curve turns to zero if and only if the curve K is locally convex. If in this case $\kappa(K) < \pi$, then the curve K, as is seen from Lemma 7.2.1, is convex.

Let K be an arbitrary one-sidedly smooth plane curve, $X(t)$, $a \leqslant t \leqslant b$, be a parametrization of this curve. Let us denote through $\tilde{\kappa}(t)$ a rotation of the arc $[X(a)X(t)]$ of the curve K. It directly follows from the definition that for K to be an f.c.t. curve, it is necessary and sufficient that the function $\tilde{\kappa}(t)$ has a finite number of maxima and minima and has no points of discontinuity at which the jump $\kappa(t)$ equals $\pm\pi$.

Let a sum of angles between the subsequent marked osculating planes at one point A_j out of the number of points realizing a canonical subdivision of the f.c.t. curve K, be termed a torsion at this point. It will be denoted by $\tau(A_j)$.

8.1.2. Let us consider a certain path of approximating a torsion of a plane curve, which employs the notion of a triple chain.

A triple chain of an arbitrary plane or a spatial curve K is a sequence x of trios of the points

$$\xi = \{(X_1, Y_1, Z_1); (X_2, Y_2, Z_2); ...; (X_m, Y_m, Z_m)\}$$

of the curve K, such that the following relations hold:

$$X_i < Y_i < Z_i \quad \text{at all} \quad i = 1, 2, ..., m,$$

$$X_1 \leqslant X_2 \leqslant \cdots \leqslant X_m \qquad Y_1 \leqslant Y_2 \leqslant \cdots \leqslant Y_m \qquad Z_1 \leqslant Z_2 \leqslant \cdots \leqslant Z_m.$$

A trio of points (X, Y, Z) of the curve K is called non-degenerate if the points X, Y and Z do not lie in one straight line. A non-degenerate trio defines a secant plane $P(X, Y, Z)$ of the curve K. If the trio (X, Y, Z) is included into the composition of a triple chain ξ of the curve K, let us say that $P(X, Y, Z)$ is a secant plane of the triple chain ξ.

Let P_1, P_2, ..., P_r all be secant planes of the triple chain ξ of the curve K, numbered in the order of the sequence of the trios of the chain defining these planes. The quantity

$$T(\xi) = \sum_{i=1}^{r-1} (P_i \widehat{} P_{i+1})$$

is termed a torsion of the triple chain ξ.

Let $L = A_0$, A_1, ..., A_p be an arbitrary plane polygonal line with no points of return. The trios (A_0, A_1, A_2); (A_1, A_2, A_3); ...; (A_{p-2}, A_{p-1}, A_p) of the vertices of the polygonal line L from a certain chain ξ. We can easily prove that

$$\tau(L) = T(\xi).$$

A multiplicity of the triple chain ξ of the curve K is the largest integer number $\nu(\xi)$, such that on the curve K there can be found at least one point belonging simultaneously to $\nu(\xi)$ and the open arcs $(X_i Z_i)$.

The largest of the diameters of the arcs $(X_i Z_{i+1})$, (AX_1), $(Z_s B)$, where A and B are the beginning and the end of the curve K is called a modulus of a triple chain

$$\xi = \{(X_1, Y_1, Z_1); \ldots; (X_s, Y_s, Z_s)\}$$

of the curve K. The modulus of a triple chain ξ will be denoted through $\lambda(\xi)$.

Let (X, Y, Z) and (X_1, Y_1, Z_1) be two arbitrary trios of the points of the curve K, in which case $X < Y < Z$, $X_1 < Y_1 < Z_1$. Let us say that (X, Y, Z) is located to the left of (X_1, Y_1, Z_1) and write $(X, Y, Z) < (X_1, Y_1, Z_1)$ if the trios (X, Y, Z) and (X_1, Y_1, Z_1) do not coincide and if the following relations are valid:

$$X \leqslant X_1, \qquad Y \leqslant Y_1, \qquad Z \leqslant Z_1.$$

THEOREM 8.1.1. *Let K be an arbitrary plane curve of a finite complete torsion. In this case for any trio of the chain ξ of the curve K the following inequality holds*:

$$T(\xi) \leqslant [\nu(\xi) + 1][\kappa(K) + \tau(K)].$$

Proof. Let $\{(X_1, Y_1, Z_1); (X_2, Y_2, Z_2); \ldots; (X_m, Y_m, Z_m)$ be an arbitrary triple chain ξ of the curve K. Without reducing the generality, all the trios comprising ξ, can be considered non-degenerate since the multiplicity of the triple chain does not increase if certain trios are excluded from it. Let $P_j = P(X_j, Y_j, Z_j)$, $j = 1, 2, \ldots$, be subsequent secant planes of the triple chain ξ. Let us prove that at all j the following inequality holds:

$$(P_j \widehat{} P_{j+1}) \leqslant \tau(X_j Z_{j+1}) + \kappa(X_j Z_{j+1}). \tag{1}$$

Let us first get

$$\tau(X_j Z_{j+1}) = 0, \qquad \kappa(X_j Z_{j+1}) < \pi. \tag{2}$$

Due to the remark in the end of Section 8.1.1, the curve K is convex and inequality (1) holds, since in this case, on the basis of Lemma 7.2.1, the planes P_j and P_{j+1} are oriented in the same way, i.e., $(P_j \frown P_{j+1})$.

If at least one of inequalities (2) is violated, then inequality (1) holds in a trivial way, since $\tau(X_j Z_{j+1}) > 0$ denotes that $\tau(X_j Z_{j+1}) \geqslant \pi$, and the angle $(P_j \frown P_{j+1})$ will never be greater than π.

From inequality (1) we get:

$$T(\xi) \leqslant \sum_{i=1}^{m-1} [\kappa(X_i Z_{i+1}) + \tau(X_i Z_{i+1})].$$

None of the points of the curve K can lie inside of more than $\nu(\xi)+1$ arcs $(X_i Z_{i+1})$. Indeed, let $X \in (X_i Z_{i+1})$ at $i = i_1, i_2, \ldots, i_k, i_1 < i_2 < \cdots < i_k$. Since $i_s > i_1$ $(s \geqslant 2)$ then $i_s \geqslant i_1+1$ and, hence, $Z_{i_s} \geqslant Z_{i_1+1}$. Therefore, $X \in (X_i Z_i)$ at $i = i_2, \ldots, i_k$. This affords that $\kappa \leqslant \nu(\xi)+1$. We thus see that a torsion at an arbitrary point out of A_j is included into the sum

$$\sum_{i=1}^{m} \tau(X_i Z_{i+1})$$

as an addend not more than $\nu(\xi)+1$ times and hence

$$\sum_{i=1}^{m} \tau(X_i Z_{i+1}) \leqslant [\nu(\xi)+1] \, \tau(K).$$

In an analogous way

$$\sum_{i=1}^{m} \kappa(X_i Z_{i+1}) \leqslant [\nu(\xi)+1] \, \kappa(K)$$

and the theorem is completely proved.

8.1.3. LEMMA 8.1.1. *If p is a number of arcs of a canonical subdivision of the curve K of a finite complete torsion, then the following inequality is valid:*

$$p \leqslant \frac{\tau(K)}{\pi} + 1. \tag{3}$$

Proof. Let $A_0, A_1, A_2, \ldots, A_p$ be the points realizing a canonical subdivision of the curve K. In this case $\tau(A_j) = \pi$ at $j = 1, 2, \ldots, p-1$. This affords

$$\tau(K) \geqslant \sum_{j=1}^{p-1} \tau(A_j) \geqslant \pi(p-1)$$

and hence

$$p \leqslant \frac{\tau(K)}{\pi} + 1,$$

which is the required proof.

Let K be an arbitrary plane f.c.t. curve, $A_0, A_1, \ldots, A_p$ be the points realizing a canonical subdivision of the curve K. Hereafter let us consider triple chains ξ of the curve K obeying som of the conditions (1) - (4) listed below.

(1) On each arc $[A_{i-1}A_i]$ not lying in one straight line there is at least one non-degenerate trio of the chain ξ.

(2) Any non-degenerate trio of the chain ξ lying on the arc $[A_{i-1}A_i]$ is oriented in the same way as the arc.

(3) If the point A_j is angular, then there is a non-degenerate trio of the chain ξ for which A_j is a mean point.

(4) If there exists a plane $P_m(A_j)$, then a secant plane of any non-degenerate trio $(X_j, Y_j, Z_j) \in \xi$, for which A_j is a mean point, coincides with $P_m(A_j)$.

THEOREM 8.1.2. *Let $\xi_1, \xi_2, \ldots, \ldots$ be an arbitrary sequence of triple chains of a plane curve K of a finite complete torsion, such that $\lambda(\xi_m) \to 0$ at $m \to \infty$ and for any chain ξ_m conditions (1) and (3) are fulfilled. In this case the following equality is valid:*

$$\lim_{m \to \infty} T(\xi_m) = \tau(K).$$

Proof. The required result is obtained by way of rather numerous considerations based on a detailed study of a great number of particular cases. The proof of the theorem in general is somewhat cumbersome, thought the considerations in each particular case are trivial.

Let us first prove that there is an m_0, such that at $m > m_0$ a triple chain ξ_m obeys conditions (2) and (4), too. Let $A_0, A_1, \ldots, A_r$ be the points realizing a canonical subdivision of the curve K. For each of the arcs $[A_j A_{j+1}]$ there can be found a $\delta_j > 0$ such that a turn of any arc $[A_j A_{j+1}]$ of the curve, the diameter of which is less than δ_j, is less than π. Let $\delta = \min_{j=1,\ldots,p} \delta_j$. If $\lambda(\xi_m) < \delta$, then condition (2) for a triple chain ξ_m is fulfilled on the basis of Lemma 7.2.1.

Let an osculating plane $P_m(A_j)$ exist at the point A_j. In this case there is a neighbourhood U_j of the point A_j, such that if $X \in U_j$, $Y = A_j$, $Z \in U_j$, $X < Y < Z$, then a secant plane of the trio (X, Y, Z) coincides with the plane $P_m(A_j)$. If $\lambda(\xi)$ is less than the minimum diameter of such neighbourhoods, then condition (4) is fulfilled.

As far as the values of the quantities $T(\xi_m)$ and $\tau(K)$ can be only numbers of type πk, where $k \geqslant 0$ is an integer, then the statement of the theorem means that, beginning from a certain value,

$$T(\xi_m) = \tau(K).$$

Let $S = (X, Y, Z)$, where $X < Y < Z$ is an arbitrary trio of the points of the curve K. Let us say that the trio S is hooked at the point C of the curve K, if $X < C < Z$.

Let us first construct for each of the points A_j, where $0 < j < r$, its certain neighbourhood $U_j = (B_j C_j)$. With this neighbourhood there will be associated a certain Cartesian orthogonal system of coordinates on the plane.

Let us assume that the point A_j is smooth. Then certain neighbourhood $U_j =$

$(B_j C_j)$ is projected in a one-to-one way onto the straight line $g = t(A_j)$. Let us assume that $A_{j-1} < B_j$, $C_j < A_{j+1}$. Let h be a straight line perpendicular to g and passing through the point A_j. On the plane let us introduce a Cartesian orthogonal system of coordinates, the axes of which are the straight lines g and h. The coordinates of an arbitrary point M will be denoted through ξ_1 and ξ_2 (ξ_1 is the coordinate along the axis g, ξ_2 is that along the axis h). The direction of the axes g and h is chosen in such a way that the arc $[B_j A_j]$ lie in the quadrant $\xi_1 \leqslant 0$, $\xi_2 \geqslant 0$. The arcs $[A_{j-1} A_j]$ and $[A_j A_{j+1}]$ are oppositely oriented in this case, and the arc $[A_j C_j]$ therefore lies in the quadrant $\xi_1 \geqslant 0$, $\xi_2 \leqslant 0$. In the constructed system of coordinates the arc $[B_j C_j]$ will be defined by the equation: $\xi_2 = f(\xi_1)$, where $\xi_1 \in [\lambda, \mu]$, $\lambda < 0 < \mu$. In this case the function f is decreasing.

If the point A_j is angular then the neighbourhood U_j is constructed in the following way. Let $a = t_l(A_j) + t_r(A_j)$, $b = \bar{t}_r(A_j) - \bar{t}_l(A_j)$, g and h be the straight lines passing through A_j, in which case g is colinear to a, h is colinear to the vector b. The straight lines g and h are orthogonal. They are the bisecting lines of the angles formed by the left-hand and right-hand tangents of the curve K at the point A_j. Let us introduce on the plane a Cartesian orthogonal system of coordinates with the origin at the point A_j, the axes of which are the straight lines g and h. The coordinates of an arbitrary point M will be denoted by ξ_1, ξ_2 (ξ_1 is the coordinate along the axis g, and ξ_2 that along the axis h). The axis g is considered to have the same direction as the vector a, and h as the vector b. The tangential unit vectors $\bar{t}_l(A_j)$ and $\bar{t}_r(A_j)$ are orthogonal to neither of the vectors a and b and, hence, there can be found points B_j and C_j, such that $B_j < A_j < C_j$ and each of the arcs $[B_j A_j]$ and $[A_j C_j]$ are projected in a one-to-one manner onto each of the straight lines g and h. It might appear that a certain left-hand semi-neighbourhood of the point A_j lies in one straight line. In this case let us assume that the arc $[B_j A_j]$ is included into this semi-neighbourhood. Analogously, if a certain right-hand semi-neighbourhood of the point A_j is rectilinear, then let us assume that the arc $[A_j C_j]$ is included into this semi-neighbourhood. The arc $[B_j A_j]$ is contained in the quadrant defined by the inequalities: $\xi_1 \leqslant 0$, $\xi_2 \geqslant 0$, and the arc $[A_j C_j]$ in the quadrant $\xi_1 \geqslant 0$, $\xi_2 \geqslant 0$. The arc $[B_j C_j]$ is defined by the equation $\xi_2 = f(\xi_1)$, where $\lambda \leqslant \xi_1 \leqslant \mu$, $\lambda < 0 < \mu$. The function f is strictly decreasing in the segment $[\lambda, 0]$ and strictly increasing in the segment $[0, \mu]$.

For each of the points A_j, $j = 1, 2, \ldots, r-1$, therefore, a certain neighbourhood $U_j = (B_j C_j)$ is defined. Let us assume that these neighbourhoods do not mutually intersect.

Let δ_j be the least of the diameters of the arcs $[B_j C_j]$ and $[A_j C_j]$ and the number $m_j \geqslant m_0$ is such that at $m \geqslant m_j$ the inequality $\lambda(\xi_m) < \delta_j/2$ holds. Let $\bar{m} = \max \{m_1, m_2, \ldots, m_{r-1}\}$. Let us then assume that $m \geqslant \bar{m}$.

Our aim is to prove that at $m \geqslant \bar{m}$ $T(\xi_m) = \tau(K)$. For the chain ξ_m, where $m \geqslant$

$\bar{m}$, let us first define a certain set of its trios which will be termed marked. Let us consider all possible cases in succession.

I. Let us assume that the arcs $[A_{j-1}A_j]$ and $[A_jA_{j+1}]$ are both non-rectilinear and their orientations are different. In this case to the point A_j there correspond two marked osculating planes $P_l(A_j)$ and $P_r(A_j)$. Let us denote through $L_j^{(m)}$ the utmost right of the non-degenerate trios of the chain ξ_m, lying on the arc $[A_{j-1}A_{j+1}]$ and such that $P(L_j^{(m)}) = P([A_{j-1}A_j])$, and let $R_j^{(m)}$ be the nearest, following $L_j^{(m)}$, non-degenerate trio lying on the arc $[A_{j-1}A_{j+1}]$. Obviously, $P(L_j^{(m)}) = -P(R_j^{(m)})$ and any trio S of the chain ξ_m, lying on the arc $[A_{j-1}A_{j+1}]$ and such that $R_j^{(m)} < S$, has the same orientation as $R_j^{(m)}$. Let us prove that any trio S of the chain ξ_m lying on the arc $[A_{j-1}A_{j+1}]$ and preceding $L_j^{(m)}$, has the same orientation as the trio $L_j^{(m)}$. Let us consider several possible cases.

(A) Let us assume that the point A_j is smooth. Let $L_j^{(m)} = (X, Y, Z)$. Let us arbitrarily choose a trio (X', Y', Z') of the chain ξ_m lying on the arc $[A_{j-1}A_{j+1}]$ and preceding $L_j^{(m)}$. The task is to prove that $P(X', Y', Z') = P(L_j^{(m)})$. If $Z' \leqslant A_j$, it is obvious. Let us assume that $Z' > A_j$. In this case $Z > A_j$, too. We have: $X < A_j$, since in the opposite case the trio $L_j^{(m)}$ would lie on the arc $[A_jA_{j+1}]$ and would have the same orientation as the arc $[A_jA_{j+1}]$. In this case also $X' \leqslant A_j$. The trios (X', Y', Z') and (X, Y, Z) are therefore both hooked by the point A_j. Due to the fact that, by the condition, $\lambda(\xi_m) < \delta_j/2$ (δ_j is the least of the diameters of the arcs $[B_jA_j]$ and $[A_jC_j]$), we conclude that the considered trios lie on the arc $[B_jC_j]$. The arc $[B_jA_j]$ is not a rectilinear segment since in the opposite case the arc $[B_jC_j]$ would be convex and oriented in the same way as the arc $[A_jA_{j+1}]$, and the trio $L_j^{(m)}$ would have the same orientation as $[A_jA_{j+1}]$, which contradicts the condition. The arc $[A_jC_j]$ is not a segment, since otherwise the arc $[A_{j-1}C_j]$ would be locally convex.

Hereafter let us use the following notation. The coordinates of the points X, X' will be denoted through (x_1, x_2), (x_1', x_2'), respectively. In an analogous way, the coordinates of the point Y will be denoted through (y_1, y_2), etc..

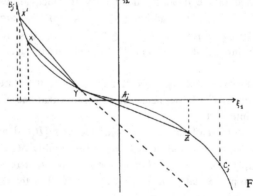

Fig. 11.

Thus, we have the trios $S = L_j^{(m)} = (X, Y, Z)$ and $S' = (X', Y', Z')$ lying on the arc $[B_j C_j]$ and such that $A_j < X$, $A_j < Z'$, $S' \leqslant S$, $P(A) = P([B_j A_j])$. In this case no left-hand and right-hand semi-neighbourhoods of the point A_j lie in one straight line. The task is to prove that $P(S') = P(S)$.

Let us assume that $Y \leqslant A_j$. Let $X' \leqslant X$, $Y' = Y$, $Z' = Z$ (Fig. 11). As far as $P(P) = P([B_j A_j])$, the point X lies above the straight line YZ. Let $\xi_2 = k\xi_1 + l$ be the equation of this straight line. We have: $y_2 = ky_1 + l$, $x_1' < x_1 < y_1$. Therefore, $x_1 = \alpha x_1' + (1-\alpha)y_1$, where $0 < \alpha < 1$. The arc $[B_j A_j]$ is convex and its convexity is downward directed. This affords: $x_2 = f(x_1) \leqslant \alpha f(x_1') + (1-\alpha)$ $f(y_1) = \alpha x_2' + (1-\alpha)y_2$. We have: $x_2 > kx_1 + l = \alpha(kx_1' + l) + (1-\alpha)(ky_1 + l) = \alpha(kx_1 + l_1) + (1-\alpha)y_2$. From this we conclude: $\alpha x_2' + (1-\alpha)y_2 > \alpha(kx_1' + l) + (1-\alpha)y_2$ and, hence, $x_2' > k x_1' + l$. We therefore come to the conclusion that the point X' lies above the straight line YZ and hence the trios (X', Y, Z) and (X, Y, Z) have the same orientation.

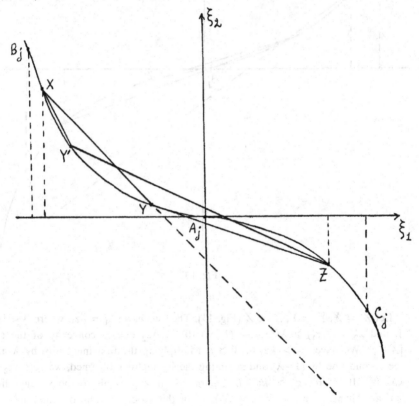

Fig. 12.

Let $X' = X$, $Y' \leqslant Y$, $Z' = Z$ (Fig. 12). The point Y lies below the straight line XZ. In this case in order to prove that $P(S') = P(S)$, it is sufficient to state that Y' also lies below the straight line XZ. Let $\xi_2 = k\xi_1 + l$ be the equation of the straight line XZ. We have: $x_1 < y_1' < y_1 \leqslant 0$. Herefrom we get: $y_1' = (1-\alpha)x_1 + \alpha y_1$, where $0 < \alpha < 1$. Due to the fact that the arc $[B_j A_j]$ is convex with its convexity turned downward, we have: $y_2' = f(y_1') \leqslant (1-\alpha)f(x_1) + \alpha(f(y_1) = (1-\alpha)x_2 + \alpha y_2$. Furthermore, $x_2 = kx_1 + l$ as far as X lies on the straight line XZ and $y_2 < ky_1 + l$. We have: $ky_1' + l = \alpha(ky_1 + l) + (1-\alpha)(kx_1 + l) > \alpha y_2 + (1-\alpha)x_2 \geqslant y_2'$, and hence Y' lies below XZ.

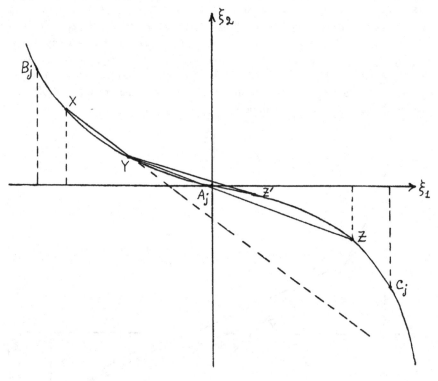

Fig. 13.

Let $X' = X$, $Y' = Y$, $Z' < Z$ (Fig. 13). Then we have: $z_1' = \lambda z_1$, where $0 < l < 1$, and $z_2' = f(z_1') \geqslant \lambda f(z_1) + (1-\lambda)f(0) = \lambda z_2$ due to convexity of the arc $[A_j C_j]$. We have: $z_2 > kz_1 + l$, $0 \geqslant l$. Multiplying the first inequality by λ and the second one by $(1-\lambda)$ and summing the inequalities obtained, we get: $\lambda z_2 > k\lambda z_1 + l$. It affords $z_2' > kz' + l$, i.e., the point Z' lies above the straight line XY and, hence, the trios S' and S have in this case the same orientation.

In line with the above proof we can conclude that the trios (X, Y, Z), (X', Y, Z), (X', Y', Z) and (X', Y', Z') have the same orientations, which is the required proof.

We asume that $Y \leqslant A_j$. Let us now consider the case when $A_j < Y$. Let $X' = X$,

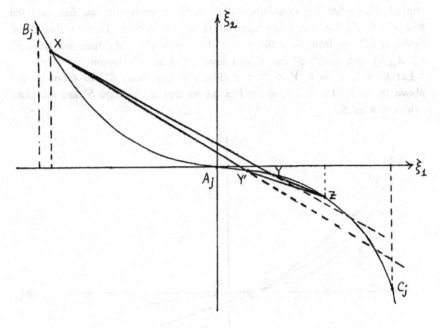

Fig. 14.

$Z' = Z$ and $A_j \leqslant Y' < Y$ (Fig. 14). Let $\xi_2 = k(\xi_1 - z_1) + z_2$ be the equation of the straight line YZ, $\xi_1 = k'(\xi_1 - z_1) + z_2$ be that of the straight line $Y'Z$. Due to the fact that the arc $[A_jC_j]$ is convex with its convexity turned upward, we have: $0 \geqslant k' \geqslant k$. In this case the point X lies above the straight line YZ. Therefore its coordinates obey the inequality $x_2 > k(x_1 - z_1) + z_2$. Hence, due to the fact that $x_1 - z_1 < 0$ and $k' \geqslant k$, we conclude that $x_2 > k'(x_1 - z_1) + z_2$, i.e., the point X lies above the straight line YZ and, thus, the

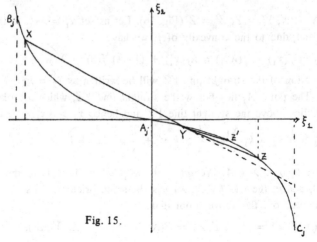

Fig. 15.

trio (X, Y', Z) has the same orientation as S. In particular, we find that the trio (X, A_j, Z) has the same orientation as the trio S. Due to the above proof, it follows from here that if for the trio S $Y' < A_j$, then the trios S', (X, A_j, X) and, hence, S' and S also have the same orientations.

Let $X' = X$, $Y' = Y$, $Y < Z' < Z$ (Fig. 15). The points Z' and Z are located above the straight line XY, which fact means that in this case S' has the same orientation as S.

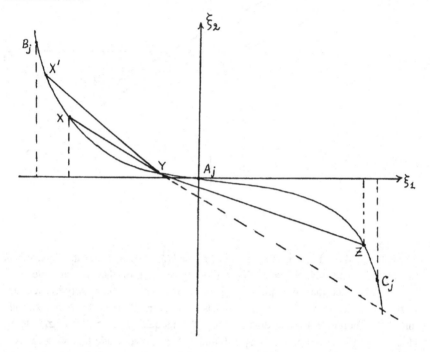

Fig. 16.

Let $X' < X$, $Y' = Y$, $Z' = Z$ (Fig. 16). Let us set $x_1 = \lambda x_1'$. In this case $0 < \lambda < 1$ and, due to the convexity of f, we have:

$$x_2 = f(x_1) = f(\lambda x_1') \leqslant \lambda f(x_1') + (1 - \lambda) f(0) = \lambda x_2'.$$

The equation of the straight line YZ will be written as $\xi_2 = k\xi_1 + l$. We have: $k < 0$. The point A_j lies below the straight line YZ, which affords $l \geqslant 0$. The point X lies above the straight line YZ and hence $x_2 > kx_1 + l$. It affords:

$$x_2' \geqslant \frac{1}{\lambda} x_2 \geqslant k_1 \frac{x_1}{\lambda} + \frac{l}{\lambda} \geqslant k_1 x' + \frac{l}{\lambda} \geqslant k_1 x' + l.$$

since $l \geqslant 0$, $0 < \lambda < 1$. We see that the point X' lies above the straight line YZ and, hence, the trio (X', Y, Z) has the same orientation as S.

It follows from the above proof that

$$P(X, Y, Z) = P(X', Y, Z) = P(X', Y', Z) = P(X', Y', Z'),$$

i.e., the trios $S = (X, Y, Z)$ and $S' = (X', Y', Z')$ have the same orientation and the required statement is proved.

(B) Let no left-hand and right-hand semi-neighbourhood of the point A_j be linear and let the point A_j be angular. In this case either $[B_jA_j]$ has its convexity upward directed, and the arc $[A_jC_j]$ has its convexity downward directed, or, vice versa, the arc $[B_jA_j]$ has its convexity downward directed and the arc $[A_jC_j]$ has its convexity upward directed. Since both cases are considered absolutely analogously, let us limit ourselves by considering only the former case.

So, let the arc $[B_jA_j]$ have its convexity turned upward, and the arc $[A_jC_j]$ have its convexity turned downward. Let $S = (X, Y, Z)$ and $S = (X', Y', Z')$ be two arbitrary non-degenerate trios of the points of the arc $[B_jC_j]$, such that $S \leqslant S'$. Let us assume that $P(S) = P_m(A_j)$. (It should be remarked that in this case the plane $P_m(A_j)$ coincides with the plane $P([A_jC_j])$.) Let us prove that in this case also $P(S') = P_m(A_j) = P([A_jC_j])$.

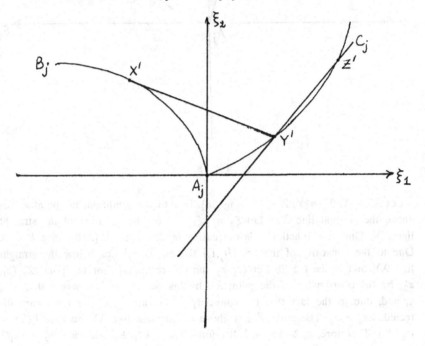

Fig. 17.

Let us first of all remark that if $Y' \in [A_jC_j]$ then $P(S') = P_m(A_j)$, whatever the location of the points X' and Z' is. Indeed, if $X' \in [A_jC_j]$, it is obvious. Let us assume that $X' < A_j$ (Fig. 17). Let $\xi_2 = k\xi_1 + l$ be the equation of the straight line $X'Z'$. Due to the convexity of the arc $[A_jC_j]$, $k > 0$. The arc $[A_jY']$ lies above the straight line $Y'Z'$. Therefore, $l < 0$. If $X' < A_j$ then x_2'

> 0, and $x_1' < 0$ and, hence, $kx_1' + l < 0$ and $x_2' > kx_1' + l$. Therefore, the point X' lies above the straight line $Y'Z'$, which shows that the trio (X', Y', Z') has the same orientation as the arc $[A_jC_j]$.

Let us assume that $Y' \leqslant A_j$. In this case also $Y \leqslant A_j$. Let us consider three possible cases.

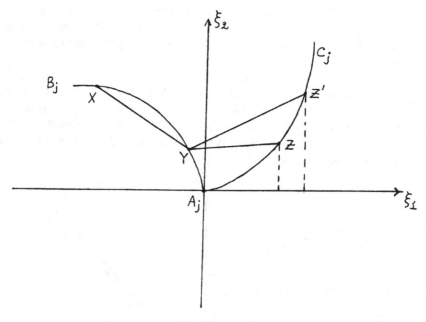

Fig. 18.

Let $X' = X$, $Y' = Y$, $Z < Z'$ (Fig. 18). Due to the condition, the point Z lies above the straight line XY. Let $\xi_2 = k\xi_1 + l$ be the equation of the straight line XY. Since the function f is decreasing in the segment $[\lambda, 0]$, then $k < 0$. Due to the convexity of the arc $[B_jA_j]$ the arc $[YA_j]$ lies below the straight line XY and hence $l \geqslant 0$. Let (z_1, z_2) be the coordinates of the point Z, (z_1', z_2') be the coordinates of the point Z'. In this case $z_1 = \lambda z_2$, where $0 \leqslant \lambda \leqslant 1$, and, due to the fact that the convexity of the arc $[A_jC_j]$ is downward directed, $\lambda z_2' \geqslant z_2$. The point Z lies above the straight line XY because $P(S) = P_m(A_j)$. Therefore, $z_2 > kz_1 + l$. It affords $\lambda z_2' > kz_1 + l$ and then: $z_2' > kz_1/\lambda + l/\lambda = kz_1' + l/\lambda \geqslant kz_1' + l$, i.e., the point Z' lies above the straight line XY. Hence, $P(S) = P(S')$

Let $X' = X$, $Y' > Y$, $Z' = Z$ (Fig. 19). The point Z lies above the straight line XY. Let us prove that Z lies above the straight line XY'. The equation of the straight line XY will be written as: $\xi_2 = k(\xi_1 - x_1) + x_2$. The equation of the straight line XY' will be presented as $\xi_2 = k'(\xi_1 - x_1) + x_2$. Since the arc $[B_jA_j]$ has its own convexity upward turned, $k' \leqslant k$. Since Z lies above the

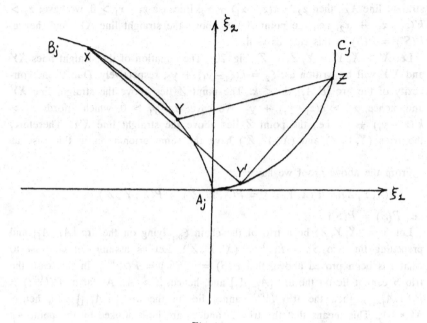

Fig. 19.

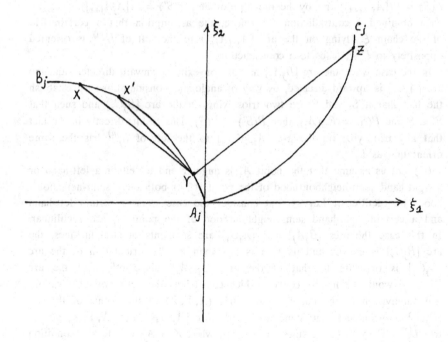

Fig. 20.

straight line XY, then $z_2 > k(z_1 - x_1) + x_2$. Because $z_1 - x_1 > 0$, we have: $z_2 > k'(z_1 - x_1) + x_2$, i.e., the point Z lies above the straight line XY' and, hence, $P(S') = P(S)$ in this case as well.

Let $X' > X$, $Y' = Y$, $Z' = Z$ (Fig. 20). The equation of the straight lines XY and $X'Y$ will be written as: $\xi_2 = k(\xi_1 - y_1) + y_2$, respectively. Due to the convexity of the arc $[B_j A_j]$, $k' < k$. The point Z lies above the straight line XY and, hence, $z_2 > k(z_1 - y_1) + y_2$. We have: $z_1 - y_1 > 0$, which affords $z_2 > k'(z_1 - y_1) + y_2$, i.e., the point Z lies above the straight line $X'Y$. Therefore, the trios (X, Y, Z) and (X', Y', Z') have the same orientation in this case as well.

From the above proof we have:

$$P(X, Y, Z) = P(X, Y, Z') = P(X, Y', Z') = P(X', Y', Z'),$$

i.e., $P(S) = P(S')$.

Let $S = (X, Y, Z)$ be a trio of the chain ξ_{m_1} lying on the arc $[A_{j-1} A_j]$ and preceding the trio $S' = L_j^{(m)} = (X', Y', Z')$. Let us assume, in contrast to what has been proved above, that $P(S) = -P(S') = P(R_j^{(m)})$. In this case the trio S cannot lie on the arc $[A_{j-1} A_j]$ and, hence, $Z > A_j$. As far as $P(L_j^{(m)}) \neq P([A_j A_{j+1}])$, then the trio $L_j^{(m)}$ cannot lie on the arc $[A_j A_{j+1}]$ and, hence, $X' < A_j$. This means that the trios S and S' are both hooked by the point A_j. Since $\lambda(\xi_m) < \delta_j/2$, this affords that S and S' lie on the arc $[B_j C_j]$. We have: $P(S) = P([A_j A_{j+1}])$ and, by the above proof, also $P(S') = P([A_j A_{j+1}])$. Thus, we have obtained a contradiction. Therefore, the assumption that a certain trio of the chain ξ_m lying on the arc $[A_{j-1} A_{j+1}]$ to the left of $L_j^{(m)}$ is oriented oppositely to $L_j^{(m)}$, leads to a contradiction.

In the case when the arc $[B_j A_j]$ has its convexity downward directed, and the arc $[A_j C_j]$ is upward directed, by way of analogous considerations we establish the fact that if S' and S are two trios lying on the arc $[B_j C_j]$ and such that $S' \leqslant S$ and $P(S) = P_m(A_j)$, then $P(S') = P(S)$. This results directly in the fact that any trio lying on the arc $[A_{j-1} A_{j+1}]$ to the left of $L_j^{(m)}$ has the same orientation as $L_j^{(m)}$.

(C) Let us assume that the point A_j is angular and let either a left-hand or a right-hand semi-neighbourhood of the point A_j, or both these semi-neighbourhoods be rectilinear. Let us first consider the case when a certain left-hand and a certain right-hand semi-neighbourhood of the point A_j are rectilinear. In this case the arcs $[B_j A_j]$ and $[A_j C_j]$ are segments of straight lines, the arc $[B_j C_j]$ is convex and its turn is less than π. The orientation of the arc $[B_j C_j]$ is opposite to that of the arc $[A_{j-1} A_j]$, since otherwise the arc $[A_{j-1} C_j]$ would be locally convex, which contradicts the definition of a canonical subdivision of the curve K. Any trio (X, Y, Z) of the points of the arc $[B_j C_j]$ is oriented in the same way as the arc $[A_j A_{j+1}]$. Let (X, Y, Z) be the trio $L_j^{(m)}$. Then $Z < A_j$, since in the case when $Z \geqslant A_j$, due to the condition $\lambda(\xi_m) < \delta_j/2$, the whole trio $L_j^{(m)}$ lies to the right of the point B_j. In this case, however, it will have the same orientation as the arc $[A_j A_{j+1}]$, which contradicts the definition of the trio $L_j^{(m)}$. Any trio S of the chain ξ_m lying on the arc $[A_{j-1}A_{j+1}]$ to the left of the trio $L_j^{(m)}$ is therefore located on the arc $[A_{j-1}A_{j+1}]$ and hence it has the same orientation as this arc.

(D) Let the point A_j be angular and a certain left-hand semi-neighbourhood of it be rectilinear, but no right-hand semi-neighbourhood of A_j lie in one straight line. If the planes $P_m(A_j)$ and $P([A_jA_{j+1}])$ coincide then the arc $[B_jC_j]$ proves to be convex and its turn here is less than π. Using considerations analogous to those employed in case (C), we come to the conclusion that the trio $L_j^{(m)}$ lies on the arc $[A_{j-1}A_j]$, and, hence, all trios of the chain ξ_m lying on the arc $[A_{j-1}A_{j+1}]$ to the left of $L_j^{(m)}$, have the same orientation as $L_j^{(m)}$.

Let us assume that $P_m = P_m(A_j) = -P([A_jA_{j+1}])$.

Let $S = (X, Y, Z)$ and (X', Y', Z') be two non-degenerate trios formed by the points of the arc $[B_jC_j]$ and such that $S' \leqslant S$. The osculating plane $P_m(A_j)$ will be denoted through P_m. Let us prove that if $P(S) = P_m$, then also $P(S') = P_m$. Indeed, as far as the trios S and S' are non-degenerate, the points Z and Z' lie on the arc $[A_jC_j]$. Then we have: $X < A_j$ since, in the opposite case, the plane $P(S)$ would coincide wiht the plane $P([A_jA_{j+1}]) = -P_m$. In the case when $Y' \leqslant A_j$, we, obviously, have $P(S') = P_m$, and let us henceforth consider that $A_j < Y'$. Let $\xi_2 = k\xi_1 + l$ be the equation of the straight line YZ. In

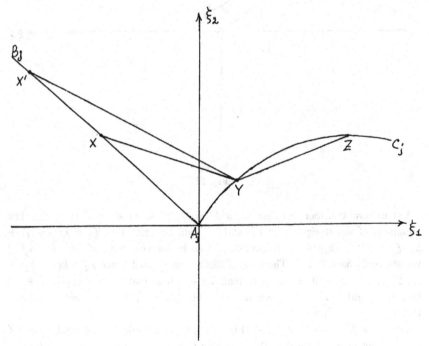

Fig. 21.

this case $k > 0$ due to the convexity of the arc $[A_jC_j]$ and $l \geqslant 0$, since the arc $[A_jY]$ is located below the straight line YZ. The point X lies above the straight line YZ and, hence, its coordinates (x_1, x_2) obey the inequality: $x_2 > k_{x1} + l$. Let $X' < X$, $Y' = Y$, $Z' = Z$ (Fig. 21). As far as the arc $[B_jA_j]$ is a segment of a straight line passing through the origin of the coordinates, then the coordinates of the point X' are expressed through x_1, x_2 in the following way: $x_1' = \lambda x_1$, $x_2' = \lambda x_2$. From the condition $X' < X$, it obviously follows that $\lambda > 1$. We have: $\lambda x_2 > k\lambda x_1 + \lambda l \geqslant k\lambda x_1 + l$. (Here we used the fact that $l \geqslant 0$.) Herefrom we get $x_2' > kx_1' + l$, i.e., the point X' also lies above the straight line YZ and hence in this case, the trios S' and S have the same orientation.

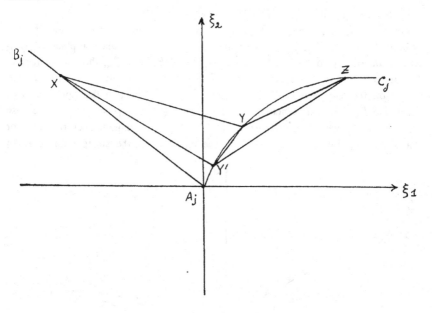

Fig. 22.

Let us now consider the case when $X' = X$, $Y' < Y$, $Z' = Z$ (Fig. 22). The equations of the straight lines YZ and $Y'Z$ will be written as: $\xi_2 = k(\xi_1 - z_1) + z_2$, $\xi_2 = k'(\xi_1 - z_1) + z_2$, respectively. Due to the convexity of the arc $[A_jC_j]$ we obviously have $k' > k$. The point X lies above YZ and hence $x_2 > k(x_1 - z_1) + z_2$. Since $x_1 - z_1 < 0$ and $k' \geqslant 0$, then we conclude that $x_2 > k'(x_1 - z_1) + z_2$, i.e., the point $Y'Z$ lies above the straight line $Y'Z$. In this case it follows that $P(S') = P(S)$ too.

Let $X' = X$, $Y' = Y$, $Z' < Z$ (Fig. 23). Let us consider the straight lines YZ and YZ', which are defined by the equations $\xi_2 = k(\xi_1 - y_1) + y_2$ and $\xi_2 = k'(\xi_1 - y_1) + y_2$, respectively. Due to the convexity of the arc $[B_jC_j]$ we have $k' \geqslant k$.

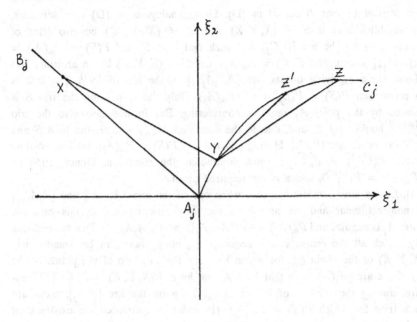

Fig. 23.

The point X lies above YZ. Therefore, $x_2 > k(x_1 - y_1) + y_2$. As far as $x_1 - y_1 < 0$ and $k' \geqslant k$, the point X is above the straight line YZ' as well. It makes it possible to conclude that $P(S') = P(S)$.

It follows from the above considerations that

$$P(S) = P(X, Y, Z) = P(X', Y, Z) = P(X', Y', Z) = P(X', Y', Z') = P(S'),$$

which is the required proof.

Let $S = L_j^{(m)}(X, Y, Z)$. If $Z \leqslant A_j$, then any trio S' of the chain ξ_m, lying on the arc $[A_{j-1}A_{j+1}]$ and preceding S, has the same orientation as S. Let us assume that $Z > A_j$. The trio $L_j^{(m)}$ cannot lie on the arc $[A_j A_{j+1}]$ (in the opposite case it would have the same orientation as the arc). Therefore, $X < A_j$, and, in line with the condition $\lambda(\xi_m) < \delta_j/2$, the trio $L_j^{(m)}$ lies on the arc $[B_j A_j]$. Let us assume that the trio $S' = (X', Y', Z')$ of the chain ξ_m precedes $L_j^{(m)}$ and lies on the arc $[A_{j-1}A_{j+1}]$. If $Z' \leqslant A_j$, then S' has the same orientation as $L_j^{(m)}$. If $Z' > A_j$, then trio S' lies on the arc $[B_j C_j]$ in line with the condition $\lambda(\xi_m) < \delta_j/2$. It follows from the above proof that in the case under discussion all trios of the chain ξ_m, lying on the arc $[A_{j-1}A_{j+1}]$ to the left of the trio $L_j^{(m)}$ have the same orientation as the arc.

(E) Let us assume that a certain right-hand semi-neighbourhood of the point A_j is rectilinear and no left-hand semi-neighbourhood of A_j lies in one straight line. It should be remarked that the plane $P_m(A_j)$ does not coincide with the plane $P([A_{j-1}A_j])$, since in the opposite case the arc $[A_{j-1}C_j]$ would be locally convex. We therefore consider here the situation which is mirror-symmetrical to that discussed in (D). Using analogous to (D) considerations, we establish that if $S = (X, Y, Z)$ and $S' = (X', Y', Z')$ be two trios of points lying on the arc $[B_j C_j]$ and such that $S \leqslant S'$ and $P(S) = P_m(A_j) = P([A_j A_{j+1}])$, then also $P(S') = P_m(A_j)$. Let $S = (X, Y, Z)$ be an arbitrary trio of the chain ξ_m lying on the arc $[A_{j-1} A_{j+1}]$ to the left of $L_j^{(m)}$. The task is to prove that $P(S) = P(L_j^{(m)}) = -P_m(A_j)$. Only the case when the trio S is hooked by the point A_j requires considering. But in this case also the trio $L_j^{(m)}$ is hooked by A_j and, due to the condition $\lambda(\xi_m) < \delta_j/2$, the trios S and $L_j^{(m)}$ lie on the arc $[B_j C_j]$. Having allowed that $P(S) = P_m(A_j)$, we immediately obtain $P(L_j^{(m)}) = P_m(A_j)$, which contradicts the condition. Hence, $P(S) = -P_m(A_j) = P(L_j^{(m)})$, which is the required proof.

II. Let us now consider the case when one of the arcs $[A_{j-1} A_j]$ and $[A_j A_{j+1}]$ is non-rectilinear and the arcs have the same orientations. In this case the point A_j is angular and $P_m(A_j) = -P([A_{j-1} A_j]) = -P([A_j A_{j+1}])$. Due to condition (3), which all the chains of the sequence ξ_m obey, there can be found a trio (X, Y, Z) of the chain ξ_m, for which $Y = A_j$. For any trio of the points (X, Y, Z) of the arc $[B_j C_j]$, such that $Y = A_j$, we have: $P(X, Y, Z) = P_m(A_j)$. Therefore, among the trios S of the chain ξ_m lying on the arc $[B_j C_j]$, there are such trios for which $P(S) = P_m(A_j)$. (In order to guarantee the existence of such trios we require condition (3). It should be noted that this is the first time when the fact that the triple chain ξ_m obeys condition (3) is used.) Let us denote by $L_j^{(m)}$ the utmost right-hand trio (X, Y, Z) of the chain ξ_m, for which $Y > A_j$ and $P(X, Y, Z) = P([A_{j-1} A_j])$. Let $R_j^{(m)}$ be the utmost left-hand trio (X, Y, Z) of the chain ξ_m, such that $A_j < Y$ and $P(X, Y, Z) = P([A_j A_{j+1}])$. Let (X', Y', Z') be a trio of ξ_m, for which $Y = A_j$. In this case $P(X', Y', Z') = P_m(A_j)$. Out of trios of the type (X, A_j, Z), belonging to the chain ξ_m, let us arbitrarily choose one and denote it by $M_j^{(m)}$. Let S be an arbitrary non-degenerate trio of the chain ξ_m lying on the arc $[A_{j-1} A_{j+1}]$. Let us prove that if $S \leqslant L_j^{(m)}$, then $P(S) = P(L_j^{(m)})$; if $S \geqslant R_j^{(m)}$, then $P(S) = P(R_j^{(m)})$; and, finally, when $L_j^{(m)} < S < R_j^{(m)}$, then the plane $P(S)$ coincides with the plane $P(M_j^{(m)}) = P_m(A_j)$. The validity of this statement results from the following.

Let $S = (X, Y, Z)$ and $S'(X', Y', Z')$ be two arbitrary trios lying on the arc $[B_j C_j]$. Let us assume that $S' \leqslant S$, $Y < A_j$ and $P(S) = P([A_{j-1} A_j])$. In this case $P(S') = P(S)$. Analogously, if $A_j < Y$, $S \leqslant S'$ and $P(S) = P([A_j A_{j+1}])$, then $P(S) = P(S')$.

The latter statement includes two points. Let us prove one of the points, referring to the case $S' \leqslant S$ and $Y < A$, since the second one is proved analogously.

So, let $S' = (X', Y', Z')$, $X' \leqslant X$, $Y' \leqslant Y$, $Z' \leqslant Z$, $Y < A_j$ and $P(S) = P([A_{j-1} A_j])$. If $Z' \leqslant A_j$, then, obviously, $P(S') = P([A_{j-1} A_j]) = P(S)$. Let us assume that $A_j < Z'$. Let us consider three possible cases in succession.

Let $X' < X$, $Y' = Y$, $Z' = Z$ (Fig. 24). Let $\xi_2 = k\xi_1 + l$ be the equation of the straight line YZ. The point X lies below the straight line YZ and, hence,

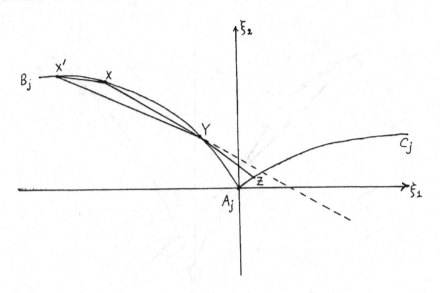

Fig. 24.

$x_2 < kx_1 + l$. We have: $x_1' < x_1 < y_1$. It affords $x_1 = \lambda x_1' + (1-\lambda)y_1$, where $0 < \lambda < 1$. As far as the arc $[B_jA_j]$ has its convexity turned upward, then $x_2 = f(x_1) \geqslant \lambda f(x_1') + (1-\lambda)f(y_1) = \lambda x_2' + (1-\lambda)y_2$. We have: $y_2 = ky_1 + l, x_2 < kx_1 + l = \lambda(kx_1' + l) + (1-\lambda)(ky_1 + l) = \lambda(kx_1' + l) + (1-\lambda)y_2$. It affords $x_2' < kx_1' + l$, i.e., the point X' lies below the straight line YZ and hence the trios (X', Y, Z) and (X, Y, Z) have the same orientations.

Let $X' = X, Y' < Y, Z' = Z$ (Fig. 25). The point Y lies above the straight line XZ. Let us show that the point Y' also lies above XZ. Let $\xi_2 = k\xi_1 + l$ be the equation of the straight line XZ. In this case $y_2 > ky_1 + l$, $x_2 = kx_1 + l$. We have: $x_1 < y_1' < y_1$, which affords $y_1' = \lambda x_1 + (1-\lambda)y_1$, where $0 < \lambda < 1$. Due to the convexity of the arc $[B_jA_j]$ we have: $y_2' = f(y_1') \geqslant \lambda f(x_1) + (1-\lambda)f(y_1) = \lambda x_2 + (1-\lambda)y_2 > \lambda(kx_1 + l) + (1-\lambda)(ky_1 + l) = k(\lambda x_1 + (1-\lambda)y_1) + l = ky_1' + l$. We conclude that the point Y' lies above the straight line XZ and hence the trios (X', Y', Z') and (X, Y, Z) have the same orientation.

Let us now consider the case when $X' = X, Y' = Y, Z' < Z$ (Fig. 26). If $Z' \leqslant A_j$, then the case is obvious. Let us, therefore, consider that $A_j < Z'$. The point Z lies below the straight line XY. Let $\xi_2 = k\xi_1 + l$ be the equation of this straight line. Since the function f decreases in the segment $[\lambda, 0]$, then $k < 0$. The point Z lies below the straight line XY and hence $z_2 < kz_1 + l$. We have: $z_1' < z_1, z_2' < z_2$. Hence, $z_2' < kz_1 + l < kz_1' + l$, due to the fact that $k < 0$. The point Z' therefore lies below the straight line XY and, hence, the trios (X', Y', Z') and (X, Y, Z) have the same orientation.

It follows from the above proved that $P(X, Y, Z) = P(X', Y, Z) = P(X', Y', Z) = P(X', Y', Z')$, i.e., that $P(S) = P(S')$, which is the required proof.

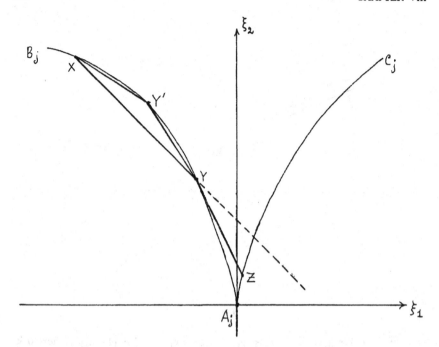

Fig. 25.

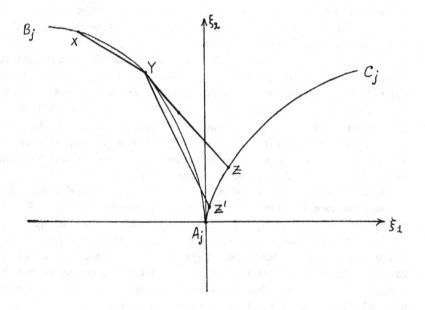

Fig. 26.

Let $S = (X', Y', Z')$ be an arbitrary non-degenerate trio of the chain ξ_m, lying on the arc $[A_{j-1}A_{j+1}]$ to the left of the trio $L_j^{(m)}$. If $Z' \leqslant A_j$, then, obviously, $P(S) = P([A_{j-1}A_j]) = P(L_j^{(m)})$. Let $Z' > A_j$. In this case the trios S and $L_j^{(m)}$ are both hooked by the point A_j. Due to the condition $\lambda(\xi_m) < \delta_j/2$, we find that these trios lie on the arc $[B_jC_j]$. In line with the above proved we conclude that $P(S) = P(L_j^{(m)})$.

In an analogous way we establish that if $R_j^{(m)} < S$, where S is a trio of the chain ξ_m lying on the arc $[A_{j-1}A_{j+1}]$, then $P(R_j^{(m)}) = P(S)$.

Let us assume that the trio S is such that $R_j^{(m)} > S > L_j^{(m)}$. Let us prove that $P(S) = P(M_j^{(m)})$. Indeed, let $L_j^{(m)} < S < M_j^{(m)}$, $S = (X, Y, Z)$. If $Y = A_j$, then $P(S) = P(M_j^{(m)})$. If $Y < A_j$, then $P(S)$ cannot coincide with $P(L_j^{(m)})$, according to the definition of the trio $L_j^{(m)}$. Therefore $P(S) = P(M_j^{(m)})$.

In the case when $A_j < Y$ the considerations are analogous. The required statement is thus proved for the case in question.

III. Let us assume that one of the arcs $[A_{j-1}A_j]$ and $[A_jA_{j+1}]$ is rectilinear. In this case its counterpart is not rectilinear and the point A_j is angular. Let us construct a neighbourhood $U_j = (B_jC_j)$ of the point A_j. In line with condition (3), which all the chains ξ_m obey, there can be found a trio S of the chain ξ_m, for which A_j is a mean point. The trio S lies on the arc $[B_jC_j]$ since $\lambda(\xi_m) < \delta_j/2$ (as usual, we assume $m \geqslant \bar{m}_j$). Out of the trios of the chain ξ_m, for which A_j is a mean point, let us arbitrarily choose one point and denote it through $M_j^{(m)}$. If the arc $[A_{j-1}A_j]$ does not lie in one straight line, then there are non-degenerate trios of the chain ξ_m on it. Let $L_j^{(m)}$ be the utmost right-hand of the trios of the chain ξ_m lying on the arc $[A_{j-1}A_{j+1}]$ and oriented in the same way as the arc $[A_{j-1}A_j]$. For the case when the arc $[A_{j-1}A_j]$ is rectilinear, let $R_j^{(m)}$ be the utmost left-hand of the trios of the chain ξ_m lying on the arc $[A_{j-1}A_{j+1}]$ and oriented in the same way as the arc $[A_jA_{j+1}]$.

Let us prove that when the arc $[A_jA_{j+1}]$ is non-rectilinear, any trio S of the chain ξ_m, following $R_j^{(m)}$ and lying on the arc $[A_{j-1}A_{j+1}]$ has the same orientation as $R_j^{(m)}$. Indeed, if the arc $[A_{j-1}A_j]$ is rectilinear, then no right-hand semi-neighbourhood of the point A_j lies in one straight line. The arc (B_jC_j), therefore, has in this case exactly the same construction as in the case I(D), and the required result is obtained from the same considerations. In the case when the arc $[A_jA_{j+1}]$ is rectilinear the situation is the same as that considered in the case I(E).

Let us now make some conclusions. At $m \geqslant \bar{m}_j$ in the triple of the chain ξ_m for each of the points A_j there are marked out trios denoted by $L_j^{(m)}$, $M_j^{(m)}$ and $R_j^{(m)}$, $j = 1, 2, \ldots, r-1$ and termed marked. The planes of the marked trios at the point A_j coincide with the marked osculating planes at this point. If the arc $[A_jA_{j+1}]$ does not lie in one straight line, $j+1 < r$, then the plane of the trios $R_j^{(m)}$ and $L_{j+1}^{(m)}$ coincide with the plane $P([A_jA_{j+1}])$ and the trios lying in between them have the same orientation as $R_j^{(m)}$ and $L_{j+1}^{(m)}$. Let $S_1, S_2, \ldots, S_q$

be marked trios numbered in the order of the sequence. In this case

$$T(\xi_m) = \sum_{i=1}^{q-1} (P(S_i), \stackrel{\frown}{P(S_{i+1})}).$$

On the other hand, $P(S_i) = P_i$, where $P_1, P_2, \ldots, P_q$ are the marked osculating planes of the curve K. We, thus, obtain

$$T(\xi_m) = \sum_{i=1}^{q-1} (P_i, \stackrel{\frown}{P_{i+1}}) = \tau(K).$$

The theorem is proved.

8.1.4. Let L be an arbitrary plane polygonal line, not lying in one straight line, $A_0, A_1, \ldots, A_m$ be vertices of the polygonal line L. Consequent trios of the vertices of the polygonal line L (A_0, A_1, A_2), (A_1, A_2, A_3), $\ldots$, (A_{m-2}, A_{m-1}, A_m) form a certain triple chain of the polygonal line L. A torsion of this triple chain will be termed a torsion of the polygonal line and denoted through $\tau(L)$. It can easily be proved that, in the case when the polygonal line L has no points of return, this definition of a complete torsion of a polygonal line gives the same result as the definition of a torsion of f.c.t. curves given above.

THEOREM 8.1.3. *Let K be an arbitrary plane f.c.t. curve, let $L_1, L_2, \ldots, L_m,$ $\ldots$ be a sequence of polygonal lines inscribed into the curve K such that $\lambda(L_m) \to 0$ at $m \to \infty$. In this case there is an m_0 such that at $m > m_0$ $\tau(L_m) = \tau(K)$.*

<u>Proof.</u> Let $A_1, A_2, \ldots, A_{r-1}$ be the points realizing a canonical subdivision of the curve K, let A_0 be its origin, and let A_r be its end. The polygonal line L_m contains r arcs $L_i^{(m)}$ inscribed into the arcs of a canonical subdivision of the curve K, each of these arcs being convex in the small, if $\lambda(L_m)$ is sufficiently small. Only osculating planes at the terminal points of these arcs can be marked osculating planes of the polygonal line, and it is these planes that are to be viewed in order to prove that $\tau(L_m) \to \tau(K)$.

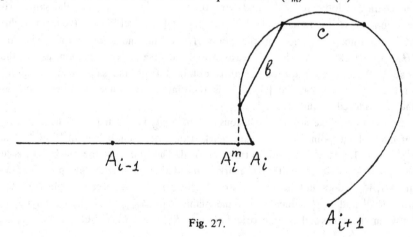

Fig. 27.

Let us dwell in detail on all possible cases here. Let $[A_{i-1}A_i]$ and $[A_iA_{i+1}]$ be arcs of a canonical subdivision of the curve K, following one another; $L_{i-1}^{(m)}$ and $L_i^{(m)}$ be inscribed into them arcs of the polygonal line L_m. Let us consider the following two cases.

1. One and only one of the arcs $[A_{i-1}A_i]$ and $[A_iA_{i+1}]$ is rectilinear.

Let it be the arc $[A_{i-1}A_i]$ (Fig. 27). (For the case when the arc $[A_iA_{i+1}]$ is rectilinear, the considerations are analogous.) Marked osculating planes of the curve at the point A_i are the planes $P_m(A_i)$ and $P_r(A_i)$. The orientations of the planes $P_m(A_i)$ and $P_r(A_i)$ are opposite, since otherwise the arc $[A_{i-1} A_{i+1}]$ would be convex in the small. Let a be a link of the polygonal line L connecting the end of the arc $L_{i-1}^{(m)}$ with the origin of the arc $L_i^{(m)}$. The link a can, in particular, degenerate into a point. Then let b and c be the utmost left-hand pair of the sequential non-colinear links of the arcs $L_i^{(m)}$. Joining, in case of necessity, the links of the polygonal line $L_i^{(m)}$, following one another, into one, we can get, as a result, that b will be the first link of the polygonal line, and c the second one. If m is sufficiently large, then the orientation of the pair of vectors (b, c) coincides with that of the plane $P_r(A_i)$, and one of the points which realizes a canonical subdivision of the polygonal line L_m, will be either the beginning of the link a or the end of the link b. Let us consider, for definitness, the first case (the second is quite analogous). Let A_i^m be the beginning of the link a. If $a \neq 0$, then marked osculating planes of the polygonal line L_m at the point A_i^m can be the planes $P_l^{Lm}(A_i^m)$, $P_r^{Lm}(A_i^m)$ and, possibly, the plane $P_m^{Lm}(A_i^m)$. But it can easily be seen that $P_l^{Lm}(A_i^m)$ coincides with the osculating plane $P_m^K(A_i)$ of the curve K, and $P_r^{Lm}(A_i^m)$ with the plane $P_r^K(A_i)$, so that the plane $P_l^{Lm}(A_i^m)$ cannot be marked. Then, if $a = 0$, then $A_i^m = A_i$ and marked osculating planes of the polygonal line L_m at the point A_i^m will be the planes $P_l^{Lm}(A_i^m) = P_m^K(A_i)$ and $P_r^{Lm}(A_i^m) = P_r^{Lm}(A_i)$.

Therefore the marked osculating planes of the polygonal line L_m at the point A_i^m coincide with the corresponding osculating planes of the curve K.

2. None of the planes $[A_{i-1}A_i]$ and $[A_iA_{i+1}]$ is rectilinear. Let c be a link of the polygonal line L_m which connects the end of the arc $L_{i-1}^{(m)}$ with the beginning of the arc $L_i^{(m)}$, a and b be the utmost right-hand pair of neighbouring non-colinear links of the polygonal line $L_{i-1}^{(m)}$ and, finally, let d and e be the utmost left-hand pair of neighbouring non-colinear links of the polygonal line $L_i^{(m)}$. Joining, if necessary, the links of the polygonal line L_m which continue one another into one, we can consider b to be the last link of the arc $L_{i-1}^{(m)}$, and let d be the first link of the arc $L_i^{(m)}$.

Let us first consider the case when the arcs $[A_{i-1}A_i]$ and $[A_iA_{i+1}]$ are oppositely oriented. The orientations of the pairs of vectors (a, b) and (d, e) coincide with those of the arcs of the curve K, into which they are inscribed. Two situations are possible here: the pairs of vectors (a, b) and (b, c) are oriented either in the same way or oppositely. In the former case the end of the link c is included into the number of points forming a canonical subdivision of the curve K, in the latter case such a point is the beginning of c. Denoting this point through A_i^m, we see that marked osculating planes at it are the planes $P_l^{Lm}(A_i^m) = P_l^K(A_i)$ and $P_r^{Lm}(A_i^m) = P_r^K(A_i)$.

Let us now consider the case when the arcs $[A_{i-1}A_i]$ and $[A_iA_{i+1}]$ have the same orientation. Here marked osculating planes of the curve K at the point A_i are the planes $P_l(A_i)$, $P_m(A_i)$ and $P_r(A_i)$. If the terminal points of the arcs $L_{i-1}^{(m)}$ and $L_i^{(m)}$ coincide with the point A_i, then at sufficiently large m the oriented planes $P_m^{Lm}(A_i^m)$ and $P_m^K(A_i)$ will coincide in the same way as the planes $P_l^K(A_i)$ and $P_l^{Lm}(A_i^m)$, $P_r^K(A_i^m)$ and $P_r^{Lm}(A_i)$, and the point A_i will be included into the number of points forming a new canonical subdivision of the polygonal line L_m.

Let now none of the terminal points of the arcs $L_{i-1}^{(m)}$ and $L_i^{(m)}$ coincide with the point A_i. In this case the three following cases are possible, depending on the orientation of the pairs of vectors (a, b), (b, c), (c, d) and (d, e) (the pairs (a, b) and (d, e) have the same orientation):

(1) orientation of the pairs (b, c) and (c, d) coincide and are opposite to those of the pairs (a, b) and (d, e);

(2) the pairs (b, c) and (c, d) are oppositely oriented;

(3) orientation of all four pair coincide.

In the first case the origin A_i^m of the link c and the end B_i^m of the link d are included into the number of points forming a canonical subdivision of the polygonal line L_m and marked osculating planes will be the planes at the point $A_i^m: P_l^{Lm}(A_i^m) = P_l^K(A_i)$ and $P_m^{Lm}(A_i^m) = P_m^K(A_i)$; at the point $B_i^m: P_m^{Lm}(A_i^m) = P_m^K(A_i)$ and $P_r^{Lm}(A_i^m) = P_r^K(A_i)$.

The second case is considered in exactly the same way as the case when the point A_i is a common canonical point of the arcs $L_{i-1}^{(m)}$ and $L_i^{(m)}$.

As to the third case, it is absolutely impossible.

It results from the above proof that $\tau(L_m) = \tau(K)$ at sufficiently large m, which the required proof.

8.1.5. THEOREM 8.1.4. *A geodesic turn of the indicatrix of the tangents Q of a plane curve K of a finite complete turn is equal to $\tau(K)$.*

Proof. Let us prove the theorem by way of induction with respect to the number p of arcs of a canonical subdivision of the curve K. For $p = 1$ the theorem is obvious. Let the statement be proved for $p = p_0$. Let us prove it for $p = p_0 + 1$. Let $A_0, A_1, \ldots, A_{p_0}, A_{p_0+1}$ be the points forming a canonical subdivision of the curve K; let Q be the indicatrix of the tangents of the arc $[A_0A_{p_0}]$; let Q' be the indicatrix of the tangents of the arc $[A_0A_{p_0+1}]$. We have:

$$\tau(A_0A_{p_0}) = |\kappa|(Q')$$

Here several cases are possible. Let us consider each in detail.

(1) The arc $[A_{p_0-1}A_{p_0}]$ is rectilinear, the arc $[A_{p_0}A_{p_0+1}]$ is not rectilinear. In this case the planes $P_m(A_p)$ and $P_r(A_p)$ are marked. The indicatrix Q'

is obtained from Q by adding two arcs $[xy]$ and $[yz]$, the orientations of which coincide with those of the planes $P_m(A_{p_0})$ and $P_r(A_{p_0})$. The plane $P_m(A_{p_0-1})$ is marked. The point y is obviously a point of return. If now $|P_m(A_{p_0-1})\overset{\frown}{,}P_m(A_{p_0})| = 0$, then x is a smooth point; if $|P_m(A_{p_0-1})\overset{\frown}{,}P_m(A_{p_0})| = \pi$, then x is a point of return. In all cases

$$|\tau|(A_0A_{p_0+1}) = |\kappa_g|(Q').$$

(2) The arc $[A_{p_0}A_{p_0+1}]$ is rectilinear, the arc $[A_{p_0-1}A_{p_0}]$ is not rectilinear. In this case the planes $P_m(A_{p_0})$ and $P_l(A_{p_0})$ are both marked. The curve Q' is obtained from Q by adding the arc $[xy]$ which has the same orientation as $P_m(A_{p_0})$.

The last link of the polygonal line Q has the same orientation as the plane $P_l(A_{p_0})$, and $|P_l(A_{p_0})\overset{\frown}{,}P_m(A_{p_0})| = \pi$, since, in the reverse case, the arc $[A_{p_0-1}A_{p_0+1}]$ would be convex in the small. Therefore, $|\tau|(A_0A_{p_0+1}) = \tau(A_0A_{p_0}) + \pi$, $|\kappa_g|(A_0A_{p_0+1}) = |\kappa|(A_0A_{p_0}) + \pi$.

(3) Neither of the arcs $[A_{p_0-1}A_{p_0}]$ and $[A_{p_0}A_{p_0+1}]$ is rectilinear, and the arcs $[A_{p_0-1}A_{p_0}]$ and $[A_{p_0}A_{p_0+1}]$ have the same orientation. In this case, obviously, there exists a plane $P_m(A_{p_0})$, $|\tau|(A_0A_{p_0+1}) = |\tau|(A_0A_{p_0}) + 2\pi$. The curve Q is obtained from Q' by adding the arcs $[xy]$ and $[yz]$ which have the same orientations as $P_m(A_{p_0})$ and $P_r(A_{p_0})$, respectively. It is clear here that x and y are points of return of the curve Q, so that

$$|\kappa_g|(Q') = |\kappa_g|(Q) + 2\pi = |\tau|(A_0A_{p_0+1}).$$

(4) The arcs $[A_0A_{p_0}]$ and $[A_0A_{p_0+1}]$ are not rectilinear and oppositely oriented. In this case $|\tau|(A_0A_{p_0+1}) = |\tau|(A_0A_{p_0}) + \pi$. But it can easily be seen that the curve Q' is obtained from Q by adding one or two arcs, in which case there arises the only new point of return, so that $|\kappa_g|(Q') = |\kappa_g|(Q) + \pi$.

8.2. Curves of a Finite Complete Torsion

8.2.1. Let L be an arbitrary polygonal line. Let us consider a triple chain ξ formed by trios of sequential vertices of the polygonal line L. Let P_1, P_2, ..., P_m be consequent secant planes of the triple chain ξ (drawn through the non-degenerate trios ξ). The sum

$$\sum_{i=1}^{m-1} (P_i \overset{\frown}{,} P_{i+1})$$

is termed a torsion of the polygonal line L and is denoted by $|\tau|(L)$.

Let K be an arbitrary curve in the space E^n, not lying in one straight line. At $\lambda(L) \to 0$ the upper limit of torsions of the polygonal lines L inscribed into the curve K, is called an absolute torsion of a curve. It will be denoted through $|\tau|(K)$.

As above, in analogous cases we will consider only the curves, the absolute torsions of which is finite.

THEOREM 8.2.1. *If for a curve K $|\tau|(K) < \infty$, then at every point X on it, no*

right-hand (left-hand) semi-neighbourhood of which is rectilinear, there
exists a right-hand (left-hand) osculating plane in the strong sense.

The proof of the theorem is omitted since it is nearly exactly the same as
that of Theorem 5.1.1.

The arc L of the curve K, which does not degenerate into a point, is termed
maximum rectilinear provided all the points of the arc L lie in one straight
line and no arc R of the curve K, which contains L and other than L, lies in
this straight line.

THEOREM 8.2.2. *If for the curve K $|\tau|(K) < \infty$, then a set of its maximum rec-*
tilinear arcs with multiple points is finite.

Proof. Let the statement of the theorem be wrong and let $\{[X_1 Y_1], [X_2 Y_2],$
$\ldots, [X_m Y_m], \ldots$ be a totality of all maximum rectilinear arcs with multiple
points. Two different maximum rectilinear arcs can have not more than one common (boundary) point. Let us choose a subsequence $X_{n_1}, X_{n_2}, \ldots, X_{n_k}, \ldots,$
converging to a certain point X_0 from one of the sides, for instance, from the
left. Obviously we have $X_{n_k} < Y_{n_k} < X_{n_k+1}$. Due to the fact that the arcs $[X_{n_k}$
$Y_{n_k}]$ are maximum, no left-hand semi-neighbourhood of the point X_0 is rectilinear and, in line with the preceding theorem, a certain left-hand semi-neighbourhood is a simple arc. This, however, contradicts the fact that the arcs
$[X_{n_k} Y_{n_k}]$ are not simple and, therefore, the theorem is proved.

From the theorem proved above it results, in particular, that if the absolute torsion of the curve K is finite and all the maximum rectilinear arcs of
the curve K have a finite turn, then a turn of the whole curve K is also finite.

To avoid considering quite cumbersome and, in essence, non-interesting
cases, let us introduce a certain limitation on the class of curves being
studied.

Let us say that the curve K is a curve of a finite complete torsion if its
absolute torsion is finite, $|\tau|(K) > \infty$, and the curve K contains no points of
return.

Therefore, we come to the conclusion that any curve of a finite complete
torsion is a curve of a finite turn.

LEMMA 8.2.1. *Let K be an arbitrary one-sidedly smooth curve in a space, α is a*
plane with the normal n. In this case, if neither vector n nor vector −n
belong to the indicatrix of the tangents of the curve K, then the orthogonal
projection of the curve K onto the plane α is a plane one-sidedly smooth
curve. In this case, if the curve K has no points of return, then a projection
of the curve K onto the plane α also has no points of return.

Proof. The first statement of the Lemma naturally results from the fact that
if the oriented straight lines l_m converge to an oriented straight line l,
which is not orthogonal to that plane α, then the projections l'_m of the
straight lines l_m onto the plane α converge to the projection l' of the

straight line l.

If the curve K has no points of return, then a projection of an angular point onto the plane α will be an angular point of the curve K, if and only if either n or $-n$ belong to the arc of the indicatrix of the tangents of the curve K, which corresponds to this angular point. This results in the second statement of the lemma.

COROLLARY. *Let K be an arbitrary curve of a finite complete torsion in E^3. In this case for almost all $\alpha \in G_2^3$ the projection K_α is a one-sidedly smooth curve without points of return.*

8.2.2. Let us establish an integro-geometrical characteristic of an absolute torsion of a curve, analogous to that proved earlier for the case of a turn.

Let P_α be a projection of an oriented plane P onto the plane $\alpha \in G_2^3$. Then, according to Theorem 6.2.1, for any two oriented planes P, Q in E^3 the following equation is valid:

$$|P, \widehat{} Q) = \int_{G_2^3} (P_\alpha, \widehat{} Q_\alpha)\, \mu_{3,2}\, (d\alpha).$$

This shows that for any polygonal line L in the space we have:

$$|\tau|(L) = \int_{G_2^3} |\tau|(L_\alpha)\, \mu_{3,2}\, (d\alpha).$$

LEMMA 8.2.2. *Let K be an arbitrary one-sidedly smooth plane curve having no points of return. Let $L_1, L_2, \ldots, L_m, \ldots$ be an arbitrary sequence of polygonal lines inscribed into the curve K, such that $\lambda(L_m) \to 0$. In this case, if there exists a constant M for whcih $|\tau|(L_m) \leqslant M$ at all m, then K is a curve of a finite complete torsion.*

Proof. In as much as the curve K is one-sidedly smooth and has no points of return, it can be divided into a finite number of arcs $K_1, K_2, \ldots, K_s$, each of which is projected in a one-to-one manner onto a certain straight line. Let $L_m^{(i)}$ be an arc of the polygonal line L_m inscribed into the arc K_i of the curve K. The arc $L_m^{(i)}$ is obviously defined only at sufficiently large m, $m \geqslant m_i$.

By the condition, $|\tau|(L_m) \leqslant M$ at all m. Therefore,

$$|\tau|(L_m^{(i)}) \leqslant M$$

at all $m \geqslant m_i$. Let K_i be projected in a one-to-one manner onto the straight line p_i. The polygonal line $L_m^{(i)}$ is also projected in a one-to-one way onto the straight line p_i and, in particular, it is a simple arc. According to Lemma 8.1.2 the arc $L_m^{(i)}$ can be divided into a finite number of convex arcs not greater than $[M/\pi] + 1$. Hence, through an evident limiting transition, we can easily prove that the arc K_i itself consists of not more than $[M/\pi] + 1$ convex arcs.

Therefore, the curve K can be divided into a finite number of convex arcs, and, hence, K is a curve of a finite complete torsion.

THEOREM 8.2.3. *Let K be a curve of a finite complete torsion. Then absolute*

torsions of the polygonal lines inscribed into K converge to $|\tau|(L)$ at $\lambda(L) \to$
0.

For almost all $\alpha \in G_2^3$ the plane curves K_α are curves of a finite complete torsion, in which case the following equality is valid:

$$|\tau|(G) = \int_{G_2^3} |\tau|(K_\alpha)\, \mu_{3,2}\,(d\alpha).$$

Proof. Let $L_1,\ L_2,\ \ldots,\ L_m,\ \ldots$ be an arbitrary sequence of polygonal lines inscribed into the curve K such that, at $m \to \infty$, $\lambda(L_m) \to 0$. Let E_1 be a set of all $\alpha \in G_2^3$ for which the plane curve K_α is not a one-sidedly smooth plane curve with no points of return, let E_2 be a set of all two-dimensional directions α, which are orthogonal to at least one secant plane passing through a trio of sequential vertices of one of the polygonal lines L_m. The set E_1 has a zero measure (as corollary to Lemma 8.2.1); obviously, the measure of E_2 is also equal to zero.

Let $\alpha \notin E_1 \cup E_2$. The polygonal lines $L_{m,\alpha}$ are inscribed into the curve K_α, in which case, obviously, at $m \to \infty$, $\lambda(L_{m,\alpha}) \to 0$.

Let us consider the quantity:

$$\tau = \lim_{m \to \infty} |\tau|(L_{m,\alpha}).$$

As far as the curve K_α is one-sidedly smooth and has no points of return ($\alpha \notin E_1$), then, in the case when $\tau < \infty$, K_α is an f.c.t. curve in line with the preceding lemma; therefore, according to Theorem 8.1.4, $|\tau|(L_{m,\alpha}) \to |\tau|(K_\alpha)$ at $m \to \infty$. In the case when $\tau = \infty$, in accord with the same theorem, K_α cannot be an f.c.t. curve and hence $|\tau|(K_\alpha) = \infty$. We have thus shown that

$$|\tau|(L_{m,\alpha}) \to |\tau|(K_\alpha)$$

at almost all $\alpha \in G_2^3$.

It follows from the above proof, in particular, that the function $|\tau|(K_\alpha)$ is measurable.

Applying Fatou's theorem and the corollary to Lemma 8.2.1, we obtain:

$$\int_{G_2^3} |\tau|(K_\alpha)\, \mu_{3,2}\,(d\alpha) \leqslant \lim_{m \to \infty} \int_{G_2^3} |\tau|(L_{m,\alpha})\, \mu_{3,2}\,(d\alpha)$$

$$= \lim_{m \to \infty} |\tau|(L_m) < \infty.$$

The function $|\tau|(K_\alpha)$ is thus summable, which fact means that the curve K_α is a plane curve of a finite complete torsion for almost all α.

Applying Theorem 8.1.1 to the triple chain of the polygonal line defined at the beginning of Section 8.1.4, we get the estimate:

$$|\tau|(L'_{m,n}) \leqslant 3[\kappa(K_n) + |\tau|(K'_n)].$$

Therefore, the functions $|\tau|(L_{m,\alpha})$ are all majorized by the summable function $\Phi(\alpha) = 3[\kappa(K_\alpha) + |\tau|(K_\alpha)]$. As far as at $m \to \infty$ $|\tau|(L_{m,\alpha}) \to |\tau|(K_\alpha)$ for almost all α, then, according to the known Lebesgue theorem on a limiting transition under the sign of the integral, we get the following relation:

$$\int_{G_2^3} |\tau|(K_\alpha)\,\mu_{3,2}\,(d\alpha) = \lim_{m\to\infty} \int_{G_2^3} |\tau|(L_{m,\alpha})\,\mu_{3,2}\,(d\alpha) =$$

$$= \lim_{m\to\infty} |\tau|(L_m).$$

We have therefore proved that there exists a finite limit $\lim_{m\to\infty} |\tau|(L_m)$. Since the sequence $L_1, L_2, \ldots, L_m, \ldots$ is arbitrary, then the limit $\lim_{\lambda(L)\to 0} |\tau|(L)$ does exist. Obviously, this limit can be equal only to an absolute torsion of the curve K. The theorem is thus proved.

REMARK 1. By analogy with the theory of a turn of spherical curves, we can try to construct a theory of curves of a finite complete torsion, having defined an absolute torsion as the lower limit of absolute torsions of the inscribed polygonal lines. This attempt, however, is doomed to failure if we do not introduce certain additional limitations on the curves, as can be shown by the example of the curve defined by the equations $x(t) = t \cos \ln t$, $y(t) = t \sin \ln t$ at $x(0) = y(0) = 0$ (a logarithmic spiral). At any $\varepsilon > 0$ and any N for this curve K we can construct the polygonal lines L_1 and L_2, such that $\lambda(L_1) < \varepsilon$, $\lambda(L_2) < \varepsilon$ and $|\tau|(L_1) = 0$, $|\tau|(L_2) \geqslant N$. The upper limit of absolute torsions of the inscribed polygonal lines is, in this case, equal to ∞ and the lower limit to 0.

REMARK 2. If the absolute torsion of the curve K is finite and the curve K is not a curve of a finite complete torsion in the sense of the definitition given above (i.e., K either has rectilinear arcs with divisible points or is a one-sidedly smooth curve with points of return), then Theorem 8.2.3 is also untrue, as is shown by the examples presented in Figs. 28 and 29. In Fig. 28

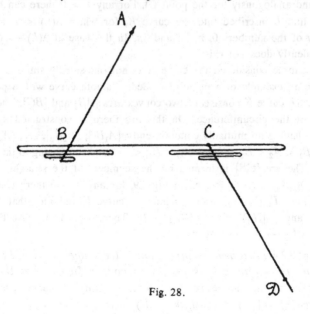

Fig. 28.

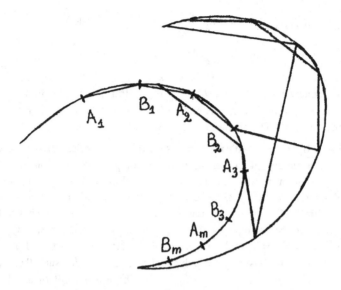

Fig. 29.

the curve K consists of three arcs $[AB]$, $[BC]$ and $[CD]$. The arc $[BC]$ lies in one straight line, $[AB]$ and $[CD]$ are rectilinear segments not lying in the straight line BC. In this case the arc $[BC]$ oscillates near the points B and C, so that any right-hand semi-neighbourhood of the point B on the curve K contains the points lying on the straight line BC to different sides of the point B (and analogously for the point C). For any $\varepsilon > 0$ there can be found a polygonal line L inscribed into the curve K, for which $\lambda(L) < \varepsilon$ and $|\tau|(L)$ equals any of the numbers 0, π, 2π and 3π. In this case at $\lambda(L) \longrightarrow 0$ the limit $|\tau|(L)$ evidently does not exist.

In the example considered the curve K is not one-sidedly smooth. In Fig. 29 we can see an example of a plane one-sidedly smooth curve with a point of return. Here the curve K consists of two convex arcs $[AB]$ and $[BC]$. The arc $[AB]$ is an arc of the circumference. On this arc there is constructed a countable set of arcs, having no mutually common ends: $[A_1B_1]$, $[A_2B_2]$, ..., $[A_mB_m]$, such that $A_1 < B_1 < A_2 < B_2 < \cdots < A_m < B_m < \cdots$, and converging to the point B at $m \longrightarrow \infty$. The arc $[CB]$ is formed by the segments of the straight lines A_1B_1, A_2B_2, ..., A_mB_m, ... As is seen in Fig. 29, for any $\varepsilon > 0$ there can be found polygonal lines L_1 and L_2, inscribed into the curve K and such that $\lambda(L_1) < \varepsilon$, $\lambda(L_2) < \varepsilon$ and $|\tau|(L_1) = \pi$, $|\tau|(L_2) = 2\pi$. Therefore, in this case the limit of $|\tau|(L)$ at $\lambda(L) \longrightarrow 0$ does not exist.

THEOREM 8.2.4. *Let K be an arbitrary one-sidedly smooth curve. For its absolute torsion to be finite it is necessary and sufficient that the absolute geodesic turn of the indicatrix of the tangents Q of the curve K were finite. In this case if $|\tau|(K) < \infty$, then $|\tau|(K) = |\kappa_g|(Q)$.*

Proof. Let first $|\tau|(K) < \infty$. In this case the curve Q is rectifiable. Let us consider a projection of the curve K onto the plane α with the normal n. Let x be the end of the unit vector n with the origin at the center of a unit sphere Ω. In this case, if neither x nor $-x$ lie on the curve Q, then the projection Q_α of the curve Q onto the great circumference $\gamma(\alpha)$ coincides with the indicatrix of the tangents of the curve K_α and, in line with Theorem 8.1.5, $|\kappa_g|(Q_\alpha)$ $= |\tau|(K_\alpha)$, which evidently affords $|\kappa_g|(Q) = |\tau|(K_\alpha)$.

Let now $|\kappa_g|(Q) < \infty$. In this case a turn of the curve Q_α is finite for nearly all α, $|\kappa_g|(Q) < \infty$. The curve Q_α is a normal indicatrix of the tangents of the plane curve K_α. If $|\kappa_g|(Q) < \infty$, then the curve K_α is evidently a curve of a finite complete torsion and $|\tau|(K_\alpha) = |\kappa_g|(Q)$.

Let $L_1, L_2, \ldots, L_m, \ldots$ be an arbitrary sequence of the inscribed into the curve K polygonal lines, such that $\lambda(L_m) \longrightarrow 0$ at $m \longrightarrow \infty$. Repeating the considerations used to prove Theorem 8.2.3, we get:

$$|\tau|(L_m) \longrightarrow \int_{G_2^3} \tau(K_\alpha)\, \mu_{3,2}\,(d\alpha) < \infty.$$

which affords $|\tau|(K) < \infty$. The theorem is thus proved.

8.3. Complete Two–Dimensional Indicatrix of a Curve of a Finite Complete Torsion

8.3.1. Let us first state some remarks concerning the curves which have an osculating plane in the strong sense.

Let us say that the curve K in E^3 is a curve with an osculating plane in the strong sense provided the following conditions are met:

(1) if no left-hand (right-hand) semi-neighbourhood of the point X of the curve K lies in one straight line, then K has at the point X a left-hand (right-hand, respectively) osculating plane in the strong sense;

(2) any rectilinear arc of the curve K is simple.

It follows from (1) and (2) that a curve with an osculating plane in the strong sense has a finite turn. Let us further assume that one more condition is also met:

(3) K has no points of return.

If K is a curve with an osculating plane in the strong sense, then, according to Theorem 7.3.2, its indicatrix of the tangents is a one-sidedly smooth curve.

Let us describe a certain construction referring to one-sidedly smooth curves on a sphere.

Let C be an oriented great circumference on the sphere S^2. Let us denote by $\mu(C)$ a point of the sphere which is the end of a unit vector orthogonal to the plane C and such that for any point $X \in C$ the trio of vectors $(OX, t_C(X),$ $\nu(C))$ is right-handed. (Here $t_C(X)$ is a tangential unit vector of the sphere C

at the point X).

Let Q be a one-sidedly smooth curve on the sphere S^2. For the point $X \in Q$ let $\nu_Q(X)$ be the shortest line on the sphere, connecting the points $\nu[C_l(X)]$ and $\nu[C_r(X)]$. We thus obtain a certain c-correspondence $X \mapsto \nu_Q(X)$. Its indicatrix is called a curve, which is polar to K, and is denoted henceforth by the symbol $\nu(Q)$. Theorem 6.2.1 readily shows that the length of the curve $\nu(Q)$ is equal to $|\kappa|(Q)$. The proof here is analogous to the proof of the fact that the length of an indicatrix equals the turn of a curve, and is thus omitted.

A two-dimensional element in E^3 will be a trio (X, l, P) consisting of a point X, an oriented straight line l and an oriented plane P passing through the straight line l. We shall call X, l and P the first, the second and the third components of the two-dimensional element $H = (X, l, P)$, respectively. Let $H_1 = (X, l, P)$ and $H_2 = (Y, m, P)$ be two-dimensional elements in E^3. Let us set

$$\rho(H_1, H_2) = |X - Y| + (l \frown m) + (P \frown Q).$$

A totality of all two-dimensional elements in E^3 will be denoted by $T_2(E^3)$. Obviously, ρ is a metric in $T_2(E^3)$.

Let there be given two-dimensional elements $H_1 = (X, l, P)$, $H_2 = (X, m, Q)$ with the same first component X. Let us say that the two-dimensional elements H_1 and H_2 are neighbouring if either $l = m$ or $P = Q$. For each pair of neighbouring two-dimensional elements H_1, H_2 in $T_2(E^3)$ we define a certain curve in $T_2(E^3)$, connecting H_1 and H_2, which will be termed the shortest curve in $T_2(E^3)$, connecting H_1 and H_2. If $l = m$, then this curve consists of two-dimensional elements which are obtained if we rotate the plane P around l, with X and l motionless, in order to superpose it with Q in the shortest possible way. If $P = Q$ and $l \neq m$, then the required curve is obtained by rotating l around the point X in the plane P, X and P remaining motionless, in order to superpose l and m in the shortest possible way.

To any curve with an osculating plane in the strong sense let us put in correspondence a certain curve in the space of two-dimensional elements $T_2(E^3)$. At the same time let us define a certain set Ω of two-dimensional planes. To each point $X \in K$ there corresponds a certain subset Ω, denoted through $\Omega(X)$. The set $\Omega(X)$ is finite and its elements are termed basic osculating planes at the point X.

Let $\tau_1(K)$ be a complete indicatrix of the tangents of the curve K. By definition, $\tau_1(K)$ is a curve in the space $T_1(E^3)$ of one-dimensional elements in E^3. To each point $Z \in \tau_1(K)$ let us put in correspondence a certain curve $\omega(Z)$ in $T_2(E^3)$.

Let $Z = (X, \lambda)$. Here several cases are possible, which we are gooing to consider in detail.

(A) X is an angular point of the curve K and λ is an intermediate tangent at this point.

In the case under discussion the intermediate osculating plane $P_m(X)$ is defined. The curve $\omega(Z)$ is considered to consist of the only point which is the two-dimensional element $(X, \lambda, P_m(X))$. The plane $P_m(X)$ is included into the number of planes from Ω referring to the point X.

(B) X is an angular point, $\lambda = t_r(X)$.

Let us assume that no right semi-neighbourhood of X lies in one straight line. In this case the right-hand osculating plane $P_r(X)$ is defined at the point X. The plane $P_r(X)$ is included into the number of planes from Ω referring to the point X. As $\omega(Z)$, let us choose the shortest arc in $T_2(E^3)$, connecting the two-dimensional elements $(X, \lambda, P_m(X))$ and $(X, \lambda, P_r(X))$. In the case when a certain right-hand semi-neighbourhood of X lies in one straight line, $\omega(Z)$ is considered to consist of the only point, which is the two-dimensional element $(X, \lambda, P_m(X))$.

(C) X is an angular point, $\lambda = t_l(X)$, and no left-hand semi-neighbourhood of X lies in one straight line.

In this case $\omega(Z)$ is the shortest line in $T_2(E^3)$, connecting the two-dimensional elements $(X, \lambda, P_l(X))$ and $(X, \lambda, P_m(X))$. The plane $P_l(X)$ is included into the number of planes from Ω referring to the point X. If X is an angular point, $\lambda = t_l(X)$ and a certain left-hand semi-neighbourhood of X is rectilinear, then the arc $\omega(Z)$ is constructed in the following way. We find the utmost left-hand point $X' < X$, such that the arc $[X'X]$ is rectilinear. If X' is the beginning of the curve K, then $\omega(Z)$ consists of the only point, which is the two-dimensional element $(X, \lambda, P_m(X))$. Let X' be not the beginning of the curve. Let $P = P_l(X')$ if X' is a smooth point, $P = P_m(X')$ if X' is an angular point of the curve K. The straight line λ lies in the plane P. In this case we set $P = P_l(X)$ and the plane P is included in the number of planes from Ω referring to the point X. As $\omega(Z)$, we choose the shortest line in $T_2(E^3)$ connecting the two-dimensional element (X, λ, P) with the two-dimensional element $(X, \lambda, P_m(X))$.

(D) X is a smooth point and no neighbourhood of X lies in one straight line.

In this case $\lambda = t_l(X) = t_r(X)$. If no left-hand and no right-hand semi-neighbourhood of X lies in one straight line, then $\omega(Z)$ is the shortest in $T_2(E^3)$ line connecting the two-dimensional element $(X, \lambda, P_l(X))$ with the two-dimensional element $(X, \lambda, P_r(X))$. Let us also set $\Omega(X) = \{P_l(X), P_r(X)\}$. If a certain right-hand semi-neighbourhood is rectilinear, then $\omega(Z)$ consists of the only point, which is the two-dimensional element $(X, \lambda, P_l(X))$, and $\Omega(X) = \{P_l(X)\}$. Let us suppose that a certain left-hand semi-neighbourhood of the point X is rectilinear. In this case we find the utmost left-hand point X', such that the arc $[X'X]$ lies in one straight line. If X' is the beginning of the curve K, then $\omega(Z)$ consists of the only point, which is the two-dimensional element $(X, \lambda, P_r(Z))$; and $\Omega(X) = \{P_r(X)\}$. If X' is not the beginning of the curve, then let $P = P_m(X')$ in the case when X' is an angular point and $P = P_l(X')$ if X' is a smooth point. The straight line λ lies in P and as $\omega(Z)$ let us choose the shortest line connecting the two-dimensional element (X, λ, P) with the two-dimensional element $(X, \lambda, P_r(X))$. Let us set $P_l(X) = P$ and $\mathfrak{Q}(X) = \{P, P_r(X)\}$.

(E) The last case: a certain neighbourhood of X is rectilinear. Let $X' \leqslant X$ be the utmost left-hand, and $Y' \geqslant X$ the utmost right-hand points of the curve K, such that the arcs $[X'X]$ and $[XY']$ are rectilinear. If X' is not the beginning of the curve, then let $P = P_m(X')$ in the case when X' is an angular point, $P = P_l(X')$ in the case when X' is a smooth point. The curve $\omega(Z)$ consists of the only point (X, λ, P), and $\mathfrak{Q}(X) = \{P\}$. If X' is the beginning of the curve K, then let us consider the point Y'. It may appear that Y' is the end of K, so that the whole curve is a segment of a straight line. In this case let us arbitrary choose a plane P containing K, and, as $\omega(Z)$, let us choose a two-dimensional element (X, λ, P) for any Z. If Y' is not the end of the curve K, then the plane $P = P_m(Y')$ is defined if Y' is an angular point, and we set $P = P_r(Y')$, if Y' is a smooth point, and $\omega(Z)$ is considered consisting of the only point, which is the two-dimensional element (X, λ, P).

Let us set $\mathfrak{Q}(X) = \{P\}$. It can be easily proved that ω is a c-correspondence in the sense defined in Section 3.5. It is the indicatrix of the given c-correspondence that is termed a two-dimensional indicatrix of the curve K.

Let us set $\mathfrak{Q} = \bigcup_{X \in K} \mathfrak{Q}(X)$. The set $\mathfrak{Q}$ is ordered, assuming that, for any point X, $P_l(X) < P_m(X)$, $P_m(X) < P_r(X)$, and at $X < Y$ let us consider that any plane $P \in \mathfrak{Q}$ precedes any plane $Q \in \mathfrak{Q}(Y)$.

8.3.2. Let K be an f.c.t. curve, and B be its complete two-dimensional indicatrix. Let (P, l, X) be an arbitrary two-dimensional element, which is a point of the curve B. For the curve B, naturally defined are certain mappings onto the one-dimensional indicatrix of the curve K and onto the curve K itself, which are obtained if to the point (P, l, X) there corresponds a point (l, X) of the one-dimensional indicatrix and, respectively, a point X of the curve K. These mappings will be termed natural projections of the two-dimensional indicatrix onto the one-dimensional indicatrix and onto the curve K, respectively.

Let us call the point X of the f.c.t. curve K an angular point of the second kind if the complete prototype of this point with respect to the natural projection of a complete two-dimensional indicatrix of the curve K onto the curve itself contains not more than one point. It follows from the very definition of the notion of an indicatrix that the complete prototype considered is always a whole arc of the curve B.

Let us see in which cases a point of an arbitrary curve of a finite complete torsion is an angular point of the second kind.

Let X be a smooth point. If a certain neighbourhood of it lies in one straight line, then X is not an angular point of the second kind.

If X is an angular point, then X is obviously also an angular point of the second kind.

If X is a smooth point, no neighbourhood of which lies in one straight line, then X will be an angular point of the second kind in the following cases: (1) no left-hand and no right-hand semi-neighbourhood of X lies in one straight line and $P_l(X) \neq P_r(X)$; (2) there exists a point $X' < X$, such that the arc $[X'X]$ is rectilinear, no left-hand semi-neighbourhood of X' lies in one straight line and the plane P, where $P = P_m(X')$ if X' is an angular point, $P = P_l(X)$ if X' is a smooth point, differs from the plane $P_l(X)$.

If X is an angular point, then X is obviously also an angular point of the second kind.

For a one-dimensional indicatrix of an f.c.t. curve K there can be defined two additive functions of the arcs which will be called a length and a turn, respectively. Let L be an arbitrary arc of the one-dimensional indicatrix B_1 of the curve K, L' be an arc corresponding to L due to natural projection of B_1 onto K. The length of the arc L' will be denoted by $s(L)$. Let us term it the length of the arc L. Therefore, a certain function of the arcs of the curve B_1 is defined. Obviously this function is continuous, being a function of the ends of the arc, and additive.

Let (X_1, l_1) be the beginning, and (X_2, l_2) be the end of the arc L of the one-dimensional indicatrix B_1 of the curve K of a finite complete torsion. We set:

$$\kappa(L) = (l_1, \widehat{} t_r(X_1)) + \kappa(X_1 X_2) + (t_l(X_2), \widehat{} l_2).$$

Let us call the defined with this equality function of the arc L a turn of the arc L of the one-dimensional indicatrix of the curve K. It is obvious from the definition that a turn, as well as a length, is an additive function of the arc.

Let now B_2 be a complete two-dimensional indicatrix of the curve K of a finite complete torsion. For the curve B_2 we can also define certain additive functions of the arcs.

Let L be an arbitrary arc of the two-dimensional indicatrix B_2 of the curve K, L' be an arc of the one-dimensional indicatrix corresponding to L under natural projection. The length and the turn of the arc L' will be termed the length and the turn, respectively, of the arc L and denoted by $s(L)$ and $\kappa(L)$.

Besides, let us define one more additive function of the arcs of the two-dimensional indicatrix − an absolute torsion. Let us first define it for those arcs of the curve B_2, the natural projection of which onto the curve K consists of a single point. Let L be such an arc. Let (X, l_1, P_1) be its beginning, and (X, l_2, P_2) be its end. Here two cases are possible. The first case is when the point X is smooth. Then the tangents l_1 and l_2 coincide and we set:

$$|\tau|(L) = |P_1, \widehat{} P_2|.$$

If the point X is angular, then the osculating planes $P_l(X)$, $P_m(X)$ and $P_r(X)$ are defined in it.

Let (X, l_0, P_m) be a two-dimensional element at the point X, containing the plane P_m. If this element lies in the arc L, we set $|\tau|(L) = |P_1 \overset{\frown}{,} P_m| + |P_m \overset{\frown}{,} P_2|$. If this element does not lie in the arc L, we set:

$$|\tau|(L) = |P_1 \overset{\frown}{,} P_2|$$

Let now L be an arbitrary arc of the two-dimensional indicatrix B. Let $\alpha_1 = (X_1, l_1, P_1)$ and $\alpha_2 = (X_2, l_2, P_2)$ be the beginning and the end of the arc L. Let us find the utmost right-hand point of the two-dimensional indicatrix B_1 and the utmost left-hand point of B_2 of the arc L, such that the natural projections of the arcs $[\alpha_1 \beta_1]$ and $[\beta_2 \alpha_2]$ are the points X_1 and X_2, respectively. Finally, let L_2' be a natural projection of the arc L onto the curve K. We set:

$$|\tau|(L) = |\tau|([\alpha_1 \beta_1]) + |\tau|(L') + |\tau|([\beta_2 \alpha_2]).$$

The function $|\tau|(L)$ is also an additive function of an arc of the two-dimensional indicatrix, as follows from Theorem 8.4.2, which will be proved below.

8.4. Continuity and Additivity of Absolute Torsion

8.4.1. Let us prove here some corollaries of Theorem 8.2.3.

THEOREM 8.4.1. *Let K be a curve of a finite complete torsion, X be an arbitrary point of the curve K. Then at $Y \to X$ the absolute torsion of the arc $[XY]$ tends to zero.*

The theorem is proved in exactly the same way as the corresponding theorem on a turn of a curve. For plane curves of a finite complete torsion it is evident. In a general case it can be readily obtained from the integral relation of Theorem 8.2.3.

Let K be an arbitrary curve of a finite complete torsion. Let us construct its complete two-dimensional indicatrix. Let X be an arbitrary point of the curve K, and let $L = p^{-1}(X)$ be its complete prototype with respect to the natural mapping of the two-dimensional indicatrix of the curve onto the curve. We set:

$$\varphi(X) = |\tau|(L).$$

The notion introduced allows one to prove a certain additive property of the absolute torsion of a curve.

THEOREM 8.4.2. *Let $K = [AB]$ be an arbitrary curve of a finite complete torsion, and let C be an arbitrary point of the curve K. In this case each of the arcs $[AC]$ and $[CB]$ of the curve K is a curve of a finite complete torsion itself, and the following equality is valid:*

$$|\tau|(K) = |\tau|(AC) + |\tau|(CB) + \varphi(C).$$

<u>Proof.</u> It is sufficient to prove the validity of the theorem for the case of plane curves, or, more exactly, the following fact. Let A', B' and X' be projections of the points A, B and X onto the plane α, in which case $K_\alpha = A'B'$ is

a curve of a finite complete torsion. In this case $|\tau|(A'B') = |\tau|(A'X') + |\varphi|$
$(X') + |\tau|(X'B')$, and the angle $|\varphi|(X')$ is expresed by the angles between the
osculating planes of the curve K_α at the point X' in exactly the same way as
$|\varphi|(\ )$ is expressed through the angles between the osculating planes of the
curve K at the point X, i.e., if, for instance, $|\varphi|(X) = 2\pi - |P_l(X)\overset{\frown}{,}P_r(X)|$
or $|\varphi|(X) = |P_l(X)\overset{\frown}{,}P_m(X)| + |P_m(X)\overset{\frown}{,}P_r(X)|$, then also $|\varphi|(X') = 2\pi -$
$|P_l(X')\overset{\frown}{,}P_r(X')|$ or, $|\varphi|(X') = |P_l(X')\overset{\frown}{,}P_m(X')| + |P_m(X')\overset{\frown}{,}P_r(X')|$, respec-
tively. As is seen from the definition of $|\varphi|(X)$, the expression $|\varphi|(X)$
through the angles between the osculating planes at the point X depends exclu-
sively on the structure of the neighbourhood of the point X. It can be easily
seen, however, that the neighbourhoods of the point $X \in K$ and the point $X' \in$
K_α will have the same structure in this respect, in any case, for almost all
α. Applying Theorem 3.2.3, we easily get the proof of the theorem.

It obviously follows from the results obtained in this paragraph that an absolute torsion is a continuous additive function of the arcs of a complete
two-dimensional indicatrix of a curve of a finite complete torsion.

8.5. Definition of an Absolute Torsion Through Triple Chains and Paratingences

8.5.1. THEOREM 8.5.1. *Let* $\xi_1, \xi_2, \ldots, \xi_m, \ldots$ *be an arbitrary sequence of
triple chains of the curve of a finite complete torsion* K, *such that* $\lambda(\xi_m) \to$
$0, \nu(\xi_m) < A < \infty$ *at all* m, *and for any angular point of the curve* K *there can
be found such an* m_0 *that at any* $m > m_0$ *this point be a mean point of a certain
trio of the points of the chain* ξ_m. *In this case, at* $m \to \infty, T(\xi_m) \to |\tau|(K)$.

Proof. Let us choose some $\alpha \in G_2^3$, such that a plane curve K_α is a curve of a
finite complete torsion. Such are, by Theorem 8.2.3, almost all α. Besides
this, let α be not orthogonal to any of the secant planes of the triple chains
$\xi_m, m = 1, 2, \ldots$.

Projecting the trios of points of the chain ξ_m onto the plane of the direc-
tion α, we obviously obtain a certain triple chain $\xi_{m,\alpha}$ of the curve K_α. It is
easily seen that

$$T(\xi_m) = \int_{G_2^3} T(\xi_{m,\alpha})\, \mu_{3,2}\, (d\alpha).$$

Omitting, if necessary, the finite number of chains, we can consider the se-
quence of triple chains of the curve K_α $\{\xi_{1,\alpha}, \xi_{2,\alpha}, \ldots, \ldots\}$ to obey conditions (1) and (3) in Section 8.1.3 and, hence, by Theorem 8.1.2, $(\xi_{m,\alpha}) \to$
$|\tau|(K_\alpha)$ at $m \to \infty$. To prove that $T(\xi_m) \to |\tau|(K)$ let us make use of the estimate of Theorem 8.1.1. We have:

$$T(\xi_{m,\alpha}) \leqslant [\nu(\xi_m) + 1]\, [|\tau|(K_\alpha) + \kappa(K_\alpha)] =$$
$$\leqslant (A + 1)\, [|\tau|(K_\alpha) + \kappa(K_\alpha)].$$

As has been proved above, $|\tau|(K_\alpha) + \kappa(K_\alpha)$ is a summable function of α. Therefore, the function $T(\xi_{m,\alpha})$ is majorized from above by a summable function, which affords $T(\xi_m) \rightarrow |\tau|(K)$ at $m \rightarrow \infty$, which is the required proof.

8.5.2. Let us consider the problem of defining an absolute torsion of a curve through paratingencies by analogy with the way we have above defined a turn through paratingencies.

Let K be an arbitrary spatial curve. A set Ω of all its paratingencies is naturally ordered. Let $\xi = \{\Pi_1 < \Pi_2 < \cdots < \Pi_m\}$ be an arbitrary sequence of the paratingencies of the curve, P_i be an arbitrary plane of the paratingency Π_i and $T(\xi) = \sup \sum_{i=1}^{m-1} (P_1 \frown P_{i+1})$, where the upper boundary is taken by all possible $P_i \in \Pi_i$, $P_{i+1} \in \Pi_{i+1}$. The exact upper boundary of the numbers $T(\xi)$ taken by all the sequences of the paratingencies will be denoted through $\nu(K)$. The equality $\nu(K) = |\tau|(K)$ is, generally speaking, not valid. As an example, we can consider a logarithmic spiral $x(t) = t \cos \ln t$, $y(t) = t \sin \ln t$, $0 \leqslant t \leqslant 1$, $x(0) = y(0) = 0$. Its paratingency at the point $(x(0), y(0))$ consists, as can be seen, of two planes which are different in their orientations. At all the other points all paratingencies of the curve consist of the only plane and coincide, so that $\nu(K)$ is in this case equal to π, while $|\tau|(K) = \infty$.

If, however, we impose on the curve K certain additional limitations, for instance, if we demand that K be one-sidedly smooth, the equality $\nu(K) = |\tau|(K)$ becomes valid. It is at this point of discussion that the theory of torsion proves less perfect as consider with the theory of turn. Here we are not going to consider the problem of minimal limitations, which are to be imposed on the curves in question for the equality $\nu(K) = |\tau|(K)$ to be valid.

LEMMA 8.5.1. *If for the curve K $\nu(K) < \infty$, then the right-hand and left-hand limits of the paratingencies at its every point consist of the only plane.*

This lemma is proved by analogy with Lemma 5.2.1. Let us assume, in contrast to what is required, that at the point X, for instance, the right-hand limit of the paratingencies consists of two different planes P and Q, $|P \frown Q| \neq 0$. Let Y be the utmost right-hand point, such that the arc $[XY]$ is rectilinear. The right-hand paratingencies of all the points of the arc $[XY]$, obviously, coincide. Let $Y_1 > Y_2 > \cdots > Y_n$ be a sequence of the points converging to Y from the right and such that at the point Y_k the paratingency Π_k of the curve contains a plane P_k which forms an angle less than $\frac{1}{3}(P \frown Q)$ with the plane P at an odd k, and with the plane Q, respectively, at an even k. Let us consider the system of paratingencies of the curve $\xi_{n+1} = \{\Pi_{n+1}, \Pi_n, \ldots, \Pi_2, \Pi_1\}$. It is easily seen that $T(\xi_{n+1}) \geqslant \sum_{j=1}^{n} (P_j \frown P_{j+1}) > \frac{1}{3}n(P \frown Q)$. Due to the arbitrariness of n, $\nu(K) = \sup T(\xi) = \infty$, which contradicts the condition.

LEMMA 8.5.2. *If the curve K in E^3 is one-sidedly smooth and has no points of return, then $\nu(K)$ equals the absolute turn of its indicatrix of the tangents.*

Proof. Between the tangential circumferences of the indicatrix of the tangents of the curve and its osculating planes there exists a one-to-one correspondence, maintaining the order.

At such a correspondence an oriented plane of each tangential circumference coincides with the corresponding to this tangent osculating plane of the curve. By the definition, $\nu(K) = \sup \sum_{i=1}^{m} |P_i \frown P_{i+1}|$ and $|\kappa_g|(Q) = \sup \sum_{i=1}^{m} |C_i \frown C_{i+1}|$. This obviously affords $\nu(K) = |\kappa_g|(Q)$.

The Lemmas proved above yield the basic theorem of the present section.

THEOREM 8.5.2. *For any one-sidedly smooth curve* $\nu(K) = |\tau|(K)$.

Indeed, if $|\tau|(K) = |\kappa_g|(Q)$, where Q is an indicatrix of the tangents, $\nu(K) = |\kappa_g|(K)$ and, according to Theorem 8.3.2, $\nu(K) = |\kappa_g|(K) = |\tau|(K)$.

8.6. Left–Hand and Right–Hand Indices of a Point. Complete Torsion of a Curve

8.6.1. Let P and Q be two oriented planes in E^3. Let us assume that the line of their intersection is oriented. In this case it is expedient to impose a certain sign to the angle between P and Q. Namely, let e be a directing vector of the line of intersection of the planes P and Q, and let a and b be the vectors lying in P and Q, respectively, and orthogonal to e. In this case we assume that $[P \frown Q) = \sigma(P \frown Q)$, where $\sigma = 1$, if the trio of vectors (a, b, e) is right-handed, and $\sigma - -1$, if this trio is left-handed. If $(P \frown Q) = \pi$, then we assume $\sigma = 1$.

Let L be an arbitrary polygonal line in the space, having no points of return, $A_0 < A_1 < \cdots < A_{r+1}$ be its sequential vertices. Let us consider that the turn L at every vertex A_i is other than zero. Let us set $P_i = P(A_{i-1}, A_i, A_{i+1})$, $i = 1, 2, \ldots, r$. A line of intersection of the planes P_i and P_{i+1} is the oriented straight line $A_i A_{i+1}$ ($\overline{A_{iA_{i+1}}}$ is its directing vector). The sum is

$$\sum_{i=1}^{r-1} [P_i \frown P_{i+1}] = (\tau)(L).$$

The quantity $(\tau)(L)$ will be termed a complete torsion of the polygonal line L.

Let K be a spatial curve of a finite complete torsion. Let us now define a certain set of two-dimensional directions $\gamma(K)$ of the zero measure, such that under projection the curve on the planes which are perpendicular to the straight lines of this directions, various kinds of unpleasant phenomena can arise.

Let E_1 be a set of all two-dimensional directions α, for which the curve K_α is not an f.c.t. curve, let E_2 be a set of all α which are orthogonal to at least one of the tangents of the curve K, let E_3 be a set of all α which are orthogonal to at least one mean osculating plane of the curve. Finally, let E_4 be a set of all α which are orthogonal to at least one plane containing a non-rectilinear plane arc of the curve K. Let us set $\gamma(K) = \bigcup_{i=1} E_i$. The set $\gamma(K)$ has a zero measure. Indeed, for E_1 this follows from Theorem 8.3.2, and for E_2 be from the rectifiability of the indicatrix of the tangents of the curve K. The sets E_3 and E_4 are the sums of a not greater than countable set of one-dimensional submanifolds G_2^3 and their measures are also zero.

Let K be a curve of a finite complete torsion. To every its point X let us put in correspondence a certain number $\varphi(X)$ in the following way. If a certain left-hand (right-hand) semi-neighbourhood of X is rectilinear, then let Y be the utmost left-hand (right-hand) point of the curve, such that the arc $[YX]$ ($[XY]$) is rectilinear. The osculating plane $P_m(Y)$ will be considered a left-hand (right-hand) osculating plane at the point.

Let us introduce the notion of the index of a point of a spatial curve $K \in E$.

Let X be an arbitrary point of K. Let us find all the osculating planes at X included in Ω. If their number proves to be three, and P, Q and R are these planes, then the left-hand index of the point X is the number $j_l^2 = \text{sign } (P \widehat{} Q)$, the right-hand index is the number $j_r^2 = \text{sign } (P \widehat{} Q)$.* If Ω contains only two of osculating planes at X, i.e., the planes P and Q, then the right-hand and the left-hand indices of the point X are considered to coincide and be equal to the number $j^2(X) = \text{sign } (P \widehat{} Q)$. Let, finally, at X there be the only osculating plane P included into Ω. Let us find a maximum rectilinear arc $[YZ]$ containing the point X, $Y \leqslant X \leqslant Z$.

The plane P is oriented. Let us call the upper side of the plane P the side towards which its normal is directed. The right-hand and left-hand indices of the point X are considered to coincide and be equal to 1 (-1) if a certain left-hand semi-neighbourhood of the point lies off the lower (upper) side of P, and a certain right-hand semi-neighbourhood of the point Z of the upper (lower) side of P, in which case any plane parallel to P intersects these semi-neighbourhoods through a point or through a whole plane arc.

Finally, the indices of all other points of the curve are considered equal to zero.

A point of the curve will be termed ordinary if its index is other than zero.

Now let K be an arbitrary spatial curve of a finite complete torsion. To each $\alpha \notin \gamma(K)$ let us put in correspondence a certain number $(\tau)(K_\alpha)$ in the following way.

If $|\tau|(K_\alpha) = 0$, let us set $(\tau)(K_\alpha) = 0$. If $|\tau|(K_\alpha) \neq 0$, then let P_1, P_2, ..., P_s be consequent marked osculating planes of the curve K_α. Let us consider two planes P_i and P_{i+1}. If they are chosen at the ends of one non-rectilinear arc of a canonical division of the curve K_α, then $|P_i \widehat{} P_{i+1}| = 0$.

If the planes P_i and P_{i+1} are chosen at the ends of a rectilinear arc $[A_j A_k]$ of the curve K_α, then the angle $|P_i \widehat{} P_{i+1}|$ multiplied by the right-hand index of the point a_j of the curve K, which is projected into the point A_j, will be

---

* When calculating the angle, we assume that the line of intersection of the planes $P_l(X)$ and $P_m(X)$ is the tangent $t_l(X)$, of the planes $P_m(X)$ and $P_r(X)$ the tangent $t_r(X)$. Finally, if X is a smooth point and the planes $P_l(X)$ and $P_r(X)$ are defined, then we assume the straight line $t(X) = t_l(X) = t_r(X)$ to be the line of their intersection.

denoted through $(P_i, \frown P_{i+1})$. If P_i and P_{i+1} are osculating planes at one point A_j of the curve K_α, then let $(P_i, \frown P_{i+1}) = \varepsilon |P_i, \frown P_{i+1}|$, where ε is the left-hand index at the point a_j, if $P_i = P_l(A_j)$, and ε is the right-hand index, if $P_i = P_m(A_j)$ (a_j is a point of the curve K projected into the point A_j).

Let us set $(\tau)(K_\alpha) = \sum_{i=1}^{s-1} (P_i, \frown P_{i+1})$.

Let K be an arbitrary curve of the class E. To its every point X let us bring in correspondence a certain number $\varphi(X)$.

If X is a smooth point, we set $\varphi(X) = (P_l(X), \frown P_r(X))$, where the angle $(P_l(X), \frown P_r(X))$ is measured under the condition that the line of intersection of the planes $P_l(X)$ and $P_r(X)$ coincides with the tangent $t(X)$. If X is an angular point, then we set $\varphi(X) = (P_l(X), \frown P_m(X)) + (P_m(X), \frown P_r(X))$. Finally, if X is a point of return with a rectilinear neighbourhood, then we set $\varphi(X) = (P_l(X), \frown P_r(X))$, where the angle $(P_l(X), \frown P_r(X))$ is measured under the condition that the line of intersection of $P_l(X)$ and $P_r(X)$ is a tangent at the utmost left-hand point Y, such that the arc $[XY]$ is rectilinear.

Let K be a spatial curve, in which case $|\tau|(K) < \infty$, $K \in E$, and let $\alpha \in \gamma(K)$. Let us choose an arbitrary point X of the curve K. By the definition, $\varphi(X)$ is equal to the sum of angles between the osculating planes at the point X, $\varphi(X)$ is equal to either $(P, \frown Q)$ or $(P, \frown Q) + (Q, \frown R)$, where P, Q and R are osculating planes at the point X. Let P_α, Q_α and R_α be projections of the planes P, Q and R onto the plane α. We set:

$$(P_\alpha, \frown Q_\alpha) = \text{sign } (P, \frown Q) |P_\alpha, \frown Q_\alpha|,$$

$$(Q_\alpha, \frown R_\alpha) = \text{sign } (Q, \frown R) |Q_\alpha, \frown R_\alpha|$$

and, finally, $\varphi_\alpha(X) = (P_\alpha, \frown Q_\alpha)$ if $\varphi(X) = (P, \frown Q)$ and $\varphi_\alpha(X) = (P_\alpha, \frown Q_\alpha) + (Q_\alpha, \frown R_\alpha)$, if $\varphi(X) = (P, \frown Q) + (Q, \frown R)$.

LEMMA 8.6.1. *Let A and B be terminal points of the curve K. In this case at any fixed X:*

$$(\tau)[(AB)_\alpha] = (\tau)[(AX)_\alpha)] + \varphi_\alpha(X) + (\tau)[(XB)_\alpha]$$

at nearly all α.

Proof. Let $\alpha \in \gamma(K)$ and, besides, α is not orthogonal to any of the osculating planes at the point X. Let us construct a canonical division of the curve K_α. If the projection X' of the point X lies inside a non-rectilinear arc of this canonical subdivision, then, as may be seen, the projections of all osculating planes at the point X coincide, so that $\varphi_\alpha(X) = 0$. If X' lies inside a rectilinear arc $[A_j A_k]$ of the canonical division, then $\varphi(X) = (P_m(a_j), \frown P_m(a_k))$, where a_j and a_k are the points of the curve K which are projected into the points A_j and A_k. Projections of the planes $P_m(a_j)$ and $P_m(a_k)$ are the planes $P_m(A_j)$ and $P_m(A_k)$. In this case we have:

$$(\tau)[(AB)_\alpha] = (\tau)[(AX)_\alpha] + (\tau)[(XB)_\alpha] + (P_m(A_j), \frown P_m(A_k)).$$

The number $(P_m(A_j), \frown P_m(A_k))$, by the definition of $\varphi_\alpha(X)$, is, in this case, equal to $\varphi_\alpha(X)$.

Now let X' coincide with one of the points realizing a canonical division of the curve K, $X' = A_j$.

First let one of the arcs $[A_{j-1}A_j]$ and $[A_jA_{j+1}]$ be rectilinear; let, for the sake of definitness, it be the arc $[A_{j-1}A_j]$. Let us find the utmost left-hand of the points A_k, for which the arc $[A_kA_j]$ is rectilinear (if it coincides with the beginning of the curve, the considerations will be simplified). In this case:

$$(\tau)[(AB)_\alpha] = (\tau)[(AX)_\alpha] + (P_m(A_k)\overset{\frown}{,}P_m(A_j)) +$$
$$+ (P_m(A_j)\overset{\frown}{,}P_r(A_j)) + (\tau)[(XB)_\alpha]$$

For the point $X = A_j$ the plane $P_m(a_k)$ is a left-hand osculating plane and $\varphi(X) = (P_m(a_k)\overset{\frown}{,}P_m(a_j)) + (P_m(a_j)\overset{\frown}{,}P_r(a_j))$, which affords

$$(P_m(A_k)\overset{\frown}{,}P_m(A_j)) + (P_m(A_k)\overset{\frown}{,}P_r(A_j)) = \varphi_\alpha(A_j).$$

The case when the arc $[A_jA_{j+1}]$ is rectilinear is considered in an analogous way.

One more case not yet considered is when the arcs $[A_{j-1}A_j]$ and $[A_jA_{j+1}]$ are not rectilinear, and marked osculating planes are the planes $P_l(A_j)$, $P_m(A_j)$ and $P_r(A_j)$. In this case:

$$(\tau)[(AB)_\alpha] = (\tau)[(AA_j)_\alpha] + (P_l(A_j)\overset{\frown}{,}P_m(A_j)) +$$
$$+ (P_m(A_j)\overset{\frown}{,}P_r(A_j)) + (\tau)[(A_jB)_\alpha]$$

in line with the definition of $\varphi_\alpha(X)$. Therefore, the lemma is completely proved.

LEMMA 8.6.2. *Let K be a curve of a finite complete torsion, in which case $K \in E$, Q is its normal indicatrix of the tangents and X is an arbitrary simple point of the curve K. Then, if Y_l and Y_r are end points of the arc of the indicatrtix Q corresponding to X, $Y_l \leqslant Y_r$, then $j_l(X) = j(Y_l)$ and $j_r()X) = j(Y_r)$.*

Proof. First let X be an angular point. In this case the points Y_l and Y_r are obviously different and the plane of the arc $[Y_lY_r]$ of the indicatrix Q connecting them is parallel to and has the same orientation as the plane $P_m(X)$. Furthermore, the planes of the great circumferences $t_l(Y_l)$ and $t_r(Y_r)$ are parallel to and have the same orientation as the planes $P_l(X)$ and $P_r(X)$, which affords $j_l(X) = j(Y_l)$ and $j_r(X) = j(Y_r)$. Considerations are analogous if X is a smooth point but $P_l(X) \neq P_r(X)$.

Now let X be a smooth ordinary point, and $P_l(X) \neq P_r(X)$. In this case $Y_l = Y_r = Y$ and Y is a smooth point of the curve. The tangents at all the points of a certain neighbourhood U of the point X, containing a maximum rectilinear arc of the curve K, to which X belongs, are directed, as follows from the definition, of one side of the plane $P_r(X)$. Therefore, a certain neighbourhood of the point $Y \in Q$ lies off one side of the tangent $t(Y)$. If $j(X) = +1$, then the tangents of the arc U are directed off the upper side of the plane $P(X)$. But in this case a certain neighbourhood of the point Y lies off the right-hand side of the tangent $t(X)$, since the right-hand side of this tangent coincides with the side to which a normal to $P(X)$ is directed. Hence, $j(Y) = +1$, i.e., $j(X) = j(Y)$. The case when $j(X) = -1$ is considered in an analogous way and, hence, the lemma is proved.

LEMMA 8.6.3. *Let K be a curve of a finite complete torsion, $K \in E$, Q be its indicatrix of the tangents. Then for nearly all $\alpha \in G_2^3$, $(\kappa_g)(Q_\alpha) = (\tau)(K_\alpha)$.*

Proof. Let $\alpha \notin \gamma(K)$. Let us prove that $(\kappa_g)(Q_\alpha) = (\tau)(K_\alpha)$. The proof will be carried out by way of induction with respect to the number p of the arcs of a canonical division of the curve K_α. For the case when $p = 1$ the lemma is obvious, since in this case $(\tau)(K_\alpha) = 0$ and $(\kappa_g)(Q_\alpha) = 0$. Let the lemma be proved for $p = p_0$, and let a canonical division of the curve K_α consist of $p_0 + 1$ arcs and be realized by the points $A_0, A_1, \ldots, A_{p_0}, A_{p_0+1}$.

Let $a_0, a_1, \ldots, a_{p_i+1}$ be the points of the curve projected onto them. Several cases are possible here. Let us consider them in detail.

(1) The arcs $[A_{p_0-1}A_{p_0}]$ and $[A_{p_0}A_{p_0+1}]$ are rectilinear. The points A_{p_0} and a_{p_0} are points of return. In this case $(\tau)(A_0A_{p+1}) = (\tau)(A_0A_{p_0})$ is a normal indicatrix of the tangents Q of the arc $[a_0a_{p_0+1}]$ which is obtained from the normal indicatrix of the tangents Q' of the arc $[a_0a_{p_0}]$ by adding the arc of a great circumference which is an extension of the tangent at the end of the arc C. This means that at the transition from Q'_α to Q_α there arise no new points of return and hence

$$(\kappa_g)(Q_\alpha) = (\kappa_g)(Q'_\alpha) = (\tau)(A_0A_{p_0}) = (\tau)(A_0A_{p_0+1}).$$

(2) The arc $[A_{p_0+1}A_{p_0}]$ is rectilinear, the arc $[A_{p_0}A_{p_0+1}]$ is not rectilinear. In this case marked are the planes $P_m(A_{p_0})$ and $P_r(A_{p_0+1})$. At the transition from Q' to Q the curve Q' is added to with an arc of a great circumference, the plane of which is parallel to and has the same orientation as the plane $P_m(a_{p_0})$ and, besides, with the arc, the plane of which is parallel to and has the same orientation as the plane $P_r(a_{p_0})$, and, finally, with the indicatrix of the tangents of the arc $[a_{p_0}a_{p_0+1}]$. As far as the projections of the planes $P_m(a_{p_0})$ and $P_r(a_{p_0})$, which, obviously, are oppositely orientated, then the projection of a common point of the point X of the first two arcs added to Q' is a point of return.

Let A_k be the utmost left-hand part of the points $A_0, A_1, \ldots, A_{p_0}$, such that the arc $[A_kA_{p_0}]$ is rectilinear. The plane $P_m(A_k)$ has either the same or opposite orientation with $P_m(A_{p_0})$ and, depending on it, the projection of the point Y of the curve Q', which is the end of the curve Q', onto the circumference $C(\alpha)$ will be either a point of return of the curve Q_α or its ordinary point. In the former case $(\kappa_g)(Q_\alpha) = (\kappa_g)(Q_\alpha) + j(X)\pi$, and, at the same time, $(\tau)(K_\alpha) = (\tau)(A_0A_{p_0}) + j_r(a_{p_0})\pi$; in the latter case $(\kappa_g)(Q_\alpha) = (\kappa_g)(Q_\alpha) + [j(X) + j(Y)]\pi$ and $(\tau)(K_\alpha) = (\tau)(A_0A_{p_0}) + j_l(a_{p_0})\pi + j_r(a_{p_0})\pi$. As far as

$j_l(a_p) = j(Y)$, and $j_r(a_p) = j(X)$, then in this case also $(\tau)(K_\alpha) = (\kappa_g)Q_\alpha$.

(3) If the arc $[A_{p_0}A_{p_0+1}]$ is rectilinear, and the arc $[A_{p_0-1}A_{p_0}]$ is not rectilinear, then Q is obtained from Q' by adding the arc of a great circumference which is parallel to and has the same orientation as the plane $P_m(a_{p_0})$. In this case we use the same considerations as in (2).

(4) None of the arcs $[A_{p_0-1}A_{p_0}]$ and $[A_{p_0}A_{p_0+1}]$ is rectilinear. Let first a_{p_0} be an angular point. In this case the indicatrix Q is obtained from Q' by adding the arcs $[XY]$ and $[YZ]$, where $[XY]$ is the arc of a great circumference, the plane of which is parallel to and has the same orientation as the plane $P_m(a_{p_0})$, and $[YZ]$ is the indicatrix of the tangents of the arc $[a_{p_0}a_{p_0+1}]$.

The plane of a right-hand tangential circumference of the curve Q at the point Y is parallel to and has the same orientation as the osculating plane $P_r^K(a_{p_0})$, while the plane of a left-hand tangential circumference at the point X is parallel to and has the same orientation as the plane $P_l^K(a_{p_0})$. The line of intersection of the planes $P_l^K(a_{p_0})$ and $P_m^K(a_{p_0})$ has the direction of the radius $\overline{OX}$, and the line of intersection of the planes $P_m^K(a_{p_0})$ and $P_r^K(a_{p_0})$ has the direction of the radius $\overline{OY}$. Therefore, for the curve Q we have:

$$(\overline{t}_l(X) \frown \overline{t}_r(X)) = (P_l^K(a_{p_0}), P_m^K(a_{p_0}))$$

and

$$(\overline{t}_l(X) \frown \overline{t}_r(X)) = (P_m^K(a_{p_0}), P_r^K(a_{p_0})),$$

and, finally,

$$(\kappa_g)(Q_\alpha) = (\kappa_g)(Q_\alpha') + (P_l^{K}\alpha(A_{p_0}), P_m^{K}\alpha(A_{p_0})) +$$
$$+ (P_l^{K}\alpha(A_{p_0}), P_r^{K}\alpha(A_{p_0})) = (\tau)(K_\alpha).$$

(5) None of the arcs $[A_{p_0-1}A_{p_0}]$ and $[A_{p_0}A_{p_0+1}]$ is rectilinear, and the point a_{p_0} is a smooth point of the curve K. If X is a corresponding to it point of the curve Q, then $Q = Q' + [XY]$, where $[XY]$ is the indicatrix of the tangents of the arc $[a_{p_0}a_{p_0+1}]$. It is easily seen that in this case

$$(\kappa_g)(Q_\alpha) = (\kappa_g)(Q_\alpha') + \pi_j(X),$$

and

$$(\tau)(K_\alpha) = (\tau)(A_{p_0+1}A_{p_0}) + \pi_j(a_{p_0}).$$

But, according to the lemma, $j(X) = j(a_{p_0})$ and therefore, the lemma is completely proved.

8.6.2. THEOREM 8.6.1. *Let K be a spatial curve of a finite complete torsion, with the angle between any two neighbouring osculating planes of this curve other than π. In this case for any sequence $L_1, L_2, \ldots, L_m, \ldots$ of polygonal lines inscribed into the curve K, such that $\lambda(L_m) \rightarrow 0$ at $m \rightarrow 0$ and every angular point of K, starting from a certain $m = m_0$, will be a vertex of the polygonal line L_m, there exists a limit of complete torsion of the inscribed polygonal lines, which is equal to*

$\int_{G_2^3} (\tau)(K_\alpha)\ \mu_{3,2}\ (d\alpha).$

Proof. Let us number all the angular points of the curve. Let L_1, L_2, ..., L_m, ... be an arbitrary sequence of the inscribed into the curve K polygonal lines, such that $\lambda(L_m) \rightarrow 0$ at $m \rightarrow \infty$, the polygonal line L_m be inscribed into L_{m+1} and the angular point numbered m be an vertex of the polygonal line L_m. Let us denote by E a set of all the directions α which are parallel to the osculating plane of at least one of the polygonal lines L_m.

Let now $\alpha \in E \cup \gamma(K)$. Since $\alpha \notin \gamma(K)$, then $\kappa(K_\alpha) < +\infty$, $|\tau|(K_\alpha) < +\infty$. The polygonal line $L_{m,\alpha}$ is inscribed into the curve K_α. Obviously, $\tau(L_\alpha) = \int_{G_2^3} (\tau)$ $(L_{m,\alpha})\ \mu_{3,2}\ (d\alpha)$. The theorem will be proved if we demonstrate that, at $\tilde{m} \rightarrow \infty$, $(\tau)(L_{m,\alpha}) \rightarrow (\tau)(K_\alpha)$.

The proof will be carried out by way of induction with respect to the number p of the arcs of a canonical division of the curve K_α. At $p = 1$ the curve K_α is convex in the small and at a sufficiently large m the polygonal line $L_{m,\alpha}$ is also convex in the small and, therefore, $(\tau)(L_{m,\alpha}) = (\tau)(K_\alpha) = 0$.

Let the theorem be proved for $p = p_0$. Let us prove it for $p = p_0 + 1$. Let A_0, A_1, ..., A_{p_0+1} be the points realizing a canonical division of the curve K_α, a_0, a_1, ..., a_{p_0}, a_{p_0+1} be the points of the curve K projected onto them. Let us consider in detail several cases which are possible here.

For simplicity let us set $L_{m,\alpha} = \Lambda_m$; let $\overline{L}_m$ denote the arc L_m inscribed into the arc $[a_0 a_{p_0}]$, Λ_m be its projection onto the plane α. Here, again, several cases are possible. Let us consider them in detail.

Let $m = m_0$ be such that at $m > m_0$ $(\tau)(\overline{L}_{m,\alpha}) = (\tau)(A_0 A_{p_0})$.

(1) The arc $[A_{p_0} A_{p_0+1}]$ is rectilinear, the arc $[A_{p_0-1} A_{p_0}]$ is not rectilinear. In this case the point a_{p_0} is angular. Let X be the utmost left-hand point of the curve K_α, such that the arc $[X A_{p_0}]$ is rectilinear, X is a point of the curve K projected into it. The arc $[X A_{p_0}]$ is obviously also rectilinear. The point a_{p_0} is an angular point of the curve and the following equality is valid:

$$(\tau)(A_0 A_{p_0+1}) = (\tau)(A_0 A_{p_0}) + (\widehat{P_l(A_{p_0})},\ P_m(A_{p_0})).$$

At $m > m_1$ the point A_{p_0} is a vertex of the polygonal line L_m. Let a_m and b_m be the links of the polygonal line $L_{m,\alpha}$ converging at the point A_{p_0}; c_m and d_m be the utmost right-hand pair of non-colinear links of the polygonal line $L_{m,\alpha}$, and, finally, let α_m, β_m, γ_m and δ_m be the links of the polygonal line L_m at projected onto them. The marked planes of the polygonal line L_m at the point A_{p_0} (they do exist provided m is sufficiently large) are $P_l(A_{p_0})$ and $P_m(A_{p_0})$. In this case the plane $P_m(A_{p_0})$ is defined by a pair of links a_m and b_m, and the plane $P_l(A_{p_0})$ by a pair of vectors c_m and d_m. Let us remark that the vectors δ_m and α_m always either coincide (if no left-hand semi-neighbourhood of the point is rectilinear) or are parallel and have the same direction. The osculating plane P_m, which is defined by the pair of vectors α_m and β_m at $m \rightarrow$

∞, has the plane $P_m(a_{p_0})$ as a limit; the plane Q_m defined by the links γ_m and δ_m converges to the plane $P_l(a_{p_0})$. A directed line of the intersection of the planes P_m and Q_m converges to the tangent $t_l(a_{p_0})$. Hence, at a sufficiently large m we have:

$$\text{sign } (P_m, \frown Q_m) = \text{sign } (P_l(a_{p_0}), \frown P_r(a_{p_0}))$$

and therefore $\tau(L_{m,\alpha}) = (\tau)(\overline{L}_{m,\alpha}) + \pi \text{ sign } (P_m, \frown Q_m)$ will be equal to $(\tau)(A'_0 A_{p_0+1})$, which is the required proof.

(2) The arc $[A_{p_0-1}A_{p_0}]$ is rectilinear, the arc $[A_{p_0}A_{p_0+1}]$ is not rectilinear. In this case A_{p_0} is also an angular point and, hence, at $m > m_1$, A_{p_0} is a vertex of the polygonal line L_m. Let X_k be the utmost left-hand point of the curve K, such that the arc $[X_k A_{p_0}]$ is rectilinear.

Let (a_m, b_m) be the first pair, counting off from the point A_{p_0} to the left, of non-colinear neighbouring links of the polygonal line $L_{m,\alpha}$; let (c_m, d_m) be a pair of links of the polygonal lie $L_{m,\alpha}$ converging at the point A_{p_0}; and, finally, let (e_m, f_m) be the first pair, counting off from A_{p_0} to the right, of the links of the polygonal line L_m not lying in one straight line. Let us also assume that $\alpha_m, \beta_m, \gamma_m, \delta_m, \varepsilon_m, \varphi_m$ are links, projected into them, of the polygonal line L_m; P_m, Q_m, R_m are the osculating planes of the polygonal line L_m defined by these pairs; P'_m, Q'_m, R'_m are their projections onto the plane P.

The line of intersection of the planes P_m and Q_m obviously has the direction of the vector γ_m, and the line of intersection of the planes Q_m and R_m has the direction of the vector δ_m. At $m \to \infty$ the limit of the plane P_m is the plane $P_m(a_k)$, that of the plane Q_m is the plane $P_m(a_{p_0})$, and that of the plane R_m is the plane $P_r(a_{p_0})$. The projections of these osculating planes are, obviously, the marked planes of the curve K_α. The lines of intersection of the planes P_m and Q_m, and Q_m and R_m converge, obviously to those of the planes $P_m(a_k)$ and $P_m(a_{p_0})$, and $P_m(a_{p_0})$, and $P_m(a_{p_0})$ and $P_r(a_{p_0})$. Therefore, at a sufficiently large m the addendum added to $(\tau)(\overline{L}_{m,\alpha})$ will be equal to $P', \frown Q') + (Q', \frown R')$, where P, Q and R are the projections of the planes $P_m(a_k)$, $P_m(a_{p_0})$ and $P_r(a_{p_0})$. In other words, at sufficiently large m, $\tau(\overline{L}_\alpha)$ will be equal to (τ) $(A_0 A_{p_0+1})$ which is the required proof.

(3) The arcs $[A_{p_0-1}A_{p_0}]$ and $[A_{p_0}A_{p_0+1}]$ are not rectilinear and have the same orientation. In this case the point A_{p_0}, by the condition, cannot be a point of return. Starting from a certain sufficiently large m the point A_{p_0} will be a vertex of one of the polygonal lines $L_{m,\alpha}$. Let a_m and b_m be the links of the polygonal line L_m converging to the point A_{p_0}, and let c_m and d_m be the first pair, counting off from the point A_{p_0} to the left, of non-colinear neighbouring links of the polygonal line $L_{m,\alpha}$; e_m and f_m be the first, counting off the point A_{p_0} to the right, pair of non-colinear neighbouring links of the polygonal line $L_{m,\alpha}$. It is seen that at sufficiently large m the pairs (c_m, d_m), (a_m, b_m) and (e_m, f_m) have the same orientations as the planes $P_l(A_{p_0})$, $P_m(A_{p_0})$ and $P_r(A_{p_0})$. (All these planes, obviously, are marked at the point

$A_{p_0}.)$

Let (γ_m, δ_m), (α_m, β_m) and $(\varepsilon_m, \varphi_m)$ be the pairs of links of the polygonal line L_m projected into (c_m, d_m), (a_m, b_m), (e_m, f_m), respectively. The defined by them osculating planes P_m, Q_m and R_m converge to the planes $P_l(a_{p_0})$, $P_m(a_{p_0})$, $P_r(a_{p_0})$. The line of intersection of P_m and Q_m, as is seen, has the same direction as the vectors d_m and a_m, the line of intersection of Q_m and R_m has the same direction as b_m and e_m. Herefrom, by analogy with the above proved, we have that at a sufficiently large m:

$$(\tau)(L_m, \alpha) = (\tau)(A_0 A_{p_0+1}).$$

(4) The arcs $[A_{p_0} A_{p_0+1}]$ and $[A_{p_0-1} A_{p_0}]$ are not rectilinear and oppositely oriented. The case when the point A_{p_0} is angular is analogous to (3) and we omit its consideration. Let us consider the case when A_{p_0} is a smooth point. Here, again, several cases are possible. Let us dwell on them in detail.

First let the osculating planes of the curve K, $P_l(a_{p_0})$ and $P_r(a_{p_0})$, not coincide at the point A_{p_0}. The angle between them, by the condition, is other than π. Let us find the utmost left-hand point A', such that the arc $[A' A_{p_0}]$ is rectilinear. A' can coincide with the point A_{p_0}. First let A' not coincide with A_{p_0}. Counting off from the point A_{p_0} to the left, let us find the first pair of links a_m and b_m of the polygonal line $L_{m,\alpha}$ not lying in one straight line, and, counting off from A' to the right, the first pair of links c_m and d_m of the polygonal line $L_{m,\alpha}$ not lying in one straight line. It is seen that at a sufficiently large m these pairs of links do exist, in which case b_m and c_m have the same direction, while the osculating planes of these links have the same orientation as $P_l(a_{p_0})$ and $P_r(a_{p_0})$.

Let α_m, β_m, γ_m and δ_m be links of the polygonal line L_m, projected into a_m, b_m, c_m and d_m, respectively. It is obvious that the osculating plane P_m, which is defined by the links α_m and β_m, converges to the plane $P_l(a_{p_0})$, while the plane Q_m, defined by the links γ_m and δ_m converges to the plane $P_r(a_{p_0})$. In this case at a sufficiently large m the line of intersection of the planes P_m and Q_m coincides with the tangent $t(a_{p_0})$, which is the line of intersection of the planes $P_l(a_{p_0})$ and $P_r(a_{p_0})$. Therefore, at a sufficiently large m the sign of the angle $(P_m, \frown Q_m)$ coincides with that of the angle $(P_l(a_{p_0}), \frown P_r(a_{p_0}))$, and, since $(P_l(a_{p_0}), \frown P_r(a_{p_0})) \neq \pi$, then

$$(\tau)(L_{m,\alpha}) = (\tau)(\overline{L}_{m,\alpha}) + \pi \, \mathrm{sign} \, (P_l(a_{p_0}), \frown P_r(a_{p_0})).$$

But in the case considered we have:

$$(\tau)(A_0 A_{p_0+1}) = (\tau)(A_0 A_{p_0}) + \pi \, \mathrm{sign} \, (P_l(a_{p_0}), \frown P_r(a_{p_0})).$$

Since at a sufficiently large m, $(\tau)(\overline{L}_{m,\alpha}) = (\tau)(A_0 A_{p_0})$, we thus have proved the required statement in this case.

Let $A' \neq A_{p_0}$, and let the planes $P_l(a_{p_0})$ and $P_r(a_{p_0})$ coincide. Let us introduce into the space an affinity coordinate system Σ, choosing the tangent $t(a_{p_0})$ as the x axis, a straight line normal to the plane P as the z axis, and any straight line not lying in the plane $P_l(a_{p_0})$ and not colinear to either x or z as the y axis. Let us project the curve K onto the plane xz, projecting along the y axis. In this case the projection of the curve in the vicinity of the points A' and A_{p_0} is convex and we orient the z axis in such a way that it be directed towards the convexity of the projection. The y axis will be oriented in such a way that the small left-hand semi-neighbourhood of the point y is in the semi-space $y \leqslant 0$. In this case the point a_{p_0} is chosen as the origin of the coordinate system. Then, by definition, $(\tau)(A_0 A_{p_0+1}) = (\tau)(A_0 A_{p_0}) + \varepsilon \pi$, where $\varepsilon = +1$, if the constructed coordinate system is right-hand, and $\varepsilon = -1$ if it is left-hand. Let us introduce, as above, the links α_m, β_m, γ_m of the polygonal line L_m. In this case $(P_m, \frown Q_m) > 0$, if the trio of vectors $(\alpha_m, \beta_m, \gamma_m)$ is right-hand, and $(P_m, \frown Q_m) < 0$, if this trio is left-hand. Let us prove that at a sufficiently large m the orientation of the trio $(\alpha_m, \beta_m, \gamma_m)$ coincides with that obtained in the coordinate system (x, y, z).

Indeed, at a sufficiently large m, $\beta_{mx} > 0$, $\beta_{my} = \beta_{mz} = 0$. Furthermore, $\gamma_{mz} > 0$, $\alpha_{my} > 0$, and, finally, $\gamma_{mx} > 0$, $\alpha_{mx} > 0$, $\alpha_{mz} > 0$, and $\gamma_{mz} > 0$. From this we have:

$$(\alpha_m, \beta_m, \gamma_m) = \begin{vmatrix} \alpha_{mx} & \alpha_{my} & \alpha_{mz} \\ \beta_{mx} & \beta_{my} & \beta_{mz} \\ \gamma_{mz} & \gamma_{my} & \gamma_{mz} \end{vmatrix} = -\beta_{mx}(\alpha_{my}\gamma_{mz} - \gamma_{my}\alpha_{mz}) > 0.$$

As far as in this case $(\tau)(L_{m,\alpha}) = (\tau)(\overline{L}_{m,\alpha}) + \pi \operatorname{sign}(P_m, \frown Q_m)$, then $(\tau)(L_{m,\alpha}) = (\tau)(A_0 A_{p_0+1})$ at a sufficiently large m.

Let us now consider the case when $A' = A_{p_0}$. By the condition, the angle between the planes $P_l(a_{p_0})$ and $P_r(a_{p_0})$ is other than α. Let us first find the utmost right-hand pair of neighbouring links a_m and b_m of the polygonal line $L_{m,\alpha}$, which has the same orientation as the arc $[A_{p_0-1} A_{p_0}]$. It is clear that at a sufficiently large m the beginning of the link a_m must belong to the arc $[A_{p_0-1} A_{p_0}]$, and the link c_m, the beginning of which is the end of the link b_m, is not colinear to the link b_m and the pair (b_m, c_m) has the same orientation as the arc $[A_{p_0} A_{p_0+1}]$.

Let α_m, β_m and γ_m be the links of the polygonal line L_m projected into a, b and c. Let us introduce an auxiliary affinity coordinate system in the space, choosing the unit vector of the tangent $(t(x)$ as a unit vector of the x axis, a straight line orthogonal to α as the z axis, and a straight line, lying in the plane perpendicular to the plane (x, z) and passing through the z axis, as the y axis. In this case we will choose the y axis in such a way that, at projecting parallel to the y axis, the projections of the planes $P_l(a_{p_0})$ and $P_r(a_{p_0})$ could coincide with the orientations. This can obviously be achieved since

$$(P_l(a_{p_0}), \frown P_r(a_{p_0})) \neq \pi.$$

At such a projecting the projection of a certain neighbourhood of the point x of the curve K onto the plane xz is obviously convex and we shall assume that the z axis is directed towards the convexity of the projection. The y axis will be directed in such a way that a certain small left-hand semi-neighbourhood of the point a_{p_0} lies in the semi-space $y < 0$. (The origin of the coordinate system is placed at the point a_{p_0}.) In this case we, obviously, have: $(\tau)(A_0 A_{p_0+1}) = (\tau)(A_0 A_{p_0}) + \varepsilon\pi$, where $\varepsilon = +1$ if the system of coordinates so constructed is right-hand, and $\varepsilon = -1$ if it is left-hand.

Along with the coordinate system Σ, let us introduce another system Σ_m, the y and z axes of which are the same as in the system Σ, while the directing unit vector of the x axis is the vector $\beta_m/|\beta_m|$. It is obvious that at a sufficiently large m the systems Σ_m and Σ have the same orientation. Furthermore, at a sufficiently large m, $(\tau)(L_{m,\alpha}) = (\tau)(\overline{L}_{m,\alpha}) + \varepsilon_m\pi$, where $\varepsilon_m = +1$, if the trio $(\alpha_m, \beta_m, \gamma_m)$ is right-hand, and $\varepsilon_m = -1$, if it is left-hand. Let us prove that at a sufficiently large m the orientation of the trio $(\alpha_m, \beta_m, \gamma_m)$ coincide with that of the system Σ_m. Let us, by analogy with the above proof, consider the coordinates of the vectors $\alpha_m, \beta_m, \gamma_m$ in the system Σ_m. We have:

$$\beta_{mx} > 0, \qquad \beta_{my} = \beta_{mz} = 0,$$
$$\alpha_{my} > 0, \qquad \alpha_{mz} > 0,$$
$$\gamma_{my} > 0, \qquad \gamma_{mz} < 0.$$

From this we get:

$$\begin{vmatrix} \alpha_{mx} & \alpha_{my} & \alpha_{mz} \\ \beta_{mx} & \beta_{my} & \beta_{mz} \\ \gamma_{mz} & \gamma_{my} & \gamma_{mz} \end{vmatrix} = -\beta_{mx}(\alpha_{my}\gamma_{mz} - \alpha_{mz}\gamma_{my}) > 0,$$

which is the required proof.

Let K be an arbitrary curve, such that $|\tau|(K) < \infty$ and $K \in E$. If $|\tau|(K) < \pi$, then a complete torsion of the curve will be a limit of the complete torsions of the polygonal lines inscribed into the curve, which exists, according to Theorem 8.6.1, in the denotations $\tau(K)$ or $\tau(AB)$, where A and B are the terminal points of the curve. If $(\tau)(K) > \pi$, then, as follows from the theorem, on the curve K we can find a sequence of the points $X_0 = A < X_1 < X_2 < \cdots X_m = B$, such that $|\tau|(X_i X_{i+1}) < \pi$ at all i, and in this case a complete torsion of the curve will be expressed by the quantity

$$\tau(K) = \sum_{i=1}^{m-1} \tau(X_i X_{i+1}) + \sum_{i=1}^{m-1} \varphi(X_i).$$

From Theorem 8.6.1, bearing in mind that

$$\varphi(X) = \int_{G_2^3} \varphi_\alpha(X)\, \mu_{3,2}\,(d\alpha),$$

we can easily conclude that

$$\tau(K) = \int_{G_2^3} (\tau)(K_\alpha)\, \mu_{3,2}\,(d\alpha).$$

The last formula shows that the quantity $\tau(K)$ is obviously independent of the choice of the points $X_1, X_2, \ldots, X_{m-1}$.

CHAPTER IX

Frenet Formulas and Theorems on Natural Parametrization

9.1. Frenet Formulas

9.1.1. Let us first prove certain results referring to spherical curves.

Let K be an arbitrary one-sidedly smooth curve on a sphere, let Q be a complete one-dimensional indicatrix of the curve K, and let $\xi(t) = (X(t), l(t))$ be the parametrization of the curve Q. Let $s(t)$ be the length of the arc $[X(a)X(t)]$ of the curve K. Let us arbitrarily choose a point X of the curve K. The function $X(t)$ is the parametrization of the curve K. Let $[\alpha, \beta]$ be a set of all the values of the parameter t which corresponds to the point X. We set $\kappa_g(\alpha) = \kappa_g[X(a)X(t)]$, and at $t \in [\alpha, \beta]$ we set: $\kappa_g(t) = \kappa_g(\alpha) + [t_l(X), \widehat{l}(t)]$, In this case the angle is chosen with a sign. In particular, $\kappa_g(\beta) = \kappa_g(\alpha) + \kappa_g(X)$. Let us call the parametrization $\xi(t) = (X(t), l(t))$, which is a complete one-dimensional indicatrix of the curve K, regular if in any segment $[\alpha, \beta]$ of the values of the parameter t which correspond to an angular point of the curve, $\kappa_g(t)$ is as follow: $\kappa_g(t) = \lambda(t - a) + \kappa_g(\alpha)$. It is obvious that a complete one-dimensional indicatrix allows a regular parametrization.

LEMMA 9.1.1. *Let K be a one-sidedly smooth curve with no points of return on the sphere S^2, let (L_m), $m = 1, 2, \ldots$, be a sequence of spherical polygonal lines inscribed into K, and such that at $m \to \infty$ $\lambda(L_m) \to 0$. In this case the complete one-dimensional indicatrices $\beta(L_m)$ of the polygonal lines L_m converge to the complete one-dimensional indicatrix $\beta(K)$ of the curve K at $m \to \infty$.*

Proof. Let $(X(t), l(t))$, $a \leqslant t \leqslant b$, be a regular parametrization of the curve K. Let us set an arbitrary $\varepsilon > 0$ and let $A_1 < A_2 < \cdots < A_r$ be all those angular points of the curve K at each of which a turn of the curve K is not less than ε. Let $A_0 = A$ be the beginning of the curve, $A_{r+1} = B$ be its end. Let $[\alpha_i, \beta_i]$ be a set of all values of the parameter t corresponding to the point A_i, where $i = 1, 2, \ldots, r$. Obviously, we have $\alpha_i < \beta_i < \alpha_{i+1} < \beta_{i+1}$ at every $i = 1, 2, \ldots, r-1$. Let us denote by δ the least of the diameters of the arcs $[A_i A_{i+1}]$, $i = 0, 1, \ldots, r$, and let $\bar{m}$ be such that at $m \geqslant \bar{m}$ $\lambda(L_m) < \delta/4$. Let us then consider that $m \geqslant \bar{m}$. Due to the condition $\lambda(L_m) < \delta/4$, at every $i = 0, 1, \ldots, r$ on the arc $[A_i A_{i+1}]$ of the curve K there are vertices of the polygonal line L_m. Let Y_i^m be the utmost left-hand, and X_{i+1}^m be the utmost

268

right-hand of the vertices of L_m lying on the arc $[A_iA_{i+1}]$. Obviously, $Y_i^m <$ X_{i+1}^m. The arc $[Y_i^m X_{i+1}^m]$ of the polygonal line L_m, denoted through $L_{i,m}$, converges to the arc $[A_iA_{i+1}]$ at $m \to \infty$. Let $Y_{i,m}(t)$, $\beta_i \leqslant t \leqslant \alpha_{i+1}$, be a parametrization of the polygonal line $L_{i,m}$, such that $Y_{i,m}(t) \to X(t)$ uniformly in the segment $[\beta_i, \alpha_{i+1}]$. (Here we assume that $\beta_0 = a$, $\alpha_{r+1} = b$.) Let us, then, construct a parametrization $X_{i,m}(t)$, $\beta_i \leqslant t \leqslant \alpha_{i+1}$ of the arc $L_{i,m}$, such that to every angular point of this arc there corresponds a set of the parameter values, which is a segment that does not degenerate into a point, and $|X_{i,m}(t) - Y_{i,m}(t)| < 1/m$ for all $t \in [\beta_i, \alpha_{i+1}]$. In this case there exists a parametrization $\xi_{i,m}(t) = (X_{i,m}(t), l_{i,m}(t))$ of the complete one-dimensional indicatrix of the polygonal line $L_{i,m}$. Let us prove that at sufficiently large m for all $t \in [\beta_i, \alpha_{i+1}]$ $\rho(\xi(t), \xi_{i,m}(t)) < 2\varepsilon$. Indeed, $|X_{i,m}(t) - X(t)| \to 0$ uniformly at $m \to \infty$. It is therefore sufficient to prove that, at sufficiently large m, $(l_{i,m}(t) \frown l(t)) < 3\varepsilon/2$ for any $t \in [\beta_i, \alpha_{i+1}]$. Let us assume that this is not the case. Then there can be found a sequence (m_k), $k = 1, 2, \ldots$, of the values m and a sequence (t_k), $k = 1, 2, \ldots$, of the points of the segment $[\beta_i, \alpha_{i+1}]$, such that $(l_{i,m_k}(t_k) \frown l(t_k)) \geqslant 3\varepsilon/2$ at every k. Without reducing the generality, we can say that at $k \to \infty$ the points t_k converge to a certain point $t_0 \in [\beta_i, \alpha_{i+1}]$, while the straight lines $l_{i,m_k}(t_k)$ converge to a certain straight line λ_0. At $k \to \infty$ $l(t_k) \to l_0 = l(t_0)$. Obviously, λ_0 passes through the point $X(t_0)$.

Let us assume that $l_{i,m_k}(t_k)$ is a tangent at an internal point of a certain link $[Z_k R_k]$ of the spherical polygonal line $L_{m,k}$. Let l_k be a secant $l(Z_k, T_k)$, p_k be a directing unit vector of the straight line l_k. At $k \to \infty$ the angle between $l_{i,m_k}(t_k)$ and l_k tends to zero. Let us assume that at all k the points Z_k and R_k lie either to one side with respect to the point $X(t_0)$, or to different sides. This can obviously be always achieved by going over to a sequence. In the former case the vector p_k has either the left-hand or the right-hand tangential unit vector at the point $X(t_0)$ of the curve K as a limit, and therefore λ_0 is either the right-hand or the left-hand tangent at the point $X(t_0)$. If t_0 is the end of the segment $[\beta_i, \alpha_{i+1}]$, then $l_0 = t_r(X(t_0))$ in the case when $t_0 = \beta_i$, $l_0 = t_l[X(t_0)]$ in the case when $t_0 = \alpha_{i+1}$. In both cases $\lambda_0 = l_0$, so that $(\lambda_0 \frown l_0) = 0$. If $\beta_i < t_0 < \alpha_{i+1}$, then $(\lambda_0 \frown l_0) < \varepsilon$, since at every angular point $X(t)$ of the curve K, where $\beta_i < t < \alpha_{i+1}$, a turn of the curve K is less than ε. It contradicts the fact that $(l_{i,m_k}(t_k), l(t_k))$ $\geqslant 3\varepsilon/2$ at all k and $l_{i,m_k}(t_k) \to \lambda_0$, $l(t_0) \to l_0$ at $k \to \infty$.

Let us consider the case when Z_k and R_k converge to $X(t_0)$ from different sides. In this case the secants $l(Z_k, X(t_0))$, $l(X(t_0), R_k)$ converge to the left-hand and right-hand tangents, respectively, at the point $X(t_0)$ of the curve K. Therefore, the straight line λ_0 to which the secants $l(Z_k, R_k)$ converge is a tangent at the point $X(t_0)$. An angle between any tangents at the point $X(t_0)$ is less than ε. Hence, $(\lambda_0 \frown l_0) < \varepsilon$. It again contradicts the fact that $(l_{i,m_k}(t_k) \frown l(t_k)) \geqslant 3\varepsilon/2$ for all k.

It follows from the above proved that there exists $\bar{m}_i$, such that, for all m $\geqslant \bar{m}_i$, $\rho(\xi_{i,m}(t), \xi(t)) < 2\varepsilon$ for all $t \in [\beta_i, \alpha_{i+1}]$ since the assumption that this is not the case results in a contradiction.

Let us now join the parametrizations $\xi_{i,m}$ of separate arcs of a complete one-dimensional indicatrix of the polygonal line L_m into one. Let us set $\xi_m(t)$ $= \xi_{i,m}(t)$, if t lies in one of the segments $[\beta_i, \alpha_{i+1}]$, $i = 1, 0, \ldots, r$. Let us additionally define the function ξ_m in the segment $[\alpha_i, b_i]$ at every $i = 1$, $2, \ldots, r$.

Let $X_i^m = Y_i^m = A_i$. In this case we set at $t \in [\alpha_i, \beta_i]$ $X_m(t) = A_i$. Let us define $l_m(t)$ in the following way. Let λ_i^m be a left-hand, μ_i^m be a right-hand tangents of the polygonal line L_m at the point X_i^m. For $t \in [\alpha_i, \beta_i]$ let us assume that $l_m(t)$ is the intermediate tangent of the polygonal line L_m at the point X_i, which forms with λ_i^m an angle equal to $\gamma_m(t - \alpha_i)$, where $\gamma = \text{const}$, $\gamma_m(\beta_i - \alpha_i) = (\lambda_i^m, \widehat{\hspace{0.3em}} \mu_i^m)$. Let us assume that the considered situation is valid for an infinitely great number of m. Going over to a limit at $m \to \infty$ with respect to a set of such m, we obtain that $\lambda_i^m \to t_l(A_i) = l(\alpha_i)$, $\mu_i^m \to t_r(A_i) = l(\beta_i)$. Since the curve K has no points of return, the point A_i is not such a one. Since the parametrization $\xi(t) = (X(t), l(t))$ is regular, then at $t \in [\alpha_i, \beta_i]$ $l(t)$ forms with $t_l(A_i)$ an angle equal to $\gamma_i[t - \alpha_i]$, $\gamma_i[\beta_i - \alpha_i] = (t_l(A_i), \widehat{\hspace{0.3em}} t_r(A_i))$. Therefore $l_m(t) \to l(t)$ uniformly when $m \to \infty$ with respect to the given set of values m.

Let us now consider the case when $X_i^m < Y_i^m$. Obviously, $X_i^m \leqslant A_i$ and $X_i^m \neq A_i$, since in the opposite case $X_i^m = Y_i^m$, since $X_i^m < A_i$. In the same way we conclude that $A_i < Y_i^m$. Let λ_i^m be a left-hand tangent of the polygonal line L_m at the point X_i^m, $\bar{\lambda}_i^m$ be a right-hand tangent of L_m at this point, μ_i^m and $\bar{\mu}_i^m$ be a left-hand and a right-hand tangents of L_m at the point Y_i^m. Let λ be a tangent of the polygonal line L_m lying between λ_i^m and μ_i^m. Let us put in correspondence to it a certain number $\theta(\lambda)$, setting $\theta(\lambda) = (\lambda_i^m, \widehat{\hspace{0.3em}} \lambda)$ if $\lambda_i^m \leqslant \lambda \leqslant \bar{\lambda}_i^m$, $\theta(\lambda) = \theta(\bar{\lambda}_i^m)$ $+ (\bar{\lambda}_i^m, \widehat{\hspace{0.3em}} \lambda)$, if $\bar{\lambda}_i^m \leqslant \lambda \leqslant \bar{\mu}_i^m$, and, finally, $\theta(\lambda) = \theta(\bar{\mu}_i^m) + (\bar{\mu}_i^m, \widehat{\hspace{0.3em}} \lambda)$ if $\bar{\mu}_i^m \leqslant \lambda \leqslant \mu_i^m$. Let $\theta_i^m = \theta(\mu_i^m)$. For $t \in [\alpha_i, \beta_i]$ let $l_m(t)$ be a tangent λ of the polygonal line L_m, such that $\lambda_i^m \leqslant \lambda \leqslant \mu_i^m$ and $\theta(\lambda) = \theta_i^m(t - \alpha_i)/(\beta_i - \alpha_i)$. Let us denote by $X_m(t)$ the point of the polygonal line L_m, a tangent of which is $l_m(t)$.

Let us consider the case when the situation in question is valid for an infinitely great number of m. At $m \to \infty$ $\lambda_i^m \to t_l(A_i)$, $\mu_i^m \to t_r(A_i)$. Let $l_m = l(X_i^m, Y_i^m)$ and λ_m be an intermediate tangent of the curve K at the point A_i, forming the smallest angle with λ_m. At m tending to ∞ with respect to the given set of m values, at $t \in [\alpha_i, \beta_i]$ $|X_m(t) - A_i| \leqslant \lambda(L_m)$ and, hence, $X_m(t) \to A_i$ at $m \to \infty$. From the way the function l_m is constructed we see that in this case $l_m(t) \to l(t)$ uniformly in the segment $[\alpha_i \beta_i]$. It is here essential that A_i is not a point of return and the parametrization $\xi(t) = (X(t), l(t))$ of the curve $\beta(K)$ is regular.

Therefore, in each of the segment $[\alpha_i, \beta_i]$ the functions $\xi_m(t) = (X_m(t), l_m(t))$ converge to $\xi(t) = (X(t), l(t))$ uniformly and in each of the segments

$[\beta_i, \alpha_{i+1}]$, at a sufficiently large m, $\rho[\xi_m(t), \xi(t)] < 2\varepsilon$. Hence, at a sufficiently large m, $\rho[\xi_m(t), \xi(t)] < 2\varepsilon$ for all $t \in [a, b]$.

We come to the conclusion that for any $\varepsilon > 0$ we can construct a parametriza-
tion $\xi_m(t)$ of the curve $\beta(L_m)$, such that $\rho(\xi_m(t), \xi(t)) < 2\varepsilon$ at all $t \in [a, b]$
for all $m > m'$, where m' is sufficiently large. Since $\varepsilon > 0$ is arbitrary, it
affords $\beta(L_m) \longrightarrow \beta(K)$, which is the required proof.

REMARK. From the above construction we can easily conclude that for any regular parametrization $\xi(t) = (X(t), l(t))$ of the curve $\beta(K)$ we can construct
regular parametrizations $\xi_m(t) = (X_m(t), l_m(t))$ of the spherical polygonal
lines L_m which uniformly converge to $\xi(t)$ in $[a, b]$.

9.1.2. Let us now establish certain integral relations which would be natural to consider as an analog of the Frenet formulas in classical differential
geometry.

A trio of vectors (p, q, r) of the space E^3 will be called an orthogonal
trihedral provided $|p| = |q| = |r| = 1$ and the vectors p, q, r are mutually
orthogonal. The space E^3 will here be considered to be oriented.

Let us first consider spherical curves. Let K be a one-sidedly smooth curve
on a unit sphere S^3. Let us construct its complete one-dimensional indicatrix
$\beta(K)$. Let $Z = (X, l)$ be an arbitrary point of the curve $\beta(K)$. Let us set:
$r_1(Z) = \overline{OX}$, where O is the centre of the sphere. Let us denote by $r_2(Z)$ a directing unit vector of the straight line l. Finally, let $r_3(Z)$ be a unit vector in E^3, such that the trio of vectors $R(Z) = (r_1(Z), r_2(Z), r_3(Z))$ is
right-handed. Let us call $R(Z)$ an accompanying orthogonal trihedral of the
curve K. If $\xi(t)$, $a \leqslant t \leqslant b$, $\eta(u)$, $c \leqslant u \leqslant d$ are two arbitrary parametrizations of the curve $\beta(K)$, then $R[\xi(t)]$, $a \leqslant t \leqslant b$, and $R[\eta(u)]$, $c \leqslant u \leqslant d$, are
equivalent paths in the space of the orthogonal trihedrals. To every spherical
curve K there, hence, corresponds a certain curve in the space of the orthogonal trihedrals.

THEOREM 9.1.1. *Let K be a curve of a finite geodesic turn on the sphere S^3,
$\xi(t) = (X(t), l(t))$, $a \leqslant t \leqslant b$, be the parametrization of the complete one-
dimensional indicatrix $\beta(K)$ of the curve K, let $(r_1(t), r_2(t), r_3(t))$ be the
corresponding parametrization of the curve of the accompanying trihedrals
of the curve K. Let $s(t) = s[X(a)X(t)]$. Let us define the function $\kappa_g(t)$ in
the following way. If $X(t)$ is a smooth point of the curve, then $\kappa_g(t) =
\kappa_g(X(a)X(t))$. If $X(t)$ is an angular point, then let $[\alpha, \beta]$ be a set of all the
values of the parameter which correspond to this point, so that $\alpha \leqslant t \leqslant \beta$. In
this case we set $\kappa_g(t) = \kappa_g[X(a)X(\alpha)] + [t_l(X(\alpha)\overset{\frown}{,} l(t)]$. The following
equalities are valid:*

$$r_1(t) = r_1(a) + \int_a^t r_2(t) \, ds(t),$$

$$r_2(t) = r_2(a) + \int_a^t [-r_1(t) \, ds(t) + r_3(t) \, d\kappa_g(t)], \tag{1}$$

$$r_3(t) = r_3(a) - \int_a^t r_2(t) \, d\kappa_g(t).$$

Proof. Let us first consider the case when the curve K is a spherical polygonal line. In this case the segment $[a, b]$ is divided into a finite number of segments by the points $\beta_0 = a < \alpha_1 < \beta_1 < \alpha_2 < \beta_2 < \cdots < \alpha_r < \beta_r < \alpha_{r+1} = b$ in such a way that in each of the segments $[\beta_i, \alpha_{i+1}]$ the function is constant, and in the segments $[\alpha_i, \beta_i]$, $i = 1, 2, \ldots, r$, the function $s(t)$ is constant. It is sufficient to prove that equalities (1) are fulfilled in each of the segments $[\alpha_i, \beta_i]$, $[\beta_i, \alpha_{i+1}]$. Let us consider the segment $[\beta_i, \alpha_{i+1}]$. Let us prove that for $t \in [\beta, \alpha_{i+1}]$:

$$r_1(t) = r_1(\beta_i) + \int_{\beta_i}^t r_2(t) \, ds(t),$$

$$r_2(t) = r_2(\beta_i) + \int_{\beta_i}^t [-r_1(t) \, ds(t) + r_3(t) \, d\kappa_g(t)], \qquad (2)$$

$$r_3(t) = r_3(\beta_i) - \int_a^t r_2(t) \, d\kappa_g(t).$$

As far as the function κ_g is constant within the segment $[\beta_i, \alpha_{i+1}]$, then its terms containing $d\kappa_g(t)$ turn to zero and the problem is reduced to proving the equalities:

$$r_1(t) = r_1(\beta_i) + \int_{\beta_0}^t r_2(t) \, ds(t),$$

$$r_2(t) = r_2(\beta_i) - \int_{\beta_0}^t r_1(t) \, ds(t), \qquad (3)$$

$$r_3(t) = r_3(\beta_i)$$

at $t \in [\beta_i, \alpha_{i+1}]$. The last equality is obvious; from the construction of the complete indicatrix it follows that the function $r_3(t)$ is constant in the segments $[\beta_i, \alpha_{i+1}]$. Let $\nu(s)$, $s_0 \leqslant s \leqslant s_1$, be a parametrization of the arc $[X(\beta_i)X(\alpha_{i+1})]$ of the polygonal line, where the parameter s is the length of the arc. In this case $\nu''(s) = -\nu(s)$ and we get:

$$\nu(s) = \nu(s_0) + \int_0^s \nu'(s) \, ds,$$

$$\nu'(s) = \nu'(s_0) - \int_0^s \nu(s) \, ds.$$

We have:

$$\nu(s(t)) = r_1(t), \qquad \nu'(s(t)) = r_2(t).$$

Herefrom, after substituting the variable, we obtain the first two equalities (3).

Let $t \in [\alpha_i, \beta_i]$. Let us prove that in this case

$$r_1(t) = r_1(\alpha_i) + \int_{\alpha_i}^t r_2(t) \, ds(t),$$

$$r_2(t) = r_2(\alpha_i) + \int_{\alpha_i}^{t} [r_1(t)\ ds(t) + r_3(t)\ d\kappa_g(t)], \tag{4}$$

$$r_3(t) = r_3(\alpha_i) - \int_{\alpha_i}^{t} r_2(t)\ d\kappa_g(t).$$

The function $s(t)$ is constant in the segment $[\alpha_i, \beta_i]$ and, hence, the terms in (4) containing $ds(t)$ turn to zero. In the segment $[\alpha_i, \beta_i]$ the function r is constant, so that the first of equalities (4) is valid in a trivial way. The aim is to prove that for all $t \in [\alpha_i, \beta_i]$:

$$r_2(t) = r_2(\alpha_i) + \int_{\alpha_i}^{t} r_3(t)\ d\kappa_g(t)],$$

$$\tag{5}$$

$$r_3(t) = r_3(\alpha_i) - \int_{\alpha_i}^{t} r_2(t)\ d\kappa_g(t).$$

The validity of these equalities is established in the way analogous to that used in (3). Let $\lambda(\varphi)$, where $-\pi < \varphi \leqslant \pi$ is a unit vector lying in the plane P, osculating the sphere S^3 at the point t_1 and such that $[r_2(\alpha_i), \lambda(\varphi)] = \varphi$ (it should be recalled that for the vectors p, q lying in the plane P $[p, q] = \sigma(p, q)$, where $\sigma = 1$ if the trio $(r_1(\alpha_i), p, q)$ is right-handed and $\sigma = -1$, if the trio is left-handed). Let $\mu(\varphi)$ be a unit vector in the plane P, orthogonal to φ and such that the trio $r(t)$, $\lambda(\varphi)$, $\mu(\varphi)$ is right-hand. In this case $\lambda'(\varphi) = \mu(\varphi)$, $\mu'(\varphi) = -\lambda(\varphi)$ and therefore

$$\lambda(\varphi) = \lambda(0) + \int_{0}^{\varphi} \mu(\psi)\ d\psi,$$

$$\mu(\varphi) = \mu(0) - \int_{0}^{\varphi} \lambda(\psi)\ d\psi.$$

From this, by an obvious change of the variables, we get (5).

From (2) and (4), by way of summing up, we come to the conclusion that equalities (1) are valid in the case considered.

Let K be an arbitrary spherical curve, such that $|\kappa_g(K) < \infty|$. Let (L_m) be a sequence of spherical polygonal lines inscribed into the curve K such that $\lambda(L_m) \to 0$ at $m \to \infty$. Let $\xi(t) = (X(t), l(t))$, $a \leqslant t \leqslant b$, be a regular parametrization of the curve $\beta(K)$, $\xi_m(t) = (X_m(t), l_m(t))$, $a \leqslant t \leqslant b$, be the parametrizations the curves $\beta(L_m)$, which uniformly converge to the function $\xi(t)$. Obviously, the parametrizations ξ can also be considered regular. Let us define the functions $s_m(t)$ and $\kappa_{g,m}(t)$, $a \leqslant t \leqslant b$. In this case, at $m \to \infty$ $s_m(t) \to s(t)$ and, due to the condition of regularity, at $m \to \infty$, $\kappa_{g,m}(t) \to \kappa_g(t)$ for all $t \in [a, b]$. At every m let define the accompanying trihedrals $(z_{1,m}(t), z_{2,m}(t), z_{3,m}(t))$ of the polygonal line L_m. At $m \to \infty$ $z_{i,m}(t) \to z_i(t)$, which follows from the construction of the accompanying trihedral. At every m, in line with what we have proved above, we have:

$$z_{1,m}(t) = z_{1,m}(a) + \int_{a}^{t} z_{2,m}(t)\ ds_m(t),$$

$$z_{2,m}(t) = z_{2,m}(a) + \int_a^t [-z_{1,m}(t) \, ds_m(t) + z_{3,m}(t) \, d\kappa_{g,m}(t)],$$

$$z_{3,m}(t) = z_{3,m}(a) - \int_a^t z_{2,m}(t) \, d\kappa_{g,m}(t).$$

From this, according to the known theorems on a limiting transition for the Stieltjes integral, we immediately obtain equalities (1) for an arbitrary spherical curve (under the supposition that the curve has no points of return). The general case is, obviously, reduced to this case. The theorem is proved.

Let K be an arbitrary curve of a finite complete torsion in E^3, $\gamma(K)$ be a complete two-dimensional indicatrix, $Z = (X, l, P)$ be an arbitrary point of the curve $\gamma(K)$. Let $\nu_1(Z)$ be a directing unit vector of the straight line l, $\nu_2(Z)$ be a unit vector perpendicular to l, lying in the plane P and such that $(\nu_1(Z), \nu_2(Z))$ be a right-hand pair of vectors in the plane P. Finally, let $\nu_3(Z)$ be a unit vector orthogonal to P and such that $(\nu_1(Z), \nu_2(Z), \nu_3(Z))$ be a right-hand trio in E^3. The orthogonal trihedral $R(Z) = (\nu_1(Z), \nu_2)Z), \nu_3(Z))$ will be called an accompanying trihedral of the curve K. We, thus, obtain a certain curve R_K in the space of orthogonal reference frames. This curve will be called a curve of the accompanying trihedrals of the curve K.

THEOREM 9.1.2. *Let K be a curve of a finite complete torsion, $\xi(t) = X(t)$, $l(t)$, $P(t)$), $a \leqslant t \leqslant b$, be the parametrizations of the two-dimensional indicatrix of the curve K, $(s(t), \kappa(t), \tau(t))$ be a corresponding to it natural parametrization of K, $(\nu_1(t), \nu_2(t), \nu_3(t))$, $a \leqslant t \leqslant b$, be the parametrization of the curve of the accompanying trihedrals. In this case the following equalities are valid:*

$$x(t) = x(a) + \int_a^t \nu_1(\sigma) \, ds(\sigma),$$

$$r_1(t) = r_1(a) + \int_a^t \nu_2(\sigma) \, d\kappa(\sigma),$$

$$r_2(t) = r_2(a) + \int_a^t [-\nu_1(\sigma) \, d\kappa(\sigma) + \nu_2(\sigma) \, d\tau(\sigma)], \qquad (6)$$

$$r_3(t) = r_3(a) - \int_a^t \nu_2(\sigma) \, d\tau(\sigma).$$

Proof. The first of equalities (6) results from the fact that $s(t)$ is the arc length, and $\nu_1(t)$ is a unit vector of a tangent of the curve K at the point $x(t)$. The last three equalities of (6) result from the fact that $(r_1(t), r_2(t), r_3(t))$ is an accompanying trihedral of the indicatrix of the tangents Q of the curve K. In this case $\kappa(t)$ is the arc length, and $\tau(t)$ is a geodesic turn. Applying Theorem 9.1.1, we immediately obtain equalities (6). The theorem is proved.

9.2. Theorems on Natural Parametrization

9.2.1. THEOREM 9.2.1. *Let $(u(t), v(t))$, $a \leqslant t \leqslant b$, be continuous functions obeying the following conditions:*
(1) $u(t)$ *is a non-decreasing function, in which case* $u(a) = 0$, $u(b) > 0$;
(2) $v(t)$ *is a function of a limited variation, in which case* $v(a) = 0$;
(3) *in any segment* $[\alpha, \beta] \subset [a, b]$, *where* $u(t)$ *is constant, the function* $v(t)$ *is monotonous, in which case* $-\pi < v(\beta) - v(\alpha) \leqslant \pi$.

In this case there exists a spherical curve K, for which the pair of functions $(u(t), v(t))$, $a \leqslant t \leqslant v$, is a natural parametrization of K.

Proof. Let the functions $u(t)$ and $v(t)$ obey all the conditions of the Theorem. Let N be a curve on the plane R^2, for which $(u(t), v(t))$, $a \leqslant t \leqslant b$, is a parametrization. This curve is rectifiable. Let $(\xi(\sigma), \eta(\sigma))$, $0 \leqslant \sigma \leqslant L$, be a parametrization of the curve N, where the parameter σ is the arc length. In this case we have: $u(t) = \xi[\sigma(t)]$, $v(t) = \eta[\sigma(t)]$, where σ is a continuous non-decreasing function, such that $\sigma(a) = 0$, $\sigma(b) = l$. Let us consider a system of common differential equations:

$$\frac{dz_1}{d\sigma}(\sigma) = \xi'(\sigma)\, z_2(\sigma),$$

$$\frac{dz_2}{d\sigma}(\sigma) = -\xi'(\sigma)\, z_1(\sigma) + \eta'(\sigma)\, z_2(\sigma), \tag{7}$$

$$\frac{dz_3}{d\sigma}(\sigma) = -\eta'(\sigma)\, z_2(\sigma).$$

where z_1, z_2, z_2 are the vector-functions with the values in V^3. The functions $\xi'(\sigma)$, $\eta'(\sigma)$ are measurable and $(\xi'(\sigma))^2 + (\eta'(\sigma))^2 = 1$ for almost all $\sigma \in [0, L]$. The trio of vector-functions (z_1, z_2, z_3) will be called a solution to system (7) provided each of them is absolutely continuous with respect to σ and for almost all $\sigma \in [0, L]$ equalities (7) hold. Due to the known results of the theory of differential equations, for any trio (p, q, r) of vectors in E^3, there exists a solution to system (7), such that $z_1(0) = p$, $z_2(0) = q$ and $z_3(0) = r$. If the trio $(z_1(0), z_2(0), z_3(0))$ is an orthogonal trihedral in E^3, then for all $\sigma \in [0, L]$, $(z_1(\sigma), z_2(\sigma), z_3(\sigma))$ is an orthogonal trihedral. The proof of this statement is here omitted since it is nearly a word-for-word repetition of the considerations found in corresponding places of the elementary course of differential geometry.

Let us arbitrarily set an orthogonal trihedral (p, q, r) and let $(z_1(\sigma), z_2(\sigma), z_3(\sigma))$, $0 \leqslant \sigma \leqslant L$, be a solution to system (7), such that $z_1(0) = p$, $z_2(0) = q$, $z_3(0) = r$. By way of integration we get from (7):

$$z_1(\sigma) = p + \int_0^\sigma z_2(\tau)\, d\xi(\tau),$$

$$z_2(\sigma) = q + \int_0^\sigma [-z_1(\tau)\, d\xi(\tau) + z_3(\tau)\, d\eta(\tau)], \tag{8}$$

$$z_3(\sigma) = r - \int_0^\sigma z_2(\tau) \; d\psi(\tau).$$

Let us set $z_i(\sigma(t)) = r_i(t)$. Substituting in (8) the variable, by the formulas $\sigma = \sigma(t) = \sigma(h)$, we get:

$$r_1(t) = r_1(a) + \int_a^t r_2(\tau) \; du(\tau),$$

$$r_2(t) = r_2(a) + \int_a^t [-r_1(\tau) \; du(\tau) + r_3(\tau) \; dv(\tau)], \tag{9}$$

$$r_3(t) = r_3(a) - \int_a^t r_2(\tau) \; dv(\tau).$$

Let K be a spherical curve for which the function $r_1(t)$ is a parametrization. The function $z_1(\sigma)$ is also a parametrization of K. The vector function $z_1(\sigma)$ is absolutely continuous. Therefore, the curve K is rectifiable and the length of the arc $[z_1(0)z_1(\sigma)]$ of this curve is equal to $\int_0^\sigma |\xi'(\tau)| \, d\tau$. The function $\xi'(\sigma)$ is non-decreasing. Therefore $|\xi'(\tau)| = \xi'(\tau)$ and we find that the arc length $[z_1(0)z_1(\sigma)]$ of the curve K equals $\xi(\sigma)$. Hence the length of the curve K, as a function of the parameter t equals $\xi(\sigma(t)) = u(t)$.

Let us arbitrarily choose a point X of the curve K and let $[\alpha, \beta]$ be a set of all the values of the parameter t which correspond to the point X. At $t > \beta$, $s(t) > s(\beta)$, we have:

$$\frac{r_1(t) - r_1(\beta)}{s(t) - s(\beta)} = \frac{1}{s(t) - s(\beta)} \int_a^t r_2(t) \; ds(t). \tag{10}$$

As far as $s(t)$ is a non-decreasing function, and the function $r_2(t)$ is continuous, then the integral in the right-hand part of (10) tends, obviously, to $r_2(\beta)$ at $t \to \beta + 0$. We, therefore, come to the conclusion that $r_2(\beta)$ is a right-hand tangent at the point X of the curve K. In an analogous way we prove that $r_2(\alpha)$ is a left-hand tangent of K at this point. At $t \in [\alpha, \beta]$ we have:

$$r_2(t) = r_2(\alpha) + \int_a^t r_3(\tau) \; dv(\tau),$$

$$r_3(t) = r_3(\alpha) - \int_a^t r_2(\tau) \; dv(\tau). \tag{11}$$

The function v is monotonous in the segment $[\alpha, \beta]$ and from equalities (11) it obviously follows that, when t runs through the segment $[\alpha, \beta]$, the function $r_2(t)$ monotonously rotates, turning at an angle equal $v(\beta) - v(\alpha)$. This means that at $t \in [\alpha, \beta]$ $r_2(t)$ is an intermediate tangent of the curve K at the point X. It makes possible to conclude that $(r_1(t), r_2(t), r_3(t))$, $a \leqslant t \leqslant b$, is an accompanying trihedral of the curve K.

The vector-function $r_2(t)$, as follows from the equality

$$r_1(t) = r_1(a) + \int_a^t [-r_1(\tau) \; du(\tau) + r_3(\tau) \; dv(\tau)],$$

is a function of a limited variation. Therefore, K is a curve of a finite turn. According to Theorem 9.1.1, we have:

$$r_1(t) = r_1(a) + \int_a^t r_2(\tau) \, ds(\tau),$$

$$r_2(t) = r_2(a) + \int_a^t [-r_1(\tau) \, ds(\tau) + r_2(\tau) \, d\kappa_g(\tau)],$$

$$r_3(t) = r_3(a) - \int_a^t r_2(\tau) \, d\kappa_g(\tau).$$

In line with the known properties of the Stieltjes integral, we have:

$$s(t) = \int_a^t <r_2(\tau), dr_1(\tau)>,$$

$$\kappa_g(t) = \int_a^t <r_2(\tau), dr_3(\tau)>.$$

Due to the same properties of the Stieltjes integral, equalities (9) afford:

$$\int_a^t <r_2(\tau), dr_1(\tau)> = u(t),$$

$$\int_a^t <r_2(\tau), dr_3(\tau)> = -v(t).$$

Herefrom we get $u(t) = s(t)$, $v(t) = \kappa_g(t)$ and hence the curve K is the one sought.

THEOREM 9.2.2. *If spherical curves of a finite turn have the same natural parametrization, they can be superposed by rotating the sphere S^3.*

Proof. Let $(s(t), \kappa_g(t))$, $a \leqslant t \leqslant b$, be a common natural parametrization of K_1 and K_2. (The pair $(s(t), \kappa_g(t))$ defines a certain curve N on the plane R^2.) Let us consider a system of differential equations:

$$\nu_1'(\sigma) = \nu_2(\sigma)\xi'(\sigma),$$

$$\nu_2'(\sigma) = -\xi'(\sigma) \nu_1(\sigma) + \eta'(\sigma) \nu_3(\sigma), \qquad (12)$$

$$\nu_3'(\sigma) = -\eta'(\sigma) \nu_2(\sigma).$$

Let (p_1, q_1, r_1) be an accompanying reference frame at the beginning of the curve K_1, let (p_2, q_2, r_2) be an accompanying reference frame at the beginning of the curve K_2, and let $(\nu_1(\sigma), \nu_2(\sigma), \nu_3(\sigma))$ be a solution to system (12), such that $\nu_1(0) = p_1$, $\nu_2(0) = q_1$, $\nu_3(0) = r_1$. In this case $\nu_1(\sigma)$ is a parametrization of the curve K_1. Let F be a rotation of the sphere, such that $Fp_1 = p_2$, $Fq_1 = q_2$, $Fr_1 = r_2$. In this case the functions $\mu_i = F\nu_i$ are a solution to system (12), which obey the initial conditions $\mu_1(0) = p_2$, $\mu_2(0) = q_2$, $\mu_3(0) = r_2$. Therefore, $\mu_1(s)$, $0 \leqslant s \leqslant L$, is a parametrization of the curve K_2 and hence the rotation F superposes the curve K_1 with the curve K_2. The theorem is proved.

9.2.2. For a complete two-dimensional indicatrix of the curve we have three functions of the arcs: a length, a turn and an absolute torsion. We can now define one more function for which an absolute torsion is a variation.

Let L be an arbitrary arc of the curve K. Let us first consider the case when a natural projection of the arc reduced to one point X of the curve K. Let $\alpha = (X_1, l_1, P_1)$ and $\beta = (X_2, l_2, P_2)$ be the beginning and the end of the arc L. If $X_1 = X_2 = X$ and the point X is smooth, then also $l_1 = l_2$. We set:

$$\tau(L) = (P_1 \frown P_2),$$

where the angle is taken with a sign, and the sign is defined from the condition that the line of intersection of P_1 and P_2 is the straight line $l = l_1 = l_2$.

Let us assume that the point X is not smooth. Let $\gamma = (X, l_0, P_m)$ be an arbitrary two-dimensional element, containing an intermediate osculating plane P_m at the point X. If γ lies on the arc L, then we set:

$$\tau(L) = (P_1 \frown P_m) + (P_m \frown P_2),$$

where the angles are calculated under the condition that the line of intersection of P_1 and P_m is considered to be the oriented straight line l_1, and the line of intersection of P_m and P_2 is the oriented straight line l_2. If γ lies outside the arc L then the planes P_1 and P_2 intersect along the line which coincides with one of the straight lines l_1 and l_2, and we set:

$$\tau(L) = (P_1 \frown P_2),$$

where the angle is calculated by an ordinary rule.

Finally, let $L = (\alpha\beta)$ be an arbitrary arc of the complete indicatrix B, such that its natural projection does not degenerate into a point. Let α_1 be the utmost right-hand, and β_1 the utmost left-hand point of the curve B, such that the natural projections of the arcs $[\alpha\alpha_1]$ and $[\beta_1\beta]$ consist of a single point each. Let us denote by L' a natural projection of the arc L. In this case we set:

$$\tau(L) = \tau([\alpha\alpha_1]) + \tau(L') + \tau([\beta_1\beta]).$$

The function $\tau(L)$ is also additive and continuous, as follows from the properties of a complete torsion proved above.

Let K be an arbitrary curve of a finite complete torsion in the space. Let B be a complete two-dimensional indicatrix of the curve K. By the construction, B is a curve in the space of two-dimensional linear elements of the curve K. Let (P, l, X) be an arbitrary point of the curve B, i.e., (P, l, X) be an arbitrary two-dimensional element of the space. In this case X is a point of the curve K, l is a tangent at the point X of the curve K, P is an osculating plane passing through the straight line l. Let $\overline{t}$ be a unit vector of a tangent of the curve K. In the plane P let us construct a unit-vector ν perpendicular to $\overline{t}$ and such that a pair of vectors $(\overline{t}, \overline{\nu})$ in the plane P is right-hand. The vector $\overline{\nu}$ will be called the main normal at the point X of the curve K. Finally, let $\overline{b}$ be a unit vector orthogonal to the vectors $\overline{t}$ and $\overline{\nu}$ and such that the trio of vectors $(\overline{t}, \overline{\nu}, \overline{b})$ has the same orientation as the trio of basic vectors of the coordinate system in the space.

Let K be a curve of a finite complete torsion in a space, Q is its indicatrix of the tangents. B is its complete two-dimensional indicatrix. Let (X, l, P) be an arbitrary two-dimensional element of the curve K, $\overline{t}, \overline{\nu}, \overline{b}$ be a unit vector of the tangent, the main normal and the bi-normal of this two-dimensional element, respectively. Let us put in correspondence to the two-dimensional element (X, l, P) a point Y of the unit sphere S^3, which is the end of the unit vector $\overline{t}$, and the straight line λ passing through Y and having the same direction as the vector $\overline{\nu}$. Obviously, (Y, λ) is a one-dimensional element of the curve Q. Therefore, to each two-dimensional element of the curve K there corresponds a certain one-dimensional element of its indicatrix of the tangents. This correspondence obviously has the same order. The results proved above give the following theorem.

THEOREM 9.2.3. *Let L be an arc of the complete two-dimensional indicatrix B of the curve of a finite complete turn K, L' be, due to the given correspondence, the corresponding arc of the one-dimensional indicatrix of a tangent of the curve K. In this case a turn of the arc L equals the length of the arc L', the absolute torsion of the arc L equals the absolute geodesic turn of the arc L', the complete torsion of the arc L equals a geodesic turn of the arc L'.*

Let $\alpha(u) = \{X(u), l(u), P(u), a \leqslant u \leqslant b\}$ be an arbitrary parametrization of the curve of a complete two-dimensional indicatrix of the curve of a finite complete torsion K. Let us denote through $s(u)$, $\kappa(u)$, $\tau(u)$ and $|\tau|(u)$ the length, turn, complete torsion and absolute torsion, respectively, of the arc $[\alpha(a)\alpha(u)]$ of the two-dimensional indicatrix B of the curve K. The trio of the functions

$$(s(u), \ \kappa(u), \ \tau(u)), \qquad a \leqslant u \leqslant b,$$

is called a natural parametrization of the curve K. It is obvious from the definition that any natural parametrization of a curve can be obtained from any of its natural parametrizations by a monotonous transformation of the parameter realized by a non-decreasing function.

From the preceding Theorem there follows an important preposition.

COROLLARY. *If $(s(u), \kappa(u), \tau(u))$, $a \leqslant u \leqslant b$, is an arbitrary natural parametrization of the curve K of a finite complete torsion, then the functions $(\kappa(u), \tau(u))$ form a natural parametrization of the indicatrix of the tangents of the curve K.*

This immediately yields the following theorem.

THEOREM 9.2.4. *If the natural parametrizations of the two curves of a finite complete torsion K_1 and K_2 coincide in the space, then the curves K_1 and K_2 can be superposed by motion. In this case by the motion we understand the proper motion, i.e., an orthogonal transformation with a positive determinant.*

Indeed, in line with Theorem 9.2.2, in this case the indicatrices of the tangents of the curves K_1 and K_2 can be superposed by motion. Let $\overline{t}(u)$ be a radius-vector of a point of the indicatrix of the tangents of the curve. In this case, as has been proved above, the radius-vector of a point of the curve is restored by its indicatrix of the tangents by the formula:

$$\overline{x}(u) = \overline{x}(a) + \int_a^u \overline{t}(\sigma) \, ds(\sigma).$$

It shows that in this case the curves themselves can be superposed by motion.

Let $[f(u), g(u)]$ be two continuous functions defined in a certain segment $[a, b]$ of a real axis. Let us say that the function $g(u)$ is subordinate to the function $f(u)$ if it obeys the following condition: in each segment (α, β), where $f(u)$ is constant, the function $g(u)$ is monotonous, in which case:

$$-\pi < g(u) - g(\beta) \leqslant \pi.$$

If the inequality $|g(\alpha) - g(\beta)| < \pi$ holds, we say that $g(u)$ is strictly subordinated to the function $f(u)$.

If $(s(u), \kappa(u), \tau(u))$ is a natural parametrization of the curve of a finite complete torsion K, then the function $\kappa(u)$ is strictly subordinate to the function $s(u)$, and the function $\tau(u)$ is strictly subordinate to the function $\kappa(u)$.

These conditions are not only necessary but also sufficient for the given trio of functions to be a natural parametrization of a certain curve in the space.

THEOREM 9.2.5. *Let $(s(u), \kappa(u), \tau(u))$, $a \leqslant u \leqslant b$, be functions, such that $s(u)$ and $\kappa(u)$ are monotonous and non-decreasing, $\tau(u)$ is the function of a bounded variation, in which case $\kappa(u)$ is strictly subordinate to $s(u)$, and $\tau(u)$ is subordinate to $\kappa(u)$. Then there exists a curve of a finite complete torsion for which $(s(u), \kappa(u), \tau(u))$, $a \leqslant t \leqslant b$, is its natural parametrization.*

Proof. In line with Theorem 9.2.1, there exists a spherical curve Q with a natural parametrization $\{(\kappa(u), \tau(u), a \leqslant u \leqslant b\}$. Let $\{\overline{t}(u), a \leqslant u \leqslant b\}$ be a corresponding parametrization of the curve K. In this case the curve K, the radius-vector of which is defined by the formula

$$\overline{x}(u) = \overline{x}(a) + \int_a^u \overline{t}(\sigma) \, ds(\sigma)$$

is the one sought. The theorem is proved.

CHAPTER X

Some Additional Remarks

10.1. For a curve in an n-dimensional Euclidean space E^n which obeys certain conditions of regularity, to each point of the curve there is put in correspondence $(n-1)$ numbers of $k_1, k_2, \ldots, k_{n-1}$ of the curvatures at the point of the curve. Let us set

$$\kappa_i(s) = \int_0^s k_i(\sigma)\,d\sigma$$

where integration is carried out with respect to the arc length. The functions obtained in this way will be called integral curvatures of the curve. There arises a problem - to give such a definition of the ith integral curvature which could be applied to an essentially wider class of curves than the class considered in differential geometry. Accordingly, the contents of Chapter V can be considered as a theory of the curves of a limited integral first curvature. Chapter IV presents a theory of curves of a limited integral first curvature in a space S^n.

Theories of curves of a generalized integral curvature of the r order are considered in the works by I.F. Majnick [13 - 18]. The class H_r of curves of a finite integral curvature is defined in the following way. First we introduce the notion of an r-dimensional osculating plane as a limit of oriented secant planes. Let us assume that $X_0 < X_1 < \cdots < X_r$ are the points of the curve K not lying in one $(r-1)$-dimensional plane. The symbol $P(X_0, X_1, \ldots, X_r)$ will denote an r-dimensional plane P passing through the given points X_i and oriented in such a way that a sequence of vectors $(\overline{X_0X_1}, \overline{X_0X_2}, \ldots, \overline{X_0X_r})$ is a right-hand reference frame in the plane P. Let X be a point of the curve K. The limit of the planes $P(X_1, X_2, \ldots, X_r, X)$, where $X_1 < X_2 < \cdots < X_r < X$ when $X_1 \rightarrow X, \ldots, X_r \rightarrow X$, if it exists, is termed a left-hand osculating plane at the point X of the curve K. Analogously, the limit of the secant planes $P(X, X_1, \ldots, X_r)$, when $X_1 \rightarrow X, \ldots, X_r \rightarrow X$ from the right, is termed a right-hand osculating plane at the point X. If the left- and the right-hand osculating r-dimensional planes at the point X of the curve K are different, then at this point we can also define intermediate r-dimensional osculating planes, which can be sufficiently numerous (but not greater than $(r-1)$). The possible variations of constructing a set of intermediate r-dimensional planes are not going to be a subject of our discussion.

281

Let us assume that the curve K is such that none of its arcs lies in one $(r-1)$-dimensional plane and at each its point X it has a left-hand (if this point is not the beginning of the curve) and a right-hand (if X is not the end of K) r-dimensional osculating planes. A set of all osculating r-dimensional planes of the curve K is ordered in a natural way. The curve K belongs to the class H_r if the quantity

$$\kappa_r(K) = \sup \sum_{i=0}^{m-1} (P_1, \widehat{} P_{i+1})$$

is finite. Here $P_0 < P_1 < \cdots < P_m$ is an arbitrary finite sequence of osculating r-dimensional planes of the curve K, and the upper boundary is chosen with respect to a set of all such sequences. The quantity $\kappa_r(K)$ is termed an r-th integral mean curvature of the curve K.

As is shown in [14], if the curve K belongs to the class H_r, then it also belongs to the class H_s for any integer s, where $1 \leqslant s \leqslant r$. For the curves obeying the conventional conditions of regularity in differential geometry, $\kappa_r(K)$ coincides with the integral from a curvature of order r with respect to the arc length. The problem on the possibility of defining an integral curvature of order r and the class of curves $H_r(K)$ by way of approximating a curve by polygonal lines remains unsolved.

10.2. Theorems on natural parametrization in Chapter IX are proved in a way that is closest to that used in the textbooks on differential geometry (instead of the term 'natural parametrization' they commonly use the term 'natural equation'). Here another approach is possible, associated with the use of the notion of a parallel displacement in an arbitrary fibre-bundled space. This approach is also of interest since it contains a certain method of introducing the curves of a limited integral curvature of order r, different from that mentioned in 10.1.

Let K be a curve on the sphere S^n in the space E^{n+1}. Let us place the space E^n in the form of a certain hyperplane into E^{n+1}. Let us assume that the sphere S^n is rolling over the plane E^n, touching E^n at every moment of time t at the points of the curve K. Formally this means that the sphere S^n moves in E^{n+1} in such a way that its instantaneous axis of rotation is parallel to E^{n+1}, and the velocity at the point of touching of S^n and E^n equals zero. In this case the point of touching of S^n and E^n draws in E^n a certain curve Q which is called a development of the curve K. It can be easily proved that if the curve K is regular, then at every $i = 1, 2, \ldots n-1$ the curvatures k_i at the corresponding points of the curves K and Q have equal values and hence the natural equations of the curve and its developments coincide.

Let K be a curve of a finite complete torsion in E^3, $(s(u), \kappa(u), \tau(u))$, $a \leqslant u \leqslant b$, be its natural parametrization. In this case $(\kappa(u), \tau(u))$, $a \leqslant u \leqslant b$, is a natural parametrization of the indicatrix of the tangents Q of the curve K. At the same time, it is a natural parametrization of the development R of the spherical curve Q. This makes it possible to employ the following method of solving the problem - to construct a curve with a given natural parametrization. Let us first construct a plane curve R with the natural parametrization $(\kappa(u), \tau(u))$. Then let us find a spherical curve Q. the development of which is the given plane curve R. This curve Q will be an indicatrix of the tangents of the sought curve K. The curve K is restored with respect to Q by way of simple integration. This method of solving the problem on constructing a curve with a given natural parametrization was suggested by A. Ya. Yusupov [35].

The construction of a development of a smooth spherical curve (as well as the construction of a spherical curve by its development) is analytically reduced to solving a certain linear system of differential equations. The coefficients of this system contain the derivatives of $x(t)$, where $x(t)$ is the parametrization of the curve. The considerations referring to the smooth case can be easily transferred to the case of rectifiable curves and therefore the notion of the development can be defined for an arbitrary rectifiable curve on the sphere.

The notion of the development is closely associated with that of a parallel displacement along a curve on the sphere. This notion was studied in a number of papers by Yu. F. Borisov [4], where, in particular, he proves that the notion of a parallel displacement can be defined for the curves of an essentially wider class than that of rectifiable curves allowing the parametrization $x(t)$, $a \leqslant t \leqslant b$, where the function x obeys the Hölder condition with the exponent $\alpha > \frac{1}{2}$, i.e., for any t', $t'' \in [a, b]$ $|x(t') - x(t'')| \leqslant L|t' - t''|^{\alpha}$. This class of curves is contained in the class of curves allowing the parametrization with the continuity modulus ω, such that

$$\int_0^1 \frac{[\omega(t)]^2}{t}\, dt < \infty$$

In [28, 29, 15] the authors construct a theory of a parallel displacement along irregular curves in an arbitrary fibre-bundled space. The only requirement on the curves is that they must allow a parametrization with the modulus of continuity obeying condition (1) and some other conditions of a 'technical character'. As shown in [29] and [15], the construction of a development spherical curve as well as the construction of a curve by its development can be reduced to that of a parallel displacement along a curve in a suitable fibre-bundled space. In [29] it is, in particular, demonstrated that the definition of a natural parametrization of a curve by the accompanying orthogonal n-hedral of the curve, as well as the construction of a curve by its natural parametrization are reduced to carrying out a parallel displacement in suitable fibrebundled spaces.

Using the notion of a development of a spherical curve, we can, by way of induction, define the class of curves of a finite integral curvature of the

r-th order. Let $H_0(E^n)$ be a class of rectifiable curves in E^n. For $K \in H_0(E^n)$ we set $\kappa_0(K) = s(K)$. Let us assume that for a certain r for any $n > r$ the class of curves $H_r(E^n)$ is defined, and to any curve $K \in H_r(E^n)$ there corresponds a certain number $\kappa_r(K) \geqslant 0$. Let Q be a curve on the sphere S^n. Let us say that $Q \in H_r(S^n)$, if the development R of the curve Q is a curve of the class $H_r(E^n)$. Let us set $\kappa_{r,g}(Q) = \kappa_r(K)$. Let us say that the curve K in E^n is a curve of the class $H_{r=1}(E^n)$ if its indicatrix of the tangents Q is a curve of the class $H_r(S^n)$. We set: $\kappa_{r+1}(K) = \kappa_{r,g}(Q)$.

10.3. An important class of spatial curves is formed by so-called monotonous curves or, in other words, by curves of a real order n. The curve K in the space E^n is called monotonous if it does not lie in one $(n-1)$-dimensional plane and the intersection of K with any $(n-1)$-dimensional plane consists of not more than n connected components. Monotonous curves are one of the basic objects of investigation in the field of geometry known as the 'theory of real orders' (see [9]). It is the theory that was used by I.F. Majnick [13] to study monotonous curves. This author has, in particular, proved that any monotonous curve has a finite integral curvature of order r for any $r \leqslant n - 1$, and that there exists a constant $C_r < \infty$, such that for any monotonous curve K, $\kappa_r(K) \leqslant C_r$.

10.4. The theory discussed in the present book has been extended on the curves in the Lobatchevsky space by D. Sh. Yusupov [36].

10.5. The theory of irregular curves presented in this book is at present still far from being complete. One of the first problems arising here is the extension of this theory on the case of curves in an arbitrary Riemann space. Of a special interest, in our opinion, is the transfer on this case the theorems on the approximation of the turn and of the integral torsion of a curve, which are analogous to Theorems 5.7.1 and 8.3.1.

References

1. Alexandrov, A. D.: 'Theory of curves based on the approximation by polygonal lines.' In: *Nautch.sess.Leningr.univer. Tesisy dokl. na sektch. matem. nauk.* Leningrad, 1946.

2. Alexandrov, A. D.: 'Theory of curves based on the approximation with polygonal lines.' *Uspekhi matem.nauk*, t. 2, vyp.4(20), 1947, pp.182-184.

3. Alexandrov, A. D.: *Internal Geometry of Convex Surfaces.* Moscow, Gostekhizdat, 1948.

4. Borisov, Yu. F.: 'Parallel displacement on a smooth surface.' *Vest.LGU, ser.mat. mekh. i astr.* I: vyp.2, N7, 1958, pp.160-171; II: vyp.4, N19, 1958, pp.45-54; III: vyp.1, N1, 1959, pp.34-50; IV: vyp.3, N13, 1959, pp. 83-92.

5. Zalgaller, V. A.: 'On the class of curves with a limited variation of a turn on a convex surface.' *Matem.sb.*, t.30(72), 1952, pp.59-72.

6. Fari, J.: 'Sur la courbure totale d'une courbe gauche faisant un noeud.' *Bull., Sos. math. France*, 77, 1949, pp.128-138.

7. Fari, J.: 'Sur la courbure totale d'une courbe gauche et surface.' *Acta sci. math. Szeged*, 19, 1951, pp.60-76.

8. Freché, M.: 'Sur quelques points de calcul fonctionnel.' *Rendic. Circ. Palermo*, 22, 1906, pp.51-69.

9. Haupt, O. and Künneth, H.: *Geometrische Organungen.* Berlin-Heidelber, Springer-Verlag, 1977.

10. Kolmogorov, A. N. and Fomin, S. V.: *Elements of the Theory of Functions and Functional Analysis.* Moscow, vyp.I, izd.MGU, 1954.

11. Kuratowsky, K.: *Topology.* Academic Press, New York, London. Panstwowe wydawnietwo naukowe, Warsawa, vol.I, 1966, vol.II, 1968.

12. Liberman, I. M.: 'Geodesic lines on a convex surface.' *Dokl. AN SSSR*, t.32, 1941, pp.310-313.

13. Majnick I. F.: *On Monotonous Curves and Integral Geometry in n-Dimensional Euclidean and Spherical Spaces.* Dep.v VINITI, N2194-76 Dep., 1976.

14. Majnick, I. F.: *On Irregular Curves in an n-Dimensional Euclidean and Spherical Spaces.* Dep.v VINITI, N 3184-76 Dep., 1976.

15. Majnick, I. F.: 'On curves in reductive spaces.' *Dokl.AN SSSR*, t.235, N.3, 1977, pp.531-533.

16. Majnick, I. F.: 'On r-inscribed curves.' *Sib.mat.zhurn.*, t.20, N 1, 1979, pp.95-108.

17. Majnick, I. F.: 'Projective criterion of the equality of spatial curves.' *Sib.Mat. zhurn.*, t.26, N 2, 1985, p.207-210.

18. Majnick, I. F.: 'On the approximation of integral curvatures of the curves in R^n and S^n. *Sib.mat.zhurn.*, t.28, N5, 1987.

19. Milnor, J.: 'On the total curvature of knots.' *Ann.of Math.*, t.52, 1950, pp.248-257.

20. Milnor, J.: 'On the total curvature of closed space curves.' *Math.Scand.*, v.1, fasc.2, 1953, pp.289-296.

21. Pogorelov, A.V.: *External Geometry of Convex Surfaces.* Moscow, Nauka, 1969.

22. Pogorelov, A.V.: *Unique Definitness of Common Convex Surfaces.* Kiev, izd. AN USSR, 1952.

23. Radon, J.: 'Über Randwertaufgaben beim logarithmischen Potential. - Sibzber.' *Akad. Wiss.Wien*, N. 128, 1919, pp.1123-1167.

24. Reshetnyak, Yu.G.: 'On the length and turn of a curve and on the area of a surface.' *Kand.diss.*, Len.univ., 1954.

25. Reshetnyak, Yu.G.: *On a Generalization of Convex Surfaces.* Matem.sb., t.40 (82), N 3, 1956, pp.381-398.

26. Reshetnyak, Yu.G.: 'Integro-geometrical method in the theory of curves.' *Trudy 3-go Vsesoyuzn.matem.s'ezda*, Moscow, t.I, 1956, p.164.

27. Reshetnyak, Yu.G.: 'Method of orthogonal projections in the theory of curves' *Vestnik LGU, ser.matem,mekh. i astron.*, vyp.3, N 13, 1957, p.22-26.

28. Reshetnyak, Yu.G.: 'On a parallel displacement along an irregular curve in the main fibre-bundle.' *Sib.mat.zhurn.*, t.13, N5, 1972, pp.1067-1090.

29. Reshetnyak, Yu.G.: 'On the notion of a rise of the irregular way in a fibrebundled manifold and its applications.' *Sib.mat.zhurn.*, t.16, N 3, 1975, pp. 588-598.

30. Reshetnyak, Yu.G.: 'Isothermic coordinates in manifolds of a limited curvature, part 1.' *Sib.mat.zhurn.*, t.I, N 1, 1960, pp.88-116.

31. Reshetnyak, Yu.G.: 'Isothermal coordinates in manifolds of a limited curvature, part 2.' *Sib.mat.zhurn.*, t.1, N 2, 1960, pp.248-276.

32. Reshetnyak, Yu.G.: 'Turn of a curve in manifolds of a limited curvature with an isothermal linear element.' *Sib.mat.zhurn.*, t.4, N 4, 1963, pp. 870-911.

33. Santalo, L.A.: 'Integral geometry and geometric probability.' *Encyclopedia of Mathematics and its Applications*, vol. 1, ed. by Gian-Carlo Rota, Addison-Wesley Publishing Company, 1976.

34. Schmidt, E.: 'Über das Extremum der Bogenlänger einer Raumkurve bei Vorgeschriebenen Einschränkungen ihrer Krümmung. *Sitz.bericht preuss.Akad.Wissensch.* Berlin, physik. math.kl., Bd.25, 1925, S.485-490.

35. Yusupov, A.Ya.: 'Theory of curves based on the approximation by polygonal lines.' *Kand.diss.*, Len.ped.inst. im A.I. Gertzen, 1949.

36. Yusupov, D.Sh.: 'On the integro-geometrical method in the theory of irregular curves.' *Izv.vys.utcheb.zaved., Matem.*, N 4, 1972, pp.120-122.

Index

Angle between two directed lines 43
  subspaces 185
Area, shortest in 33
  surface 87

Chain 23
  in a metric space 16
  on a curve 16
  double 148
  limits of a function of 17
  modulus of 16
  system of (in a metric space) 24
  triple 219
  vertices of 16
Complete indicatrix of tangents of a curve 57
  in a differentiable manifold 71
Cone
  convex 97
  dual 98
  support hyperplane of 99
Contingency
  Geodesic, of a curve 197
  right/left-handed 49
Curve
  absolute torsion of 243
  c-correspondence for a curve 57
  closed 14
  closed (open) arc of 10
  closed turn of 125
  complete torsion of 267
  degenerate parametrized 7
  diameter of 11
  family of 21
  length of 29
  monotonous 284
  multiple point of 12
  non-oriented 8
  normal parametrization of 15
  obtained by changing orientation 8
  one-sidedly smooth 44
  parametrized 6

  parametrization of 7
  point of 10
  rectifiable 29
  reduced length of 38
  secant of 43
  secant plane of 209
  smooth 20
  smooth angular point of 44
  support of 7
  tangent at a point of 43
  uniformly divisible family of 21

Direction, k-dimensional, in E^n 76
Displacement, parallel, along a curve 283
Distance between curves 18

Element, two-dimensional, in E^3 250

Function, one-sidedly smooth 54

Great circumference 175

Indicatrix
  of correspondence 57
  of tangents of a curve 56
  complete two-dimensional 252
Invariant measure 86

Line, directed straight in E^n 43

Manifold
  chart or coordinate system in 41
  Lipschitz 41
  measure on 85
  n-dimensional 40
  reduced tangential 67
Mapping
  classes of 40
  strictly differentiable at a point 45
Metric spaces, isometry of 32
Modulus
  of a curve in a metric space 23
  of a double chain 148
  of an inscribed polygonal line 16
  of a triple chain 220

287

Multiplicity
  of a double chain 148
  of a triple chain 220

Ordered sets
  anti-similarity of 7
  similarity of 7

Parametrization
  natural, of a curve 279
  natural, of a spherical curve 275
  consistent 61

Parametrized curves
  analytical criterion of equivalence of 12
  equivalence of 7

Path (parametrical curve)
  point of 6
  support of a point of 6

Plane
  middle oscillating 209
  orientation of 208
  parallel to the k-dimensional direction 77
  right/left-handed 209
  right/left-handed oscillating 209

Plane curve
  complete torsion of 219
  natural parametrization of 171
  rotation of 165
  of a finite complete torsion 217

Polygonal line
  absolute torsion of 243
  inscribed 16
  inscribed spherical 17
  spherical 17
  turn (integral curvature) of 118
  vertices of 16

Rotation
  Geodesic, of a spherical curve 206
  Geodesic, of a spherical polygonal line
    203

Set
  Barrelian, union of 85
  canonical correspondence of 77
  (linearly) ordered 6
  relatively compact 21
Sequence of curves, convergence of 20
Spherical curve, development of 282
Space
  complete metric 20
  n-dimensional Euclidean 33
  (n − &)-dimensional spherical 33
  tangential, of a manifold at a point 65
Sphere
  great 175
  k-dimensional 175
  net of 48

Tangent in the strong sense 44
Trihedral, accompanying 271
Turn
  of a curve lying in a straight line 123
  of an arbitrary curve 120
  Geodesic, of a curve in a spherical space
    191
  Geodesic, of a polygonal line in a spherical
    space 177

Vector
  support 97
  support in the strict sense 98
  tangent right/left-handed unit 43